SEA-LEVEL RISE AND COASTAL SUBSIDENCE

Coastal Systems and Continental Margins

VOLUME 2

The titles published in this series are listed at the end of this volume.

Sea-Level Rise and Coastal Subsidence

Causes, Consequences, and Strategies

Edited by

John D. Milliman

School of Marine Science,
College of William and Mary,
Gloucester Point, Virginia, U.S.A.

and

Bilal U. Haq

Bethesda, Maryland, U.S.A.

KLUWER ACADEMIC PUBLISHERS

DORDRECHT / BOSTON / LONDON

A C.I.P. Catalogue record for this book is available from the Library of Congress.

ISBN 0-7923-3933-9

Published by Kluwer Academic Publishers,
P.O. Box 17, 3300 AA Dordrecht, The Netherlands.

Kluwer Academic Publishers incorporates
the publishing programmes of
D. Reidel, Martinus Nijhoff, Dr W. Junk and MTP Press.

Sold and distributed in the U.S.A. and Canada
by Kluwer Academic Publishers,
101 Philip Drive, Norwell, MA 02061, U.S.A.

In all other countries, sold and distributed
by Kluwer Academic Publishers Group,
P.O. Box 322, 3300 AH Dordrecht, The Netherlands.

Printed on acid-free paper

Printed in the Netherlands

Dedicated to the Memory of our Friend and Colleague,
James M. Broadus, 1947–1994

Table of contents

IN MEMORIAM

Dr. James M. Broadus, Senior Scientist and Director of the Marine Policy Center of the Woods Hole Oceanographic Institution, died accidentally on September 28, 1994 while swimming with his daughter in Hawaii. Jim was born in Mobile, Alabama and grew up there and in Lexington, Kentucky. He was educated at Oberlin College and at Yale University, where he received his PhD in Economics in 1976. Before coming to Woods Hole in 1981, Jim taught at the University of Kentucky and worked in the Antitrust Division of the U.S. Department of Justice. He was appointed Director of the Marine Policy Center in 1985.

Perhaps Jim's greatest contribution to the field of marine policy was to raise the level of economic rigor. His research interests covered a range of topics in the use and conservation of marine resources. In the early 1980's, his work focused on the economics of seabed mining. This work was summarized in a paper entitled "Seabed Materials" published in *Science* in 1987 (Broadus, 1987). During this same period, Jim and others at the Marine Policy Center were invited by the Government of Ecuador to participate in the planning of a marine reserve in the Galapagos Islands. This project culminated in the establishment in 1986 of the Galapagos Marine Resources Reserve (Broadus and Gaines, 1987).

After 1985, Jim increasingly turned his attention to global environmental issues — primarily, sea level rise and the conservation of biological diversity. In a much-cited collaboration with John Milliman and others, Jim examined the potential economic effects of different sea level rise scenarios on Bangladesh and Egypt (Milliman, Broadus, and Gable, 1989). In some of the earliest work of its kind, Jim collaborated with Andrew Solow and Stephen Polasky on defining and measuring biological diversity (Solow, Polasky, and Broadus, 1993).

In addition to his work on environmental issues, Jim directed and participated in studies of the organization of the marine electronics industry and of the effectiveness of multilateral agreements to reduce land-based sources of marine pollution. At the time of his death, Jim was working on a number of projects, including a study of historical adaptation to coastal storm risks.

Jim received many professional honors during his career. He served as a member of the National Research Council's Marine Board and on the Marine and Coastal Ecosystems Directorate of the U.S. Department of State Program on Man and Biosphere. He was also invited to organize and participate in the White House Conference on Global Environmental Change in 1990.

In addition to his professional interests, Jim was a shell collector, scholar of Biblical archaeology, and avid (if moderate) Bourbon drinker. He was an outstanding colleague, known for his gentility, generosity, and humor. Jim leaves his wife and three young children.

Andrew Solow
Woods Hole, MA

References

Broadus, J.M. (1987). Seabed materials. *Science*, **235**, 853–860.

Broadus, J.M. and Gaines, A.G. (1987). Coastal and marine area management in the Galapagos Islands. *Coastal Management*, **15**, 75–88.

Milliman, J.D., Broadus, J.M., and Gable, F. (1989). Environmental and economic impact of rising sea level and subsiding deltas: The Nile and Bengal Examples. *Ambio*, **18**, 340–345.

Solow, A.R., Polasky, S., and Broadus, J.M. (1993). On the measurement of biological diversity. *Journal of Environmental Economics and Management*, **24**, 60–68.

Sea-Level Rise and Coastal Subsidence: Towards Meaningful Strategies

John D. Milliman and Bilal U. Haq

Global Climate Change

No environmental issue in the past 20 years has raised as much scientific and public interest and debate as the greenhouse effect. Because the problem is global, no reaction of a single nation or group of nations can necessarily solve its many ensuing problems.

The basic principle of the greenhouse effect, first discussed 100 years ago by Arhhenius, is relatively simple: Water vapor and some of the gases in the atmosphere admit the Sun's incoming radiation and inhibit the Earth's outgoing radiation. Any increase in atmospheric greenhouse gases therefore should cause enhanced warming of the Earth's surface. Amongst the effective greenhouse gases are CO_2, methane, nitrous oxide, and the chloroflouorcarbons (CFCs), all of which have been enhanced (or, in the case of CFCs, entirely produced) by the emissions of human activities.

At the current rate of emissions, CO_2 contributes about 80% of the total greenhouse forcing (Lashof and Ahuja, 1990). The intergovernmental Panel on Climate Change (IPCC) has estimated that CO_2 concentrations in the atmosphere have increased steadily from a pre-industrial level of about 280 ppm (by volume) to 353 ppm in 1990, with a particularly rapid increase since the 1950's (Houghton, 1990). Over half of the total enhanced greenhouse-forcing during this time can be attributed to CO_2 levels. While the other greenhouse gases have atmospheric concentrations that are orders of magnitude smaller than CO_2, their radiative capacity and greenhouse potential are much larger, ranging from 10× (methane) to 3700× (CFC-12) (Lashof and Ahuja, 1990). Ozone in the troposphere also is a contributor to greenhouse forcing. The cumulative impact of each greenhouse gas is determined by its residence time in the atmosphere. Retention of CO_2 in the atmosphere is particularly complex because of its other reservoirs in the biota, ocean and lithosphere.

Temperature records suggest that during the 20th century the Earth has warmed between 0.5 and 1.0°C (Wigley, 1992), with greatest changes in mid and higher latitudes (e.g. Jones et al., 1991). It is more difficult to generalize changes in global precipitation, as this is subject to regional variability. Even less clear is the degree to which these global changes may reflect natural perturbations in weather patterns.

J.D. Milliman and B.U. Haq (eds.), Sea-Level Rise and Coastal Subsidence, 1-9.

Predicted Change

General Circulation Models (GCMs) forecast a mean global warming of 1.2 to 2.0° to as much as 3.5 to 5.2°C (see Raper et al., this volume) over the next 40–50 years, but such predictions are subject to so many variables that the predictions almost certainly will change as the climate and greenhouse systems are understood further. Most models, for example, do not adequately take into account sea-atmosphere interactions, changes in ocean circulation patterns, changes in snow cover (and hence albedo) or, most importantly, changes in cloud cover and precipitation patterns. Even a few percent increase in cloud cover, for instance, could offset most, if not all, of the effect of increased greenhouse gases (e.g. Dickinson, 1989), and change in snow cover could have a dramatic effect on global climate (Barnett et al., 1988).

Climate change by the greenhouse effect also will be manifested in other ways than temperature warming. Precipitation patterns almost certainly will shift, and increased air temperatures will lead to increased evapotranspiration. On the other hand, increased concentrations of CO_2 may stimulate plant growth (Warrick et al., 1988). At least locally, storm intensity may increase (Emanuel, 1987), particularly affecting low-lying areas subject to coastal flooding.

The Rise of Relative Mean Sea Level

Global sea level, as measured at various tide gauges around the world, appears to have risen by 12–15 cm this century (Houghton, 1990; Emery and Aubrey, 1991), largely due to the thermal expansion of sea water in response to increasing global temperature. However, considerable uncertainty about the veracity of these numbers still remains, primarily because any change in local reflects movement of land as well as the rise (or fall) of global sea level. Said another way, a relative mean sea-level rise could indicate subsiding land as well as rising ocean level.

Assuming increased global warming, most researchers expect accelerated rise of global sea level. In a "business-as-usual" scenario (which assumes continued increase in greenhouse emissions at current rates) IPCC (Houghton, 1990) projected a low, high, and best estimate value of sea level by the end of the next century to be 0.3, 1.1 and 0.66 m. Even if the greenhouse emissions were stabilized, however, sea level is expected to continue to rise due to the lag in response of the atmospheric and (particularly) the oceanic systems.

One particular limiting factor in any meaningful prediction of global sea level change is the uncertainty of the response of Antarctic ice sheets. The West Antarctic ice sheet appears most susceptible to breakup since it is largely grounded at or below sea level. A relatively small rise in sea level could unhitch this ice shelf (Thomas and Bentley, 1978), and increased ocean temperature could accelerate the melting, which, if totally melted, could raise sea level by 8.5 m. Collapse of the West Antarctic ice sheet may have caused the 6-m highstand of sea level 120,000 years ago (Emery and Aubrey, 1991). Convective heat models, however, suggest that the Antarctic ice sheets probably will remain relatively stable for at least the next century, implying that in that period sea-level rise will be primarily due to

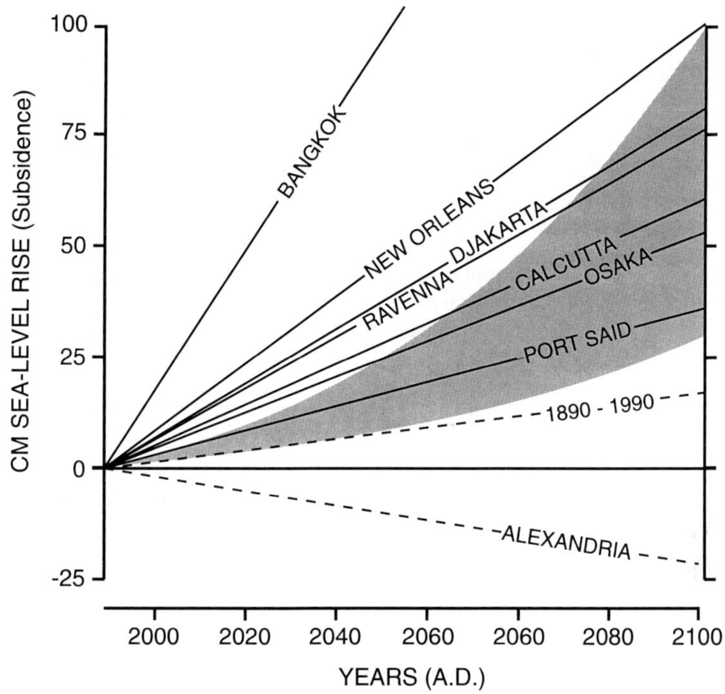

Fig. 1. Comparison of past and predicted sea-level rise with subsidence rates in some low-lying coastal areas discussed in this volume. Subsidence (or, in the case of Alexandria, uplift) is determined by relative change in mean sea level minus measured by local tide gauges, minus eustatic sea-level rise, here assumed to be 1.5 mm/year. Data are from Emery and Aubrey (1991). Note that in most areas predicted subsidence over the next 50–100 years will equal or exceed the predicted range of eustatic sea-level rise; accelerated subsidence because of groundwater withdrawal will only increase this difference. At Alexandria, sea level will continue to fall, although at Port Said, it will rise substantially.

steric expansion. In fact, re-evaluation of polar and alpine ice melting led Meier (1990) to lower his estimate of future sea-level rise to 15–25 cm over the next 50 years.

Subsiding Coastal Areas

No matter whose predictive model of sea-level rise is most accurate, the eustatic rise of sea level may be small compared to the relative sea-level rise in rapidly subsiding coastal areas. Whether the ocean is rising or land is subsiding makes little difference to a person who is up to his waist in water.

Nowhere is the problem of subsidence greater than on low-lying deltas, where populations tend to be concentrated and subsidence rates greater than the rise of global mean sea level (Fig. 1). Under natural conditions, delta subsidence is offset by the deposition of river-borne sediment, particularly the sediment deposited by

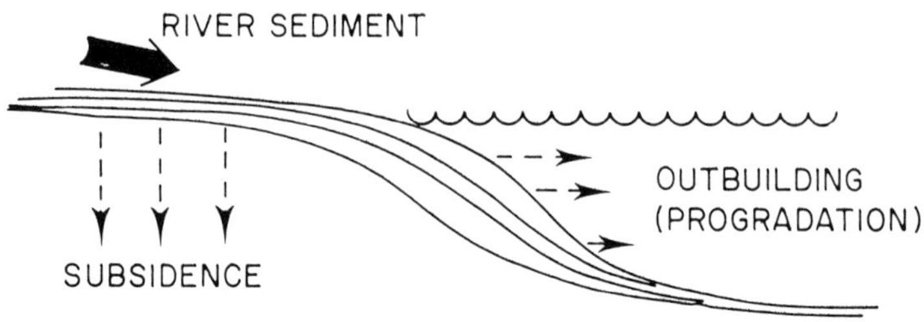

Fig. 2. Schematic vertical and horizontal accretion of sediment within a delta (after Milliman, 1992). Flooding provides a mechanism by which river sediment can escape laterally over the subaerial delta, offsetting subsidence. Sedimentation on the delta front will result in delta progradation.

flood overspill of river banks (Fig. 2). Sediment reaching the coastal environment can accumulate in a seaward progression of the shore front and/or delta front. Some low-lying areas experience natural subsidence rates as great as several cm/year, more than 10 times the present-day rise of global sea level. These subsidence rates reflect regional and local tectonic effects as well as the consolidation (dewatering) of sedimentary strata underlying the delta (Dolan and Goodell, 1986). In addition, channeling, diverting or damming a river can prevent fluvial sediment from reaching the delta, and thus subsidence may not be compensated by new sediment accumulation.

Natural subsidence can be accelerated through the removal of ground water or petroleum from the underlying strata, thereby accelerating sediment compaction. For example, a rapid increase in Bangkok's population in the 1960–80's led to an increased pumping of ground water, a corresponding drop in the water table, resulting in an accelerated subsidence that locally exceeded 10 cm/year (Nutalaya et al., this volume), nearly two orders of magnitude greater than the global rise of sea level.

Effect of Sea-Level Rise on Subsiding Coastal Areas

Rivers provide a basis for human habitation, not only in terms of water and transportation, but also as sources for fertile soils, often renewed by annual floods. As deltaic environments become more densely populated, the need for river modification increases: annual flooding must be contained, channels stabilized, irrigation increased, and hydroelectric power utilized.

The problem of river change and delta subsidence is especially critical in southern Asia, where more than half of the megapolises in the world (urban populations greater than 5 million) are located. Of these 13 cites, all but one (Dehli) are located at least partly on deltas or coastal lowlands: Shanghai, Bangkok, Tokyo, Manilla, Calcutta, Dhaka and Bombay are obvious examples. In fact, as seen in Figure 3, many of the greatest population densities in southeast Asia are located in coastal lowlands.

One reason for the concentration of delta cities in southern Asia is the predominance of large rivers. As much 75 percent of the sediment carried by rivers to the

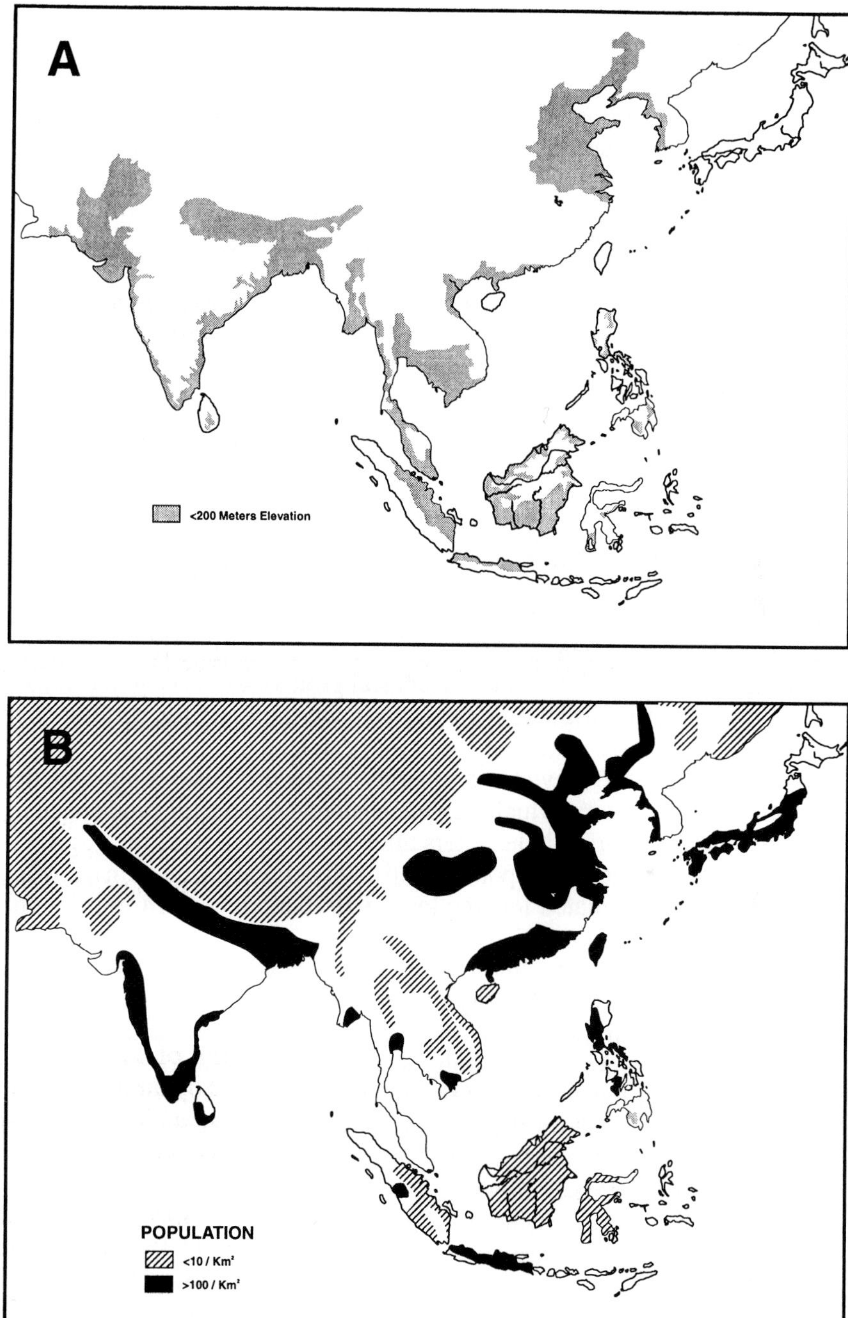

Fig. 3. Comparison of coastal lowlands (shaded areas in A) and population densities in southeast Asia (B). Coastal lowlands in this map are defined by the 200-m elevation contour, but disproportionate amounts of these various lowlands have elevations lower than 5 m above sea level, as discussed many of the papers in this volume. Note that many coastal lowlands are densely populated, exceptions being areas that are arid (e.g., southeastern Pakistan) or densely forested (e.g., Burma and Borneo).

ocean is discharged from southern Asia and the adjacent oceanic islands (Milliman and Meade, 1983; Milliman and Syvitski, 1992). The mountainous drainage basins are especially susceptible to sediment erosion, particularly when exposed to heavy monsoon rains (e.g. Walling and Webb, 1983). Moreover, human land use has resulted in very high sediment yield for many of the Asian rivers (e.g Saunders and Young, 1983; Milliman and Syvitski, 1992). One result of the increased sediment discharge over the past several millenia has been accelerated delta growth. The city of Shanghai, which presently has a metropolitan population of nearly 20 million, for example, was tidal flat as recently as 2–3 thousand years ago.

River containment can result in a series of potential problems for the deltaic environment. Annual floods by the Ganges and Brahmaputra Rivers, for instance, can inundate as much as 35 percent of the total land area of Bangladesh, and storm surges from the Bay of Bengal can greatly exacerbate the problem (e.g., Murty et al., 1986). For much of the year, however, most of the Bengal Delta is rain-deficient. Containment and diversion of these rivers therefore can serve agricultural activities and economic development (e.g., increased hydroelectric power), irrigation and flood control in Bangladesh; but construction of embankments along river also may prevent the annual renourishment of the soil and therefore assure non-compensated subsidence of the delta (Rogers et al., 1989).

Stemming river discharge by damming or diversion also can have a major impact upon the coastal zone. For example, decreased or diverted river flow can lead to increased saltwater intrusion into coastal ground waters (e.g. Kishk, 1986), and salinization and desertification can accentuate the need for drilling shallow and deep wells for ground water, which in turn can lead to accelerated subsidence (see below). Decreased freshwater discharge means diminished nutrient flux to estuarine and coastal waters, which adversely affects biological productivity. The lack of influx of nutrient-rich Nile waters after the completion of the High Aswan Dam in 1964, for example, corresponded with a 95 percent decrease in the offshore sardine fisheries; the estimated income loss in 1970 alone was 14 million dollars (Abdel-Aal, 1985; Wahby and Bishara, 1981).

Damming of rivers means that the necessary brackish water conditions no longer are present; for instance, mangrove communities in the Indus River Delta have declined since damming of the river in the early 1960's and probably will continue to decline as more of the Indus's water is diverted for agricultural use (Snedaker, 1984). This has resulted in the displacement of many of the indigenous peoples who had depended on the forests for their food, shelter and fuel. Near Bangkok (Thailand) accelerated regional subsidence has effectively drowned the seaward portion of a formerly productive mangrove system; in places the mangroves have retreated more than 500 m in less than 30 years. The decreased flux of river-borne sediment also can result in increased shoreline erosion. The short-term effect of decreased sediment flux is seen in locally accelerated erosion of the Nile Delta since construction of the Aswan Dam.

The possible environmental and economic impact of river damming and the ensuing coastal subsidence (natural plus accelerated) and sea-level rise could affect as much as 19 percent of the Nile Delta (16 percent of Egypt's current population and GDP) by the year 2050; coastal erosion could cause a 1-km retreat of the shoreline, but locally it could be greater (Broadus et al., 1986; Milliman et al., 1989). For the Bengal Delta, by 2050 coastal loss by local sea-level rise could be up to 18 percent of land that currently supports about 15 percent of the nation's population and activities representing 13 percent of the total GDP. If natural and accelerated subsidence rates are higher than assumed, the impact could be even greater (Broadus et al., 1986; Milliman et al., 1989). Such projected impacts indicate the potential for disaster in societies that already are stressed by overpopulation, economic underdevelopment and environmental degradation.

Implicit in such calculations, however, is the assumption that regional subsidence and river modifications will have far greater impact on the relative mean sea level on most deltas than will any global rise in sea level. The environmental and economic effects of river diversion/damming and the local rise of sea level are sufficiently grave to warrant their consideration in future river and coastal planning, not only in Egypt and Bangladesh but also in other areas with low-lying deltas and large rivers, particularly those in southern Asia, which contains many large and generally untrained rivers and where recent economic development has accelerated the need for large-scale river development projects.

The Bangkok Workshop and This Volume

Based on the above arguments, a SCOPE-sponsored workshop was held in Bangkok in late 1988 to discuss the effects of sea-level rise and subsidence on low-lying coastal areas. Approximately 50 scientists representing more than a dozen countries participated.

The meeting (and this book) was broken down into three broad areas — a general discussion of processes and problems regarding sea-level rise and subsidence; case studies; and strategies as to how the resulting environmental and economic impacts can be estimated and/or mitigated. Based on preprints and talks delivered at the meeting, the papers included in this volume were prepared and modified to address the three broad topics.

The processes and problems revolve around four broad papers that deal with global sea-level rise (Raper et al.), land subsidence (Jelgersma), river sediment input to sea (Meade) and coastal erosion (Bird). The case studies include a number of papers dealing with subsidence, coastal inundation and remedial actions in Thailand, as well as the Bengal Delta, India, the Yangtze Delta, the Nile Delta, Venice, the Niger Delta, The Netherlands, and the Mississippi Delta). In the third section, strategies for human response and mitigating the effects of sea-level rise and subsidence are discussed in terms of both economics (Broadus) and engineering strategies (de Ronde; Sabhasri et al.), and a final chapter (Haq and Milliman) summarizes possible options open to scientists, managers and planners.

The purpose of this volume is not attempt to answer the problem of future inundation of low-lying coastal areas. Nor is it to repeat discussion on the impact on coastal cities (Frassetto, 1991; Eisma, 1995) or global change *per se* (e.g., Warrick et al., 1993). Rather, by presenting the problems and case histories, specific problems and possible solutions for individual areas can be delineated and considered when discussing such seemingly diverse topics as river diversion projects and the future of coastal lowlands: What are the annual water and sediment discharges of the river(s) draining onto the coastal lowlands; where is the sediment deposited; how much river-borne sediment escapes the coastal area? What are local rates of subsidence, and how would increased ground water/petroleum removal affect these rates? What are the ecological and economic benefits of coastal mangrove forests and other coastal ecotomes, and what would be the environmental and economic consequences of their decline? With these answers, can one begin to make cost-benefit and engineering analyses that may facilitate discussion of both the economic and social effects of such changes and ways in which the negative impacts can be minimized.

Acknowledgments

SCOPE funded the Bangkok workshop, which served as the preliminary vehicle by which this book was prepared. We thank Wanda Cohen, Harold Burrell and Kay Stubblefield (VIMS, College of William and Mary) for their help in the preparation of the manuscripts and illustrations.

Any opinions, findings or conclusions are those of the authors and do not necessarily reflect the views of the National Science Foundation.

References

Abdel-Aal, F.M., 1985. Erosion of the Nile Delta coast. Proc. Coastal Zone '85, Amer. Soc. Civil Eng., 1601-1611.

Barnett, T.P., 1988. Global sea level. in NCPO, Climate Variations over the past century and the greenhouse effect. A report based on the First Climate Trends Workshop, Sept., 1988, Washington D.C., National Climate Program Office/NOAA, Rockville, Maryland.

Broadus, J.M., Milliman, J.D., Edwards, S., Aubrey, D.G., and Gable, F., 1986. Rising sea level and damming of rivers: possible effects in Egypt and Bangladesh. In: J.G. Titus (ed.), *Effects of Changes in Stratospheric Ozone and Global Climate*. UNEP/EPA, 4, 165-189.

Dickinson, R.E., 1989. Long-term climatic and environmental records from Antarctic ice. In: A. Berger and R. Dickinson (eds.), *Understanding Climate Change*. Geophys. Monogr., 52.

Dolan, R. and Gooddell, H.G., 1986. Sinking cities. *Amer. Sci.*, 74, 38-47.

Eisma, D. (ed.), 1995. *Climate Change: Impact on Coastal Habitation.* CRC Press, Boca Raton (Florida), 260 pp.

Emanuel, K.A., 1987. The dependence of hurricane intensity on climate. *Nature*, 326, 483-485.

Emery, K.O. and Aubrey, D.G., 1991. *Sea Levels, Land Levels and Tide Gauges*. Springer-Verlag, New York, 237 pp.

Frassetto, R. (ed.), 1991. *Impact of Sea Level Rise on Cities and Regions*. Centro Intern. Citta d'Acqua. Venice.

Houghton, J.T., 1990. *Climate Change. The IPCC Scientific Assessment*. Cambridge Univ. Press, Cambridge, 365 pp.

Jones, P.D., Wigley, T.M.L., and Farmer, G., 1991. Marine and land temperature data sets: A comparison and look at recent trends. In: M.E. Schlesinger (ed.), *Greenhouse*

Gas-induced Climate Change: A Critical Appraisal of Simulations and Observations. Elsevier, Amsterdam, 153-172.

Lashof, D. A. and Ahuja, D.R. ,1990. Relative contribution of greenhouse gases to global warming. *Nature*, 344, 529-531.

Kishk, M.A., 1986. Land degradation in the Nile valley. *Ambio*, 15, 226-230.

Meier, M.F., 1990. Reduced rise in sea level. *Nature*, 343, 115-116.

Milliman, J.D., 1992. Management of the coastal zone: Impact of onshore activities on the coastal environment. In: K.J. Hsü and J. Theide (eds.), *Use and Misuse of the Seafloor.* John Wiley and Sons, Ltd., Chichester, 213-227.

Milliman, J.D., Broadus, J.M. and Gable, F., 1989. Environmental and economic implications of rising sea level and subsiding deltas - the Nile and Bengal examples. *Ambio*, 18, 340-345.

Milliman, J.D. and Meade, R.H., 1983. Worldwide delivery of river sediment to the oceans. *J. Geol.*, 91, 1-21.

Milliman, J.D. and Ren, M.-e, 1995. River flux to the sea: Impact of human intervention on river systems and adjacent coastal areas. In: E. Eisma (ed.), *Climate Change: Impact on Coastal Habitation.* CRC Press, Boca Raton (Florida), 57-83.

Milliman, J.D. and Syvitski, J.P.M., 1992. Geomorphic/tectonic control of sediment dischrge to the ocean: the importance of small mountainous rivers. *J. Geol.*, 100, 525-544.

Murty, T.S., Flather, R.A., and Henry, R.F., 1986. Storm surges in the Bay of Bengal. *Prog. Oceanogr.*, 16, 195-233.

Nutalaya, O., Yong, R.N., Chumnankit, T. and Buapeng, S., 1995. Land subsidence in Bangkok during 1978-1988. (This volume).

Rogers, P., Lydon, P. and Seckler, D., 1989. Eastern Water Study: Strategies to Manage Flood and Drought in the Ganges-Brahmaputra Basin. Rept. to US AID by the Irrigation Support Project for Asia and Far East, Washington D.C., 83 pp.

Saunders, I and Young, A., 1983. Rates of surface processes on slopes, slope retreat, and denudation. *Earth Surficial Processes and Landforms*, 8, 473-501.

Snedaker, S.C., 1984. Mangroves: a summary of knowledge with emphasis on Pakistan. In: B.U. Haq and J.D. Milliman (eds.), *Marine Geology and Oceanography of the Arabian Sea and Coastal Pakistan.* Van Nostrand-Reinhold, New York, 255-262.

Thomas, R.H. and Bentley, C.R., 1978. Thickening of Ross Ice Shelf and equilibrium state of the West Antarctic Ice Sheet. *Nature*, 259, 180-183.

Wahby, S.D. and Bishara, N.F., 1981. The effect of the River Nile on Mediterranean water, before and after the construction on the High Dam at Aswan. *Proceedings of a Review Workshop on River Inputs to Ocean Systems.* United Nations, New York, 311-318.

Walling, D. and Webb, B.W., 1983. Patterns of sediment yield. In: K.J. Gregory (ed.), *Background to Palaeohydrology.* Wiley, Chicester, 69-100.

Warrick, R.A., Gifford, R.M., and Parry, M.L., 1988. CO_2, climate and argiculture. In: B. Bolin, B. Doos, J. Jager, and R. Warrick (eds.), *The Greenhouse Effect: Climate Change and Ecosystems.* J. Wiley, Chichester, 363-392.

Warrick, R.A., Barrow, E.M., and Wigley, T.M.L. (eds.), 1993. *Climate and Sea-Level Change: Observations, Projections and Implications.* Cambridge University Press, Cambridge, 424 pp.

Wigley, T.M.L., 1992. Future climate change in the Mediterranean Basin with particular empahsis on changes in precipitation. In: L. Jeftic, J. Milliman and G. Sestini (eds.), *Climate Change and the Mediterranean.* Edward Arnold, London, 15-44.

CHAPTER 1

Global Sea-level Rise: Past and Future

S.C.B. Raper, T.M.L. Wigley, R.A. Warrick

ABSTRACT: This paper reviews observationally based estimates of past global-mean temperature change and sea-level rise and compares them with model-based estimates. The climate model used is a simple upwelling-diffusion, energy-balance model, which is coupled to a set of simple ice-melt models to give total sea-level change. For best-guess model parameter values there is reasonable agreement between observed and modelled results. The same models are used to estimate future temperature changes and sea-level rise for the standard IPCC95 set of emissions scenarios, updating earlier work. Projected warming over 1990–2100 ranges between 1.4 and 2.9°C for the central emissions scenario (IS95a), while sea-level rise ranges between 20 and 86 cm. Mid-value estimates for a climate sensitivity of 2.5°C for a CO_2 doubling are 2.0°C and 49 cm. Temperature and sea-level rise estimates are also given (out to 2500) for five standard (IPCC) CO_2 concentration scenarios in which CO_2 levels stabilize at 350, 450, 550, 650 and 750 ppmv. The sea-level-rise commitment after stabilization is very large: for stabilization levels of 550 ppmv or above, sea-level rise continues for many centuries at rates similar to those occurring at the stabilization point in spite of the constancy of radiative forcing. Finally, the sensitivity of these results to changes in the ocean's thermohaline circulation is examined. The effects of a thermohaline slowdown are reduced warming rate and increased rate of sea-level rise.

Introduction

On the decade-to-century timescale, changes in global mean temperature result in changes in sea-level mainly through thermal expansion of the ocean and the melting of land ice. Over the past 100 years the world has warmed by about 0.5°C. Due to increasing atmospheric concentrations of carbon dioxide and other greenhouse gases, the prospect for the future is an even warmer world. Has sea-level also risen over the last 100 years? If so, is there evidence that the rise is climate-related? What are the projected rates of global warming due to anthropogenic forcing and how fast is sea level likely to rise? These are the questions that we address in this paper, paying particular attention to the uncertainties attached to all estimates.

J.D. Milliman and B.U. Haq (eds.), Sea-Level Rise and Coastal Subsidence, 11-45.

We begin by briefly reviewing both the evidence for past sea-level change and estimates of global temperature change based on observational data. An energy-balance climate model forced with greenhouse-gas concentration and aerosol changes is used to simulate past and future temperature changes. Estimates of oceanic thermal expansion and land ice contributions to both past and future sea-level change are then considered. Finally, we offer estimates of future global-mean sea-level rise, including the range of uncertainty arising from both uncertainties in future forcing changes and modelling uncertainties.

The paper concludes that sea level has been rising at the average rate of about 1.0–2.0 mm/year over the last 100 years and that this rise can be explained by climate-related factors. For future greenhouse-gas-induced warming, our best estimate is a temperature rise of 2.0°C from 1990 to 2100, with an attendant rise in sea level of 49 mm. If the various uncertainties are taken into account, estimates of sea-level rise over this period range from 20–86 cm but no definite probabilities can be assigned to these values.

Observed Changes in Global Mean Sea Level

Mean sea-level changes, as estimated from tide gauge data, reflect both water level changes relative to the land surface and vice versa. In projecting future sea-level rise we are concerned with both aspects of sea-level change. In relating these projections to those of future climatic change, however, and in elucidating the climate-related component of past sea-level changes our primary concern is with the global-mean eustatic rise, i.e., the change that results from a change in the volume of ocean water.

At any individual location, vertical movement of the land relative to the geoid must appear as a sea-level change. Before information on eustatic change can be obtained, such land movement effects must be removed from the tide gauge data. Various approaches have been used to accomplish this; data analysis techniques (Woodworth, 1990), geological information (Gornitz and Lebedeff, 1987), and geophysical models (Peltier and Tushingham, 1989) have all been employed to factor out the land movement term. All of these methods have their strengths and weaknesses. For example, the modelling approach only accounts for the long-term isostatic component of land movement, while geological analyses suffer from data interpretation and dating problems. Purely empirical methods are deficient in that they do not address specific physical processes and so may produce results that are mere statistical artefacts.

Almost all estimates of sea-level change over the past century or so (Table 1) are based on the same tide gauge data set, that maintained by the Permanent Service for Mean Sea Level (PSMSL) at the Bidston Observatory, Birkenhead, England. There are currently about 420 tide gauge stations with record lengths exceeding 20 years in this data set. These show a geographical bias towards the continental coastal margins of the Northern Hemisphere, in particular, North America, Europe and Japan (Fig. 1). The data show large variability on all time scales. On time scales less than one year, variations in mean sea level occur due to tidal and atmospheric

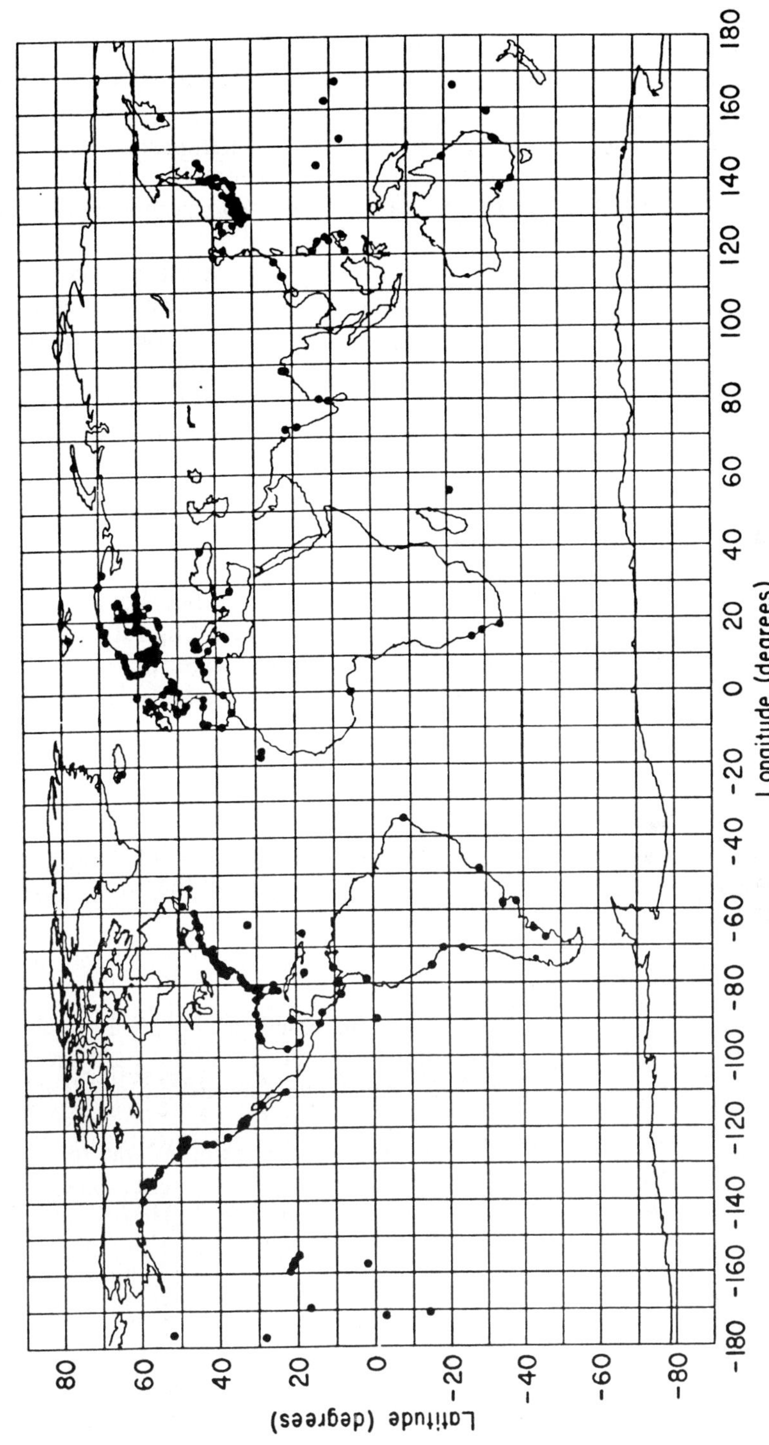

Fig. 1. Map showing the location of tide-gauge stations with record lengths ≥20 years (from Gornitz and Lebedeff, 1987).

TABLE 1

Estimates of global sea-level change from tide gauge data (updated from Barnett, 1985; Robin, 1986; Warrick and Oerlemans, 1990; Gornitz, 1995).

Rate (mm/yr)	Comments	References
1.2	Many stations >regions, 1880-1980	Gornitz et al. (1982)
1.5	Many stations, 1900-1975	Klige (1982)
1.5 ± 0.15*	Selected stations, 1903-1969	Barnett (1983)
1.4 ± 0.14*	Many stations >regions, 1881-1980	Barnett (1984)
2.3 ± 0.23*	Many stations >regions, 1930-1980	Barnett (1984)
1.2 ± 0.3*	130 stations, 1880-1982	Gornitz & Lebedeff (1987)
1.0 ± 0.1*	130 stations, >11 regions, 1880-1982	Gornitz & Lebedeff (1987)
1.15	155 stations, 1880-1986	Barnett (1988)
2.4 ± 0.9**	40 stations, 1920-1970	Peltier & Tushingham (1989; 1991)
1.7 ± 0.13**	84 stations, 1900-1980	Trupin and Wahr (1990)
1.8 ± 0.1	21 stations, 1880-1980	Douglas (1991)

* Value plus 95% confidence interval

** Mean and standard deviation

changes (see Pugh, 1987, for a comprehensive discussion). On the interannual to decadal time scale, annual average data are also quite variable. They show spatially coherent fluctuations which reflect oceanographic and atmospheric conditions such as the El Niño/Southern Oscillation phenomenon (Korma and Enfield, 1987) and the North Atlantic Oscillation (Maul and Hanson, 1991). The periodic fluctuations in tidal amplitude associated with the 18.6 year lunar nodal cycle also produce trends in short records (Lisitzin, 1974).

Even when these basic data have been processed in some way to attempt to remove the unwanted noise, uncertainties remain. These are partly removed by area-averaging to produce a global mean value for eustatic sea-level rise. However, this is a non-trivial task because of the very uneven spatial distribution of data points, and there is still some dispute over the optimum way to obtain a global-mean value for the sea-level trend. These analysis differences are of some importance; indeed Barnett (1984a) has noted that different methods can cause variations in the estimates of sea-level rise of up to ±50%. Peltier and Tushingham (1991) find that the sea-level trends depend on the choice of minimum record length and on the time period. Thus, Gornitz (1994) notes that analyses that have used fewer but longer records generally get higher estimates. She also notes that the components of SLR due to vertical land movements estimated from Peltier's glacier rebound model (Peltier and Tushingham, 1989, 1991; Douglas 1991) are smaller than estimates based on local geological data (Gornitz and Lebedeff, 1987), making the corrected tide gauge estimates greater for the former.

Despite the uncertainties, global mean sea level is thought to have risen substantially over the last 100 years. Estimates of the average rate of rise range from 0.5–3.0 mm/year, as shown in Table 1. We note that we are looking for the average rate over the last one hundred years or so which is compatible with higher or lower rates over subperiods. There appears to be no sound evidence for an acceleration in the rate of rise over the past 100 years, although there is some indication that the rate of rise was less in the preceding century (Woodworth, 1990). It is difficult to put realistic error bars on these estimates. There is, however, as noted by Warrick and Oerlemans (1990), the possibility that the results of all the studies are biased in the same direction. For instance, Pirazzoli et al. (1987) assert that there may have been a systematic downwarping of the world's coastal regions due to the effects of sedimentation and the extraction of groundwater and oil. Such a bias, if it exists, has been judged to be small, and an estimate in the range 10–20 cm rise over the past 100 years, represents our best judgement based on available data.

Observed Changes in Global Mean Temperature

Global mean surface air temperature variations have been reviewed recently by Jones and Wigley (1990) and Folland et al. (1990, 1992). Arguably the best global mean surface air-temperature series available is that produced by the Climatic Research Unit (Jones and Briffa, 1992; see also Parker et al., 1994) (Fig. 2). There are larger uncertainties in the data before around 1900 than subsequently. To account for different interpretations of these early data, a composite series has been adopted by the Intergovernmental Panel on Climate Change (IPCC). This differs slightly from the Climatic Research Unit series shown here (see Folland et al., 1990, for details). Both Figure 2 and the IPCC equivalent are homogenised temperature data series covering the land and ocean areas of both hemispheres. The main feature to note is the overall warming over the last 100 years of about 0.5°C. Note also the marked warming to 1940, the period of relatively little change from then to the mid 1970s, and the renewed rapid warming since the late 1970s. These trend changes are more apparent in the northern hemisphere, with more gradual warming over the century in the southern hemisphere.

The reliability of this and other equivalent temperature series can be questioned for the following reasons. Spatial coverage is at best less than 75% and it changes with time (for example, Antarctica is included only after 1957). Even though the land data used (those of Jones, 1994) are carefully homogenised, it is still possible that some residual urban warming signal remains. Recent work with rural data series over USA, USSR, Australia and China, however, indicates that residual urban influences are minimal (Jones et al., 1989, 1990). The upper limit for this effect in the Jones global land temperature series is thought to be less than 0.05°C. In another study, where the urban warming effect has *not* been removed, Hansen and Lebedeff (1987) estimate a warming of 0.7°C between 1880–1885 and 1980–

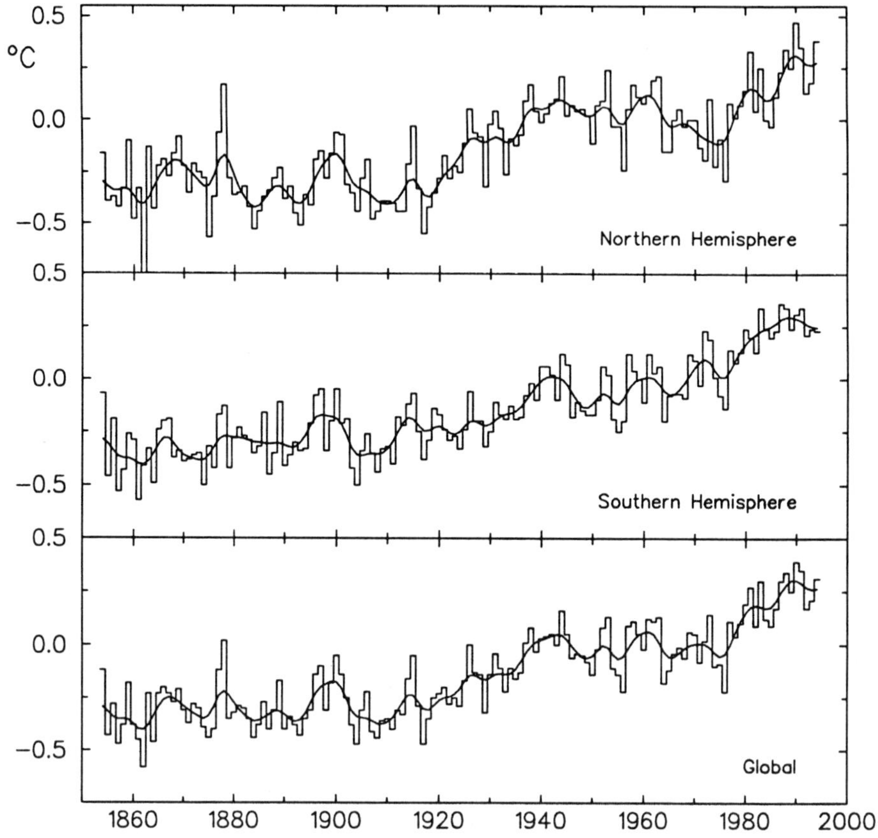

Fig. 2. Annual mean surface air temperatures for the Northern Hemisphere (upper panel), Southern Hemisphere (middle panel) and Globe (lower panel). The series are expressed as departures from the reference period 1950–1979. Smooth curves show 10-year Gaussian filtered values. From Jones and Briffa (1992).

1985 based only on land stations. They conclude that this could include an urban warming bias of up to 0.2°C.

Ocean temperature data also have their problems due to changes in measurement techniques. For example, sea surface temperatures are affected by the change from bucket to engine intake measurements, and marine air temperatures are affected by exposure changes (Jones and Wigley, 1990; Jones et al., 1991, Parker et al., 1994; Folland and Parker, 1995). However, both of the primary homogenized marine data sets (viz. the Climatic Research Unit data set of Jones and Briffa, 1992, and the UK Meteorological Office data set of Parker et al., 1994) have addressed these problems.

For this study we assume a temperature increase of 0.5°C over the last century, but acknowledge that there is a range of uncertainty of around ±0.15°C (Folland et al., 1990, give the range of uncertainty as 0.3–0.6°C). The extent to which changing atmospheric concentrations of the greenhouse gases have been responsible for this global warming over the last century is uncertain (Wigley and Barnett, 1990). Other possible contributing factors include natural internal variability (Wigley and Raper, 1990, 1991a, 1991b), the effects of sulphate aerosol changes (Wigley, 1989; Charlson et al., 1992) and natural factors external to the climate system such as solar variability (Kelly and Wigley, 1990, 1992; Schlesinger and Ramankutty, 1992) and volcanic eruptions (e.g. Sear et al., 1987; Bradley, 1988).

The Climate Model

As will be explained below, our projections of future sea-level change will be made using a set of relatively simple models. The primary model is an upwelling-diffusion, energy-balance climate model that estimates global mean temperature changes in response to external radiative forcing changes. This model also produces estimates of the thermal expansion component of sea-level change. Simple box-diffusion energy-balance climate models have been used in previous studies of both temperature and sea-level change (e.g. Hoffert et al., 1980; Gornitz et al., 1982; Harvey and Schneider, 1985; Hoffert and Flannery, 1985). The model we use here is a modified form of the upwelling-diffusion climate model (UD-model) of Wigley and Raper (1987, 1992). The base version of the model differs from other similar models in that it differentiates the hemispheres, and land and ocean regions in each hemisphere. This model was used for making the main temperature and sea-level projections given by IPCC Working Group 1 in 1990 (Bretherton et al., 1990; Warrick and Oerlemans, 1990). It has also been used to investigate the implications of the new science and emissions scenarios contained in the 1992 IPCC update (Mitchell and Gregory, 1992; Wigley and Raper, 1992). The modifications we have made to the UD-model and the reasons for them are described below.

It has been suggested that the results from transient experiments with coupled ocean/atmosphere general circulation models (O/AGCMS) give lower thermal expansion than that indicated from the UD-model (Cubasch et al., 1992; Hasselmann et al., 1993; Gregory, 1993). A detailed comparison between the MPI O/AGCM (ECHAM-1/LSG) transient runs and UD-model results shows that lower expansion results in this particular model may arise partly through the substantial land/ocean temperature change contrasts leading to lower temperature changes over the ocean (Raper and Cubasch, 1995). There is evidence that these land/ocean temperature contrasts could be caused by differences in the land/ocean climate sensitivities (Murphy, 1995). In the absence of evidence for a land/ocean forcing contrast, we have therefore revised the UD-model to incorporate different land/ocean sensitivities. We do this by defining an additional parameter R, which specifies the ratio of the equilibrium temperature changes over land versus ocean. Here we use $R = 1.3$, based on results from the MPI and Hadley Centre models.

Another factor contributing to differences between standard UD models and coupled O/AGCMs is the variability of the thermohaline circulation, whose strength is parameterized by the upwelling rate (w) in the UD-model. Previous simulations with the UD-model were performed with a constant rate of upwelling

($w = 4$ m/year), as has generally been assumed in climate simulations using simple models of this type. All current transient O/AGCM simulations, however, show a reduction in upwelling rate as greenhouse-gas-induced warming increases. In the scenario A transient run of the ECHAM1-LSG model, the upwelling rate is estimated to have declined by about 30% over the 100 years of the simulation (Santer et al., 1994). We have therefore modified our UD-model to have a variable upwelling rate, using a simple relationship between w and the ocean surface-temperature change given by

$$w(t) = w_o(1 - \Delta T/\Delta T^+),$$

where w_o is the initial upwelling rate and ΔT is the hemispheric ocean surface-temperature change. We use $\Delta T^+ = 7.0°C$ and a lower limit for $w(t)$ of 0.1 m/year based on the MPI work and on results given by Manabe and Stouffer (1994).

A decrease in w also tends to produce a land/ocean temperature change contrast by slowing the ocean surface warming. For any given ocean surface warming, however, decreasing w tends to *increase* the thermal expansion because the ocean tends to move towards an isothermal state (which it would eventually reach if w dropped to zero) so causing a greater warming at depth, greater overall heat retention by the ocean and, hence, greater expansion. The combined effect on thermal expansion of land/ocean sensitivity differences and changes in w depends on the forcing history: over long time periods, the effects of changes in the thermohaline circulation become relatively more important.

Details of the anthropogenic forcing are given in the Appendix. The 1990–2100 future forcings are based on the IPCC92 emissions scenarios (Leggett et al., 1992), except that scenario d (IS92d) emissions for the halocarbons are used throughout, consistent with the Copenhagen amendment of the Montreal Protocol. We account only for anthropogenic forcing; past and future changes in the "natural forcing" upon which the anthropogenic forcing change is imposed introduce uncertainties in our results.

Apart from the imposed forcing changes, the factor that most affects the modelled temperatures is the global mean climate sensitivity, defined here as the equilibrium global mean warming due to a doubling of atmospheric carbon dioxide concentration ($\Delta T_{2\times}$). The estimated range for ($\Delta T_{2\times}$) based on General Circulation Model (GCM) experiments is 1.5–4.5°C with a best estimate of 2.5°C (these are the IPCC-recommended values; Mitchell et al., 1990). Other uncertainties in our model temperature changes arise due to uncertainties in the magnitude of the heat flux into the ocean (here we use the best-guess parameter values given by Wigley and Raper, 1991b). The value of $\Delta T_{2\times}$ which gives a best-fit between our model temperature change results and observed temperature changes over 1861–1994 is ~6°C (see Appendix), which differs noticeably from the IPCC central value of 2.5°C. It is, however, impossible to use this empirical result to justify a revision to the IPCC values, because of the large uncertainties surrounding any empirical sensitivity estimate (see Wigley and Raper, 1991b) arising mainly through uncertainties in aerosol forcing and the possible effects of natural variability. We therefore use the GCM-based values for $\Delta T_{2\times}$.

Factors Contributing to Sea-Level Rise

Global warming is thought to change sea level mainly through thermal expansion of the oceans and by changing the net mass balance of glaciers and ice sheets. In addition, the stability of the West Antarctic Ice Sheet and uncertainties due to the pre-industrial state of the mass balance of Greenland and Antarctica must be considered.

Thermal Expansion

The coefficient of thermal expansion is a non-linear function of temperature; for a given temperature change, cold water expands less than warm water. In the UD-model, the initial vertical temperature profiles are based on the observations from Levitus (1982). For each hemisphere, time-dependent vertical profiles of the thermal expansion coefficient are calculated using the equations of state given in Gill (1982), assuming a salinity of 35‰.

The combination of expansion coefficient non-linearities and spatial differences in the changes in temperature could lead to inaccuracies in the thermal expansion estimates made with the UD-model, since this uses only a mean temperature profile for each hemisphere. The size of any such inaccuracy would depend on the geographical distribution of the ocean temperature changes. However, coupled O/AGCM results may be used to show that this is unimportant: comparisons of the thermal expansion calculated from three dimensional grid point data and from global mean profiles of temperature change for four ECHAM1/LSG experiments give negligible differences (Raper and Cubasch, 1995).

Glacier and Small Icecap (GSIC) Contribution

Over the last 100 years, there has been a general worldwide retreat in small glaciers (Röthlisberger, 1986; Grove, 1988). Estimates of the contribution that this has made to sea-level change have been made over the period 1900–1961. The methods are different and the results vary greatly and have large uncertainties; for example, for their central estimates Meier (1984) gets 28 mm, Oerlemans and Fortuin (1992) give results corresponding to 6.2 mm, and Wigley et al. (1995) get 6.6 mm. Wigley et al. (1995) show that the reason Meier's estimate is larger than the other two is at least in part due to the fact that it includes the effect of a likely initial (1900) negative mass balance of glaciers. The latter two studies consider only the sensitivity of glacier mass balance to temperature change over 1900–1961. For these latter studies, although the global mean estimates are similar, the contributions from different regions are different, so the global results hide an implied additional uncertainty. Since the model for estimating the contribution to sea-level rise from GSICs over the last 100 years and for the future must, in part, be calibrated using these estimates, the various uncertainties and inter-study differences constitute an impediment to accuracy.

There is also an uncertainty associated with estimates of the total volume of ice bound up in the GSICs. The estimate of the pre-industrial ice volume of 50 cm sea-level equivalent used by IPCC (Warrick and Oerlemans, 1990) and in our earlier work (Wigley and Raper, 1993), is revised downwards in the present analyses to around 30 cm (M. Meier, personal communication). This number is important because it represents the maximum potential contribution that the GSICs could make to sea-level rise.

For the calculation of the GSIC contribution to sea-level rise we use a simple model that relates glacier volume change to temperature changes. A similar type of model has been used by Oerlemans (1989) and Wigley and Raper (1993). The model used here, and its calibration, is described in detail and compared with the Wigley and Raper (1993) model in Wigley and Raper (1995). A summary is given below.

We start with a single glacier/ice cap ice volume equation

$$\mathrm{d}V/\mathrm{d}t = (V_\infty - V)/\tau, \tag{1}$$

where t is time (y), V is the ice volume (m^3), V_∞ is the equilibrium value of V, and τ is a glacier response time (y). Expressions for τ and V_∞ are derived in Raper et al. (1995). Here we assume that the response time, τ, is a constant and assume V_∞ decreases linearly with increasing temperature change, so that

$$V_\infty = V_{\infty,o}(1 - \Delta T/\Delta T^*), \tag{2}$$

where $\Delta T = T - T_o$ (T_o is the temperature at $t = 0$) and ΔT^* is the minimum temperature increase at which, if it were maintained, the glacier would eventually disappear.

The equations can be written in terms of sea-level equivalent (s.l.e) by substituting (2) into (1) to give

$$\mathrm{d}z/\mathrm{d}t = (-z + Z_o - Z_{\infty,o}(1 - \Delta T/\Delta T^*))/\tau, \tag{3}$$

where z is the sea-level change (positive for sea-level rise), Z_o is the initial ice volume in s.l.e., and $Z_{\infty,o}$ is the equilibrium ice volume in s.l.e. at $t = 0$.

We take into account the regional variations in the altitudinal range of the world's glaciers and hence the fact that some glaciers will disappear before others with global warming. We do this by solving equation (3) for a set of values of ΔT^* and τ. We assume that in all cases $Z_o = Z_{\infty,o}$ in 1880 and that the ice volume is evenly distributed over the various glacier types represented by the ΔT^*s and corresponding τs. For the temperature forcing, we use the UD-model global mean temperature results. There is clearly scope for the further development of this GSIC model regarding the functional form of the equation and the use of regional glacier and temperature data. The parameter values used for the low, medium and high ice melt estimates are given in Table 2.

TABLE 2
Ice-melt model parameter values used to obtain the high, medium and low sea-level rise estimates.

Estimate	GSIC		Greenland		Antarctica		
	τ	ΔT^*	ΔB_0	β	ΔB_0	β_1	β_2
	yr	°C	mm/yr	mm/yr/°C	mm/yr	mm/yr/°C	mm/yr/°C
High	35–65	0.6–2.5	0.0	0.5	0.6	-0.15	0.2
Medium	70–130	0.7–3.0	0.0	0.3	0.1	-0.30	0.1
Low	105–195	0.9–4.5	0.0	0.1	-0.4	-0.45	0.0

THE GREENLAND AND ANTARCTIC ICE SHEETS

For Greenland and Antarctica, our method is essentially that used by Warrick and Oerlemans (1990) and Wigley and Raper (1993). The response time of Greenland and Antarctica is long compared to the time scale considered here, so that as long as the temperature change is moderate it is reasonable to assume that the area of the ice sheets is constant. We can then divide the mass balance of the ice sheets into two components.

The first component, denoted ΔB_0, is a constant with units mm/year, and represents gain or loss of ice due to the initial state of the ice sheet. We take the initial time to be 1880. Thus, if the ice sheet is in equilibrium with the climate in 1880, ΔB_0 is zero. If, however, the ice sheet is not in equilibrium but still reacting to a previous temperature change, then ΔB_0 will be non zero. The second component, ΔB_1, is assumed to be linearly dependent on the temperature change relative to the initial state (1880) and also has units mm/year. Thus, the change in sea level due to the mass balance of the ice sheets may be expressed as

$$\mathrm{d}z/\mathrm{d}t = \Delta B_0 + \Delta B_1. \tag{4}$$

For Greenland, the ice sheet is assumed to be in equilibrium in 1880 so that ΔB_0 is zero. We do not investigate the effects of uncertainties in this parameter here. The temperature-dependent term is

$$\Delta B_1 = \beta \Delta T, \tag{5}$$

where β is the sensitivity value for the mass balance (we use $\beta = 0.3\pm0.2$ mm/year/°C).

For Antarctica, we assume $\Delta B_0 = 0.1 \pm 0.5$ mm/year. The temperature-dependent term is assumed to have two sensitivities so that

$$\Delta B_1 = (\beta_1 + \beta_2)\Delta T, \tag{6}$$

where β_1 is the sensitivity value for the mass balance (we use $\beta_1 = -0.3 \pm 0.15$ mm/year/°C), and β_2 represents the influence of a possible instability of the West Antarctic ice sheet (we use $\beta_2 = 0.1 \pm 0.1$ mm/year/°C).

TABLE 3
Model results for temperature change (ΔT) and sea-level rise (Δz) over 1880–1990, using the Mid I estimate for sulphate aerosol forcing (see Appendix). L, M and H refer to the low, mid and high, Δz estimates. The parameter values are as in Table 2.

ΔT_{2x} °C	ΔT °C	Δz (M) cm	Δz(L/M/H) cm
1.5	0.21	5.3	−0.9 (L)
2.5	0.29	6.8	6.8 (M)
4.5	0.40	9.0	16.6 (H)

We use these formulae to predict both past and future contributions to sea-level change from Greenland and Antarctica. The parameter values for the high, medium and low sea-level rise estimates are given in Table 2. For the past, the temperature used is the model-based global mean temperature. We have not attempted here to use observationally based regional data, nor have we weighted the global mean temperature. For the future, however, based on general circulation model results, Greenland is assumed to warm at a rate 1.5 times the global mean. For Antarctica, the unweighted global mean temperature change is used.

Temperature and Sea-Level Rise Results

Table 3 shows the model results for temperature change and sea-level rise over 1880 to 1990 for three climate sensitivities, 1.5°C, 2.5°C and 4.5°C. The modelled temperature change range is 0.21–0.40°C, which is noticably below the observed range of 0.35–0.65°C. These ranges are compatible considering that the model results only give the anthropogenic effect, while the observations include, additionally, the effects of natural variability.

We have used the past global mean model temperature change results to drive the sea-level models. The results range from −1 to 17 cm sea-level rise over 1880–1990 with a mid estimate of about 7 cm.

If sea level rise results are interpolated to correspond to a 1880–1990 warming of 0.5°C the range is 2–19 cm, with a central estimate of around 10 cm. These results are in satisfactory agreement with the observed range of 10–20 cm given uncertainties in the amount of warming and in the ice melt models.

One factor that substantially affects the past modelled net ice melt is the state of balance of the ice masses in 1880. This is represented in our model equations for Greenland and Antarctica as the parameter ΔB_0, and in the GSIC model as the difference between the initial ice volume in s.l.e (Z_0) and the initial equilibrium

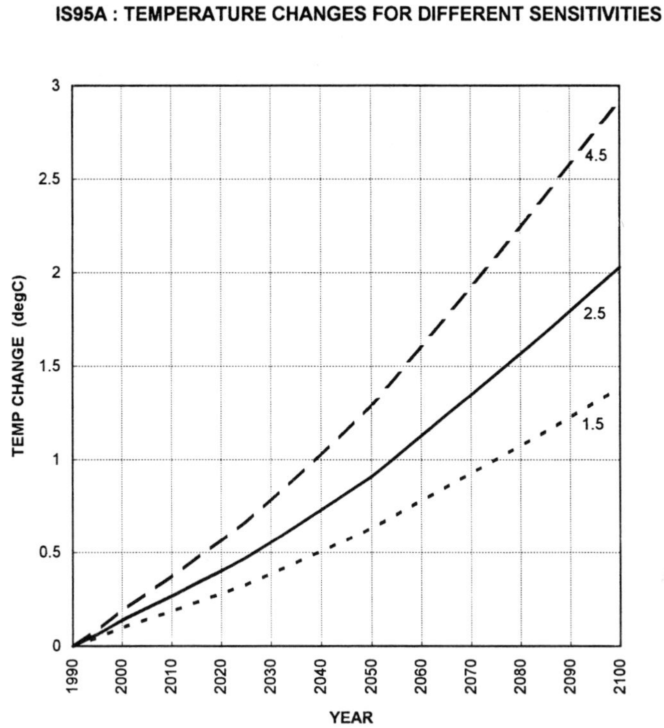

Fig. 3. Projections of global mean temperature change for emissions scenario IS95a. The three lines are for $\Delta T_{2\times} = 1.5$, 2.5 and 4.5°C.

ice volume in s.l.e ($Z_{\infty,o}$). Given the uncertainties in the initial mass balances, there is certainly scope for improving the past modelled sea-level rise estimates. However, we do not feel that the evidence is sufficiently compelling to do so at this time.

We now show results for a range of future emissions scenarios, and for scenarios in which CO_2 concentrations stabilize at levels from 350 ppmv to 750 ppmv. First, we show results for the central emissions scenario given by IPCC in 1992 (Leggett et al., 1992), both to illustrate the general magnitude of future global mean warming and sea-level rise and to give an idea of the uncertainties in these projections. This scenario, which is generally referred to as IS92a we call IS95a because minor modifications have been made as described in the Appendix. The uncertainties considered are the most important ones by far, viz. those arising from uncertainties in the climate sensitivity and in the various ice-melt parameters.

Figure 3 shows temperature projections ($\Delta T(t)$) for sensitivity values of $\Delta T_{2\times} = 1.5$, 2.5 and 4.5°C, with other climate model parameters set to the "best guess" values used by Wigley and Raper (1992). The warming amounts are generally a little less than in Wigley and Raper (1992) - henceforth WR92. For example, for $\Delta T_{2\times} = 2.5$°C, WR92 give a 1990–2100 ΔT of 2.5°C compared with 2.0°C here. The lower value is the result of a combination of factors, the

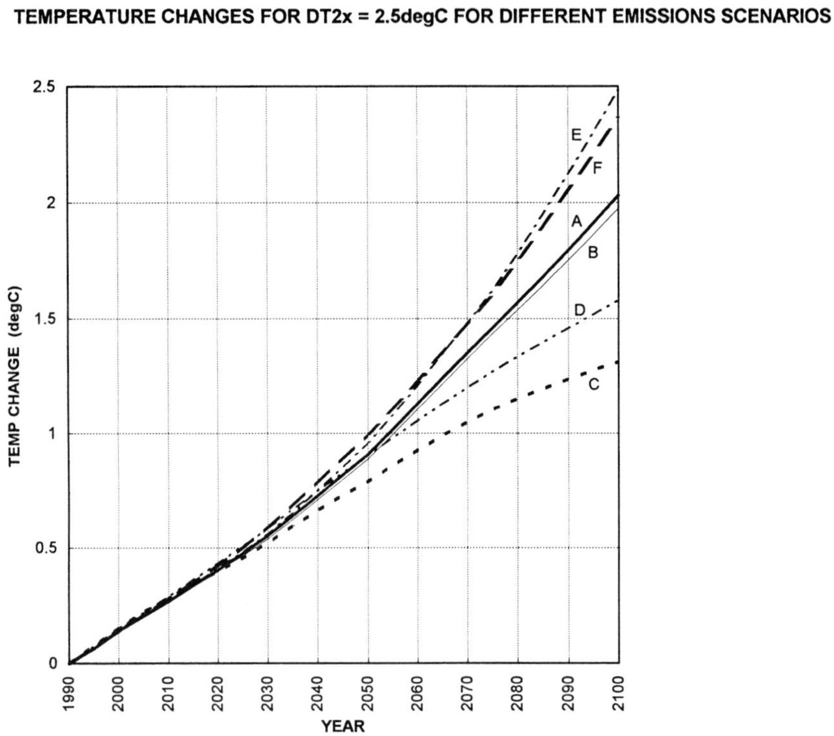

Fig. 4. Projections of global mean temperature change for emissions scenarios IS95a-f. Mid-value model parameters are used, in particular $\Delta T_{2\times} = 2.5°C$.

most important of which are: different forcing, both up to 1990 (which influences future temperature changes through the different level of disequilibrium of the climate system in 1990) and subsequently — see Appendix; and the effect of the decreasing intensity of the thermohaline circulation (which slows down the response as it progressively increases the effective thermal inertia of the ocean). The uncertainty range due to $\Delta T_{2\times}$ uncertainties ($-0.6/+0.9°C$) is similar to that given in WR92. Uncertainties due to other factors are relatively small (see Wigley and Raper, 1993).

Figure 4 shows the effect of emissions scenario on the temperature projections for the standard case of $\Delta T_{2\times} = 2.5°C$. The uncertainty range in the year 2100 ($-0.7/+0.6°C$) is similar to the uncertainty due to $\Delta T_{2\times}$. The time-dependence of this uncertainty, however, differs radically from that due to $\Delta T_{2\times}$. In the year 2050, the full emissions-related uncertainty range is only 0.3°C, one half of that due to $\Delta T_{2\times}$. The slow response of CO_2 concentration level to CO_2 emissions changes, and the additional lag in the response of the climate system to external forcing changes, mean that, even for quite large emissions changes, it takes many decades before their impacts on climate become obvious. This effect is even more pronounced for sea level — see below.

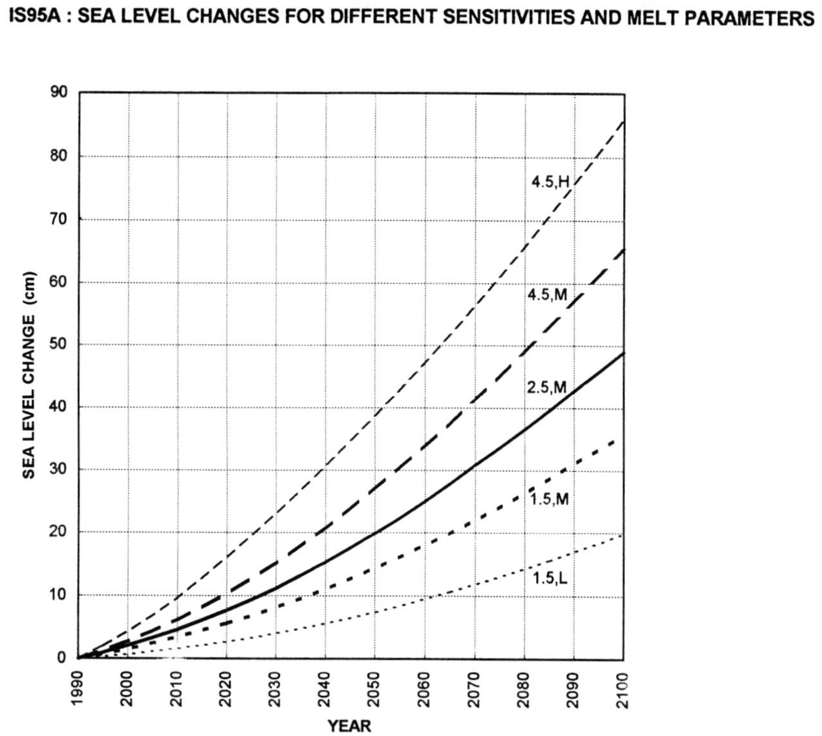

Fig. 5. Projections of global mean sea-level rise for emissions scenario IS95a for a mid-estimate and for a range of ice melt parameter values and for $\Delta T_{2\times} = 1.5, 2.5$ and 4.5°C. Curves are labelled by $\Delta T_{2\times}$ and L, M, H to indicate low, mid, high ice melt parameter values.

Figure 5 is analogous to Figure 3, but for sea level. It shows uncertainties for IS95a projections due to both $\Delta T_{2\times}$ and ice-melt model parameter uncertainties. The central value is around 49 cm for the 1990–2100 change. This is similar to, but a little higher than WR92. As for temperature, this is the combined effect of forcing differences and changes in the thermohaline circulation; but the use of a differential land/ocean climate sensitivity and the different forcing effects of aerosols and tropospheric ozone over land and ocean also have an important effect.

The effect of decreasing upwelling rate is to cause more heat to be injected into the deeper ocean, at the expense of the surface layers. This has compensating effects on sea-level rise: greater ocean heat penetration gives greater expansion (an effect that is partly offset by the influence of differential sensitivity), but the concomitant reduced global mean surface warming leads to less ice melt. Note, however, that since the present calculations use a different GSIC model from that used in WR92, one would not expect the ice melt component here to equal WR92 even if the surface warming were the same. For the same warming, the new model produces slightly larger GSIC melt than the old model (see Wigley and Raper, 1995).

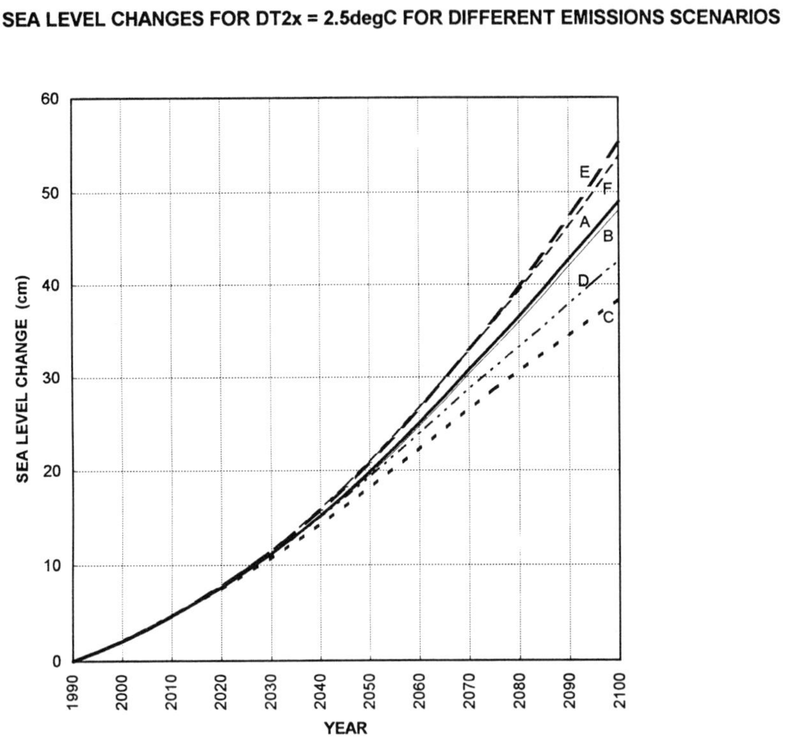

Fig. 6. Projections of global mean sea-level rise for emissions scenarios IS95 a-f. Mid-value model parameters are used.

The inner three curves of Figure 5 show the effect of $\Delta T_{2\times}$ uncertainties on sea-level rise for the mid ice-melt parameter values ($-13/+17$ cm). Ice-melt parameter uncertainties are illustrated by comparing the upper and lower pairs of curves. Changing from mid to high ice-melt parameters increases sea-level rise by some 20 cm for $\Delta T_{2\times} = 4.5$°C, while changing from mid to low ice-melt parameters decreases sea-level rise by around 16 cm for $\Delta T_{2\times} = 1.5$°C. Ice-melt parameter uncertainties are thus comparable in magnitude to those arising from uncertainties in $\Delta T_{2\times}$.

Figure 6 is the sea-level equivalent to Figure 4. It shows that the effect of emissions changes on the course of future sea-level rise is very slow acting — even in the year 2100 the range of sea-level rise (IS95e minus IS95c) is only 19 cm, noticeably less than either $\Delta T_{2\times}$ or ice-melt-parameter related uncertainties. A full assessment of these uncertainties would require projections beyond 2100.

A breakdown of the contributions to sea-level rise is given in Table 4 for IS95a (the percentage breakdown is very similar for all scenarios). It can be seen that thermal expansion is nearly always the dominant contributor, followed by GSIC melt for the mid and high ice-melt-parameter cases. For the low ice-melt cases, Antarctica has a large negative contribution, effectively offsetting the positive melt terms from GSIC and Greenland. In absolute terms, the full uncertainty ranges are

TABLE 4
Percentage contributions to sea-level rise relative to 1990 for IS95a for different $\Delta T_{2\times}$ and low (L), mid (M) and high (H) ice-melt parameter values (ΔT in °C, Δmsl in cm). Not all numbers sum to 100% due to rounding.

Year	ΔT_{2x} °C	L, M or H	ΔT °C	Δmsl cm	expn %	GSIC %	Greenland %	Antarctic %
2020	1.5	L	0.28	2.7	122	35	6	–62
	2.5	M	0.40	7.6	59	33	9	0
	4.5	H	0.56	16.0	38	41	10	12
2050	1.5	L	0.63	7.4	110	35	6	–51
	2.5	M	0.91	19.8	57	34	10	–1
	4.5	H	1.29	38.6	40	38	12	10
2100	1.5	L	1.39	19.8	100	36	7	–43
	2.5	M	2.03	49.0	58	33	12	–3
	4.5	H	2.93	86.0	47	29	16	9

comparable for all components: in 2100 they are, expansion 20 cm, GSIC 18 cm, Greenland 12 cm, Antarctica 16 cm; with an overall uncertainty range of 66 cm (see Fig. 5).

We next show much longer time scale temperature and sea-level changes (out to 2500) for cases where CO_2 concentrations stabilize at different levels. The reason for these calculations arises from the following statement in Article 2 of the Framework Convention on Climate Change, which gives as an ultimate aim of this Convention:

> *... to achieve stabilization of greenhouse gas concentrations ... at a level that would prevent dangerous anthropogenic interference with the climate system ... within a time frame sufficient to allow ecosystems to adapt naturally to climate change, to ensure that food production is not threatened and to enable economic development to proceed in a sustainable manner.*

To understand the implications of this statement, it is necessary both to estimate emissions requirements for concentration stabilization and the attendant climate and sea-level consequences. In order to do this, one of the present authors (Wigley) and Dr. Ian Enting of CSIRO Division of Atmospheric Physics, Australia, were charged by IPCC Working Group I with deriving a set of future concentration profiles in which CO_2 levels eventually stabilized at 350, 450, 550, 650 and 750 ppmv. Enting and Wigley then coordinated a multi-institution carbon cycle modelling exercise to determine the required emissions trajectories and their uncertainties. Full details are given in Enting et al. (1994). Preliminary results for the implied temperature and sea-level changes have been given by Wigley (1995) — here, we update these results.

As part of the 1995/96 IPCC assessment, some of the details underlying these stabilization carbon-cycle calculations have been modified to account for more recent information. In particular, the details of the 1980s-mean carbon budget used to initialise the calculations have been changed as follows: 1980s-mean net deforestation reduced from 1.6 GtC/year to 1.1 GtC/year; and 1980s-mean atmospheric mass build-up reduced from 3.4 GtC/year to 3.3 GtC/year. The latter also involved a revision to the recent part of the standard IPCC CO_2 concentration history, and this in turn required modifications to all the future concentration profiles in order to maintain a smooth transition around 1990. It is these new data that are used here. The revised concentration profiles (Figure 7) differ only marginally from those given in Enting et al. (1994) and used by Wigley (1995). We note, however, that the implied emissions are unrealistic in economic terms and that IPCC is, at the time of writing, devising a new set of stabilization profiles that differs noticeably from those used here. We have carried out sensitivity studies that show that the subsequent effects on future temperature and sea level are small — in other words, it is the stabilization level rather than the pathway to stabilization that is the primary control on the temperature and sea level results.

Other important changes have been made relative to the calculations presented in Wigley (1995). First, the results we now give are for CO_2 alone — i.e., they use the full best-guess history of radiative forcing to 1990, and then assume that CO_2 changes are the only post-1990 forcing agent. In Wigley (1995), based on Wigley and Raper (1992), the post-1990 CO_2 forcing was scaled up by 23% to account for the net effect of other gases and aerosols. The present calculations, therefore, isolate the CO_2 effect. We do not attempt to predict the future forcing contributions from other sources — although we note that these are likely to be positive at least out to 2100.

Second, the climate model has been changed to include land/sea differences in the climate sensitivity and a temperature-related slow down in the thermohaline circulation, as described above. Third, the GSIC model has been revised radically from the Wigley and Raper (1993) model used in Wigley (1995). Finally, although there are five different stabilization profiles, we only give full results for two profiles, those stabilizing at 450 ppmv in the year 2100 (S450) and at 650 ppmv in the year 2200 (S650). A selection of other results is given in tabular form.

Figure 8 shows global mean temperature variations for S450 and S650 for different values of the climate sensitivity ($\Delta T_{2\times} = 1.5, 2.5$ and $4.5°C$). Note that temperatures continue to rise after the points of stabilization (dots on the Figure), and that equilibrium is only reached well after 2500. This continued warming occurs because, at the point of stabilization (after which there are no radiative forcing changes) the climate system is generally quite far removed from equilibrium due to oceanic lag effects. Subsequent warming represents the (slow) approach to equilibrium. This lag effect or "warming commitment" is more pronounced for higher stabilization levels and higher climate sensitivities.

Both the lag between forcing and response and, consequently, the warming commitment are noticeably enhanced by the decreasing strength of the thermohaline circulation, as shown for S650 in Figure 9. This effect is particularly pronounced at high climate sensitivities: for $\Delta T_{2\times} = 2.5°C$, the variable upwelling rate lags behind the constant upwelling rate case in the year 2200 by 0.25°C, while

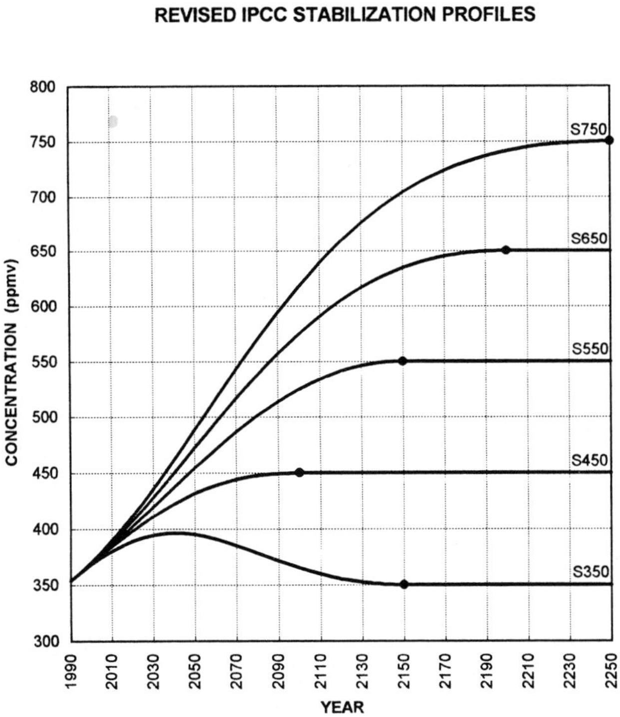

Fig. 7. CO_2 concentration profiles for stabilization at 350, 450, 550, 650 and 750 ppmv. Dots show the stabilization dates.

for $\Delta T_{2\times} = 4.5°C$ this difference is 0.72°C. We know of no other calculations comparable to these, but we note that a similar slow and prolonged approach towards equilibrium for the decreasing upwelling rate case has been obtained by Manabe and Stouffer (1994) using the GFDL coupled ocean-atmosphere GCM.

Results for sea level equivalent to Figures 8 and 9 are shown in Figures 10 and 11. These are quite startling results — they show that, for sensitivities of around 2°C and higher, halting the increase in radiative forcing has only a very small effect on the sea-level rise trend. Sea level continues to rise at almost the same rate for at least three centuries beyond the stabilization point, mainly because of the very long timescales associated with the responses of oceanic thermal expansion (which involves the whole ocean column) and the Greenland ice sheet (see Wigley, 1995). In other words, the sea-level rise commitment after stabilization of CO_2 concentration (i.e., of radiative forcing) is dramatically greater than the corresponding warming commitment, a point noted previously by Warrick and Oerlemans (1990), Wigley and Raper (1993) and Wigley (1995).

Figure 10 shows that there are enormous uncertainties in the projected sea-level changes for stabilization arising from both climate sensitivity and ice-melt parameter uncertainties. At the high end of possibilities, the global mean sea-level rises by over 3 m over 1990–2500 for S650 (around 2 m for S450). At the low

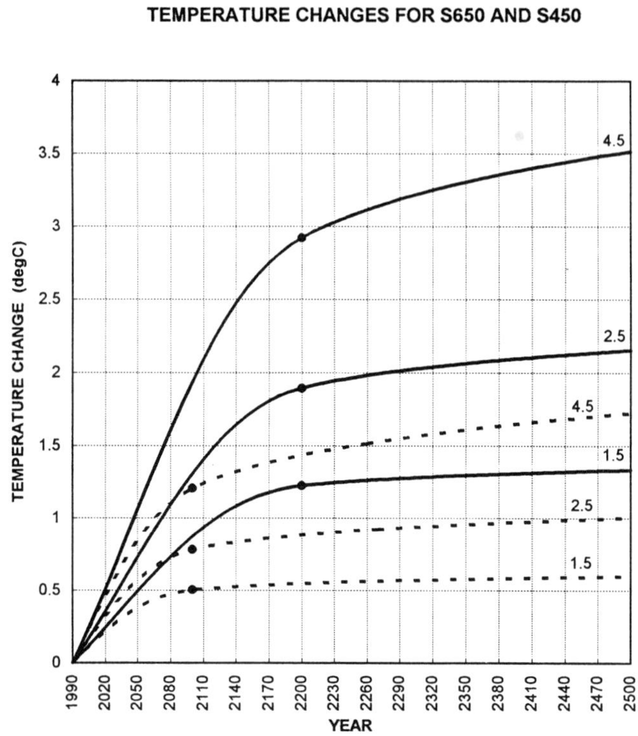

Fig. 8. Projections of global mean temperature change for profiles of CO_2 concentration stabilizing at 450 ppmv (dashed lines) and 650 ppmv (solid lines). Dots denote the dates of CO_2 concentration stabilization. Results are for $\Delta T_{2\times} = 1.5$, 2.5 and 4.5°C.

end, the sea-level rise over this period is only 30 cm for S650, less for S450. Figure 11 shows that these results are affected by the assumption of a decreasing thermohaline circulation, but this influence is relatively small.

Finally, we give results for the other stabilization profiles (S350, S550 and S750), and compare these with the results of Wigley (1995) in Table 5. For almost all cases, the new results give higher sea-level rise projections than the earlier results, in spite of the fact that the post-1990 forcing in these earlier calculations was enhanced by 23% relative to that used here — although the changes are still small relative to the overall uncertainties. The main reasons for the larger values

TABLE 5

Uncertainties in global mean sea-level projections (cm) arising from model parameter uncertainties. Low, mid and high values assume climate sensitivities of 1.5, 2.5, and 4.5°C, respectively, best-guess values for other climate model parameters, and low, mid and high values for ice-melt model parameters. Results in brackets are as given in Wigley (1995). Revised 2 October 1995.

Stabilization profile	**Year**	**Low**		**Mid**		**High**	
S350	2100	4	(0)	20	(16)	44	(39)
	2200	1	(-7)	25	(14)	60	(49)
	2300	-3	(-13)	30	(17)	76	(63)
	2400	-6	(-18)	34	(19)	93	(77)
	2500	-10	(-23)	38	(21)	109	(92)
S450	2100	8	(4)	29	(26)	58	(56)
	2200	11	(1)	48	(37)	98	(92)
	2300	11	(-4)	62	(45)	134	(125)
	2400	10	(-10)	74	(52)	169	(157)
	2500	7	(-16)	84	(59)	202	(189)
S550	2100	11	(7)	34	(32)	65	(65)
	2200	19	(7)	64	(52)	123	(121)
	2300	22	(2)	84	(65)	173	(169)
	2400	22	(-3)	101	(77)	221	(216)
	2500	20	(-10)	117	(87)	269	(260)
S650	2100	13	(9)	37	(36)	70	(72)
	2200	25	(12)	74	(63)	139	(142)
	2300	30	(8)	101	(81)	202	(203)
	2400	31	(2)	123	(96)	263	(261)
	2500	30	(-5)	142	(109)	323	(318)
S750	2100	14	(11)	41	(40)	75	(78)
	2200	30	(15)	82	(72)	152	(158)
	2300	37	(12)	114	(94)	227	(231)
	2400	39	(7)	140	(112)	299	(300)
	2500	38	(0)	163	(128)	371	(366)

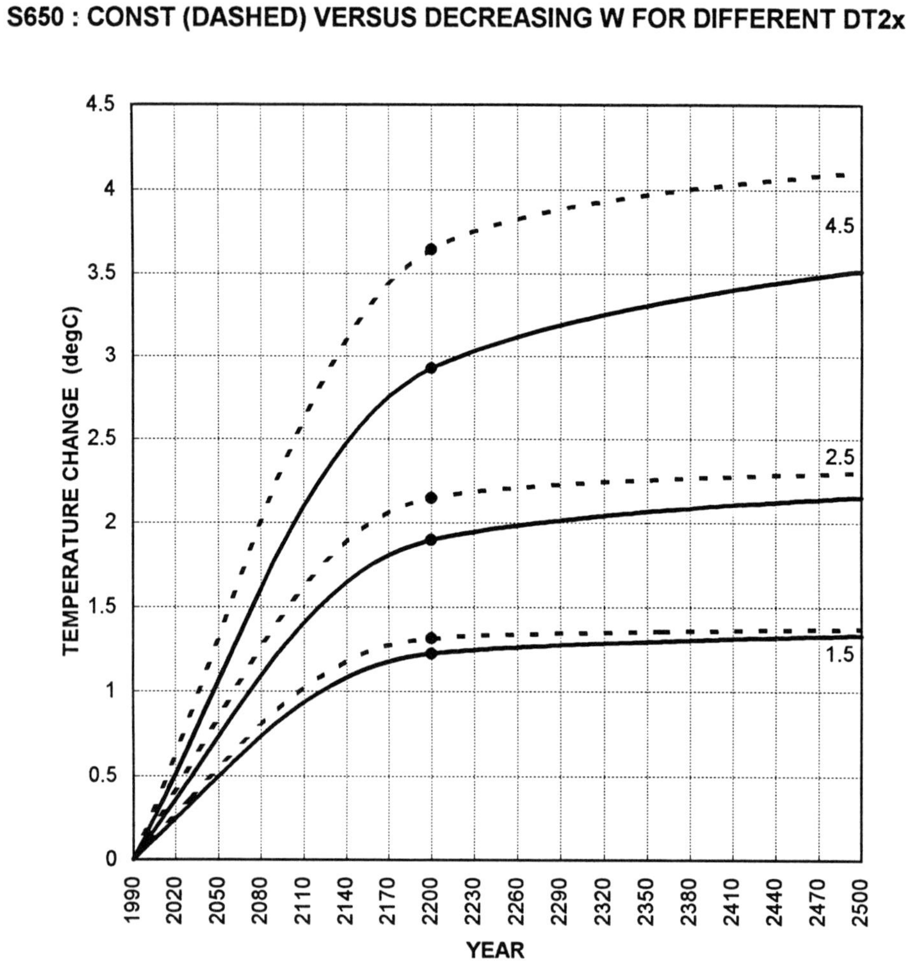

Fig. 9. Projections of global mean temperature change for the S650 stabilization profile, comparing the effect of constant (dashed lines) and declining (solid lines) thermohaline circulation. Results are shown for $\Delta T_{2\times} = 1.5$, 2.5 and 4.5°C.

are the changed ice-melt and climate model details, particularly, in the latter, the effect of the decreasing intensity of the thermohaline circulation.

Summary and Conclusions

We have reviewed observationally based estimates of global mean temperature change and sea-level rise over 1880 to 1990. We conclude that the temperature has risen by 0.5±0.15°C and sea level has risen by about 10–20 cm.

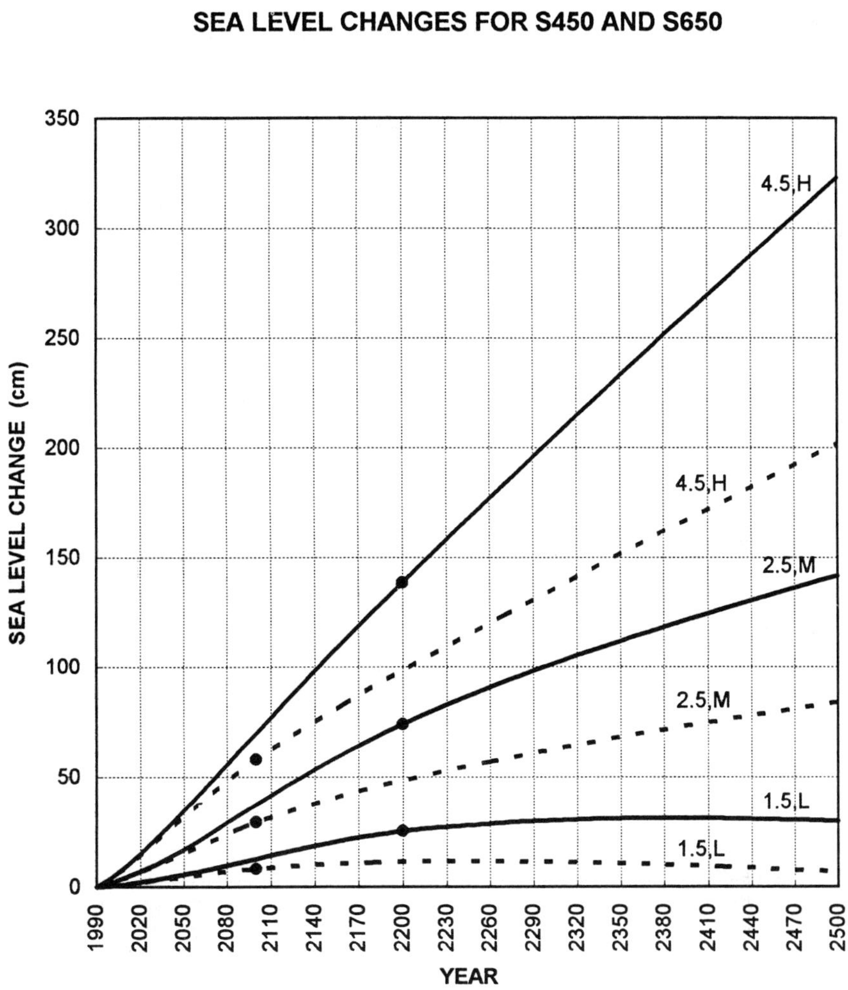

Fig. 10. As in Figure 8 but for sea-level rise. S450 results are shown dashed, S650 by full lines. Curves are labelled by $T_{2\times}$ (1.5, 2.5 or 4.5°C) and ice-melt parameter values (L, M or H).

Model-based estimates of past and future global mean temperature change and sea-level rise are presented. The climate model has been changed from earlier versions to include land/ocean differences in the climate sensitivity and a temperature change-related slow down in the thermohaline circulation. The glacier and small ice cap melt model has been revised radically from the Wigley and Raper (1993) model. The estimates for change over 1880–1990 are a warming of 0.21–0.40°C and a sea-level rise of −1–17 cm. Note that the ice melt component of the sea-level rise is driven by global mean temperature change; regional changes over the ice masses will differ from this. Considering this, and the fact that the model results only give the anthropogenic effect, while the observations include, additionally, the

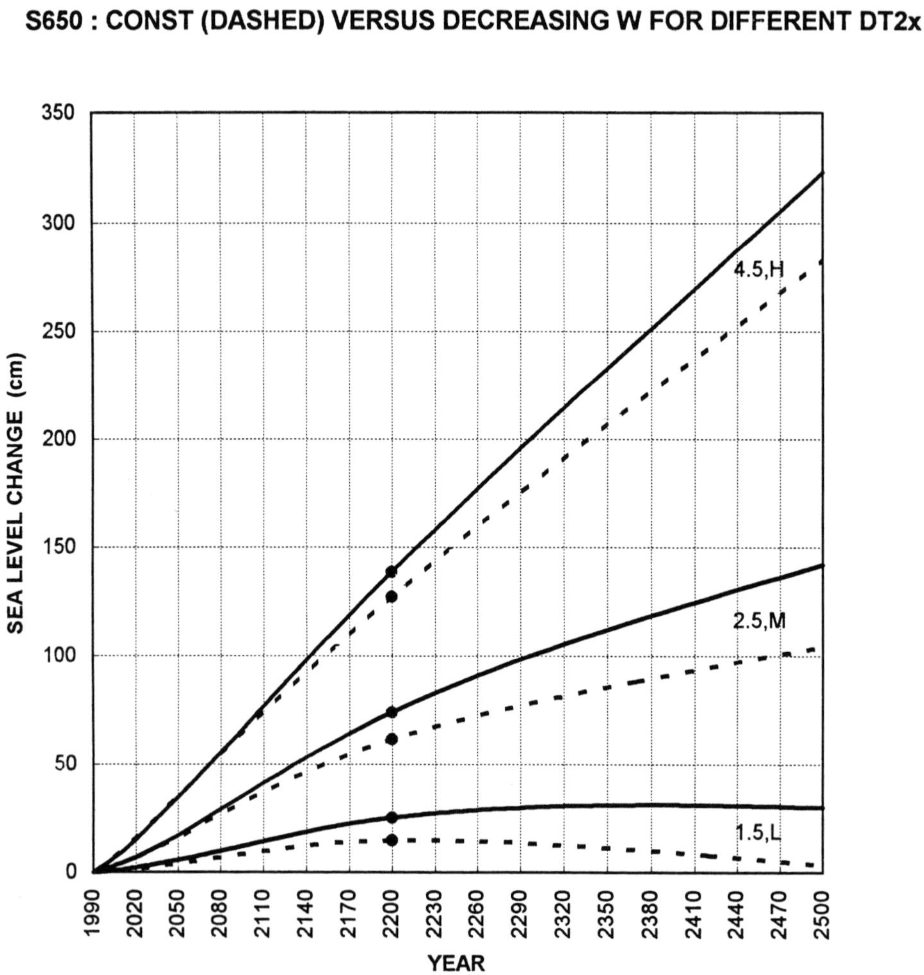

Fig. 11. As in Figure 9 but for sea-level rise. Dashed lines indicate constant upwelling rate, full lines indicate variable upwalling rate. Results are shown for low, mid and high $\Delta T_{2\times}$, ice-melt parameter combinations.

effects of natural variability, we conclude that the agreement between the observed and modelled results is satisfactory.

Results are given for a range of future emissions scenarios over 1990–2100. The warming amounts are generally a little less than in Wigley and Raper (1992). For example, for scenario IS95a with $\Delta T_{2\times} = 2.5°C$, the 1990–2100 temperature change is 2.0°C compared with the previous 2.5°C. This is due to different forcing as well as the changes to the climate and sea-level models. Uncertainty ranges due to $\Delta T_{2\times}$ uncertainties and emission scenarios are, respectively, $-0.6/+0.9°C$ and $-0.7/+0.6°C$. The central estimate for sea-level rise over 1990–2100 is 49 cm. This is similar to, but a little higher than that given in Wigley and Raper (1992).

The uncertainty range in 2100, due to $\Delta T_{2\times}$ uncertainties together with ice melt parameter uncertainties, is 20–86 cm for the IS95a forcing.

These results are for an assumed aerosol-forcing magnitude in 1990 that follows the latest IPCC recommendations. It is possible that this aerosol-forcing is too large, since it leads to a warming over 1880–1990 that is noticably less than observed. However, part of the observed warming may be due to natural factors. To account for aerosol uncertainties, results for a range of forcing possibilities are given in the Appendix.

Finally, we show much longer time scale temperature and sea-level changes (out to 2500) for cases where CO_2 concentrations stabilize at different levels. These results differ from those given in Wigley (1995) because of both forcing and model differences. We note that the 'warming commitment' after stabilization of CO_2 concentration is more pronounced for higher stabilization levels and higher climate sensitivities. Also, the 'warming commitment' is enhanced noticeably by the decreasing strength of the thermohaline circulation.

Sea-level rise results for scenarios of CO_2 concentration stabilization show that, for $\Delta T_{2\times}$ values of around 2°C and higher, halting the increase in radiative forcing has only a very small effect on the sea-level rise trend. Sea level continues to rise at almost the same rate for at least three centuries beyond the stabilization date.

Acknowledgements

This work has been supported by the European Community under Grant No. EV5V-CT91-0051 and by the U.S. Department of Energy, under Grant No. DE-FG02-86ER60397.

References

Barnett, T.P., 1983. Recent changes in sea level and their possible causes. *Climatic Change*, 5, 15-38

Barnett, T.P., 1984. Estimation of "global" sea-level change: a problem of uniqueness. *Geophys. Res.* , 89, 7980-7988.

Barnett, T.P., 1988. Global sea level. In: NCPO, Climate Variations over the past century and the Greenhouse effect. A report based on the first Climate trends Workshop, 7-9 Sept., 1988, Washington D.C., National Climate Program Office/NOAA, Rockville, Maryland.

Bradley, R.S., 1988. The explosive volcanic eruption signal in northern hemisphere continental temperature records. *Climate Change* , 12, 221-243.

Bretherton, F.P., Bryan, K. and Woods, J.D., 1990. Time-dependant greenhouse-gas-induced climate change. In: J.T. Houghton, G.J. Jenkins and J.J. Ephraums (eds.), *Climate Change. The IPCC Scientific Assessment* , Cambridge Univ. Press, 175-193.

Charlson, R.J., Schwartz, S.E., Hales, J.M., Cess, R.D., Coakley, J.A.Jr., Hansen, J.E., Hofman, D.J., 1992. Climate forcing by anthropogenic aerosols. *Science*, 255, 423-430.

Cubasch, U., Hasselmann, K., Höck, H., Maier-Reimer, E., Kolajewiczu, M., Santer, B.D., Sausen, R., 1992. Time-dependent greenhouse warming computations with a coupled ocean-atmosphere model. *Climate Dynamics*, 8, 55-69.

Douglas, B., 1991. Global Sea Level Rise. *J. Geophys. Res.*, 96, 6981-6992.

Enting, I.G., Wigley, T.M.L. and Heiman, M., 1994. Future emissions and concentrations of carbon dioxide: key ocean/atmosphere/land analyses. CSIRO Aust. Div. Atmos. Res. Tech. Pap. No. 31, 1-118.

Folland, C.K., Karl, T.R. and Vinnikov, K. Ya., 1990. Observed climate variations and change. In: J.T. Houghton, G.J. Jenkins and J.J. Ephraums (eds.)*Climate Change. The IPCC Scientific Assessment*, Cambridge Univ. Press, 195-238.

Folland, C.K., Nicholls, N., Nyenzi, B.S., Parker, D.E., Vinnikov, K.Ya., 1992. Observed climate variability and change. In: J.T. Houghton, B.A. Callander and S.K. Varney (eds.), *Climate change 1992: The Supplementary Report to the IPCC Scientific Assessment.* Cambridge Univ. Press, 135-170.

Folland, C.K. and Parker, D.E., 1995. Correction of instrumental biases in historical sea surface temperature data. *Q.J.R. Meteorol. Soc.*, 121, 319-367.

Gill, A.E., 1982. Atmosphere-Ocean Dynamics. Academic Press, New York, 662pp.

Gornitz, V., 1994. Monitoring sea level changes. *Climate Change* (submitted).

Gornitz, V., Lebedeff, S. and Hansen, J., 1982. Global sea level trends in the past century. *Science*, 215, 1611-1614.

Gornitz, V. and Lebedeff, S., 1987. Global sea level changes during the past century. In: D. Nummedal, O.H. Pilkey and J.D. Howard (eds.), *Sea-level fluctuation and coastal evolution.* Soc. of Econ. Paleont. Mineral. Spec. Publ. 41.

Gregory, J.M., 1993. Sea level changes under increasing atmospheric CO_2 in a transient coupled ocean-atmosphere GCM experiment. *J. Climate*, 6, 2247-2262.

Grove, J.M., 1988. *The Little Ice Age.* Methuen, London.

Hansen, J. and Lebedeff, S., 1987. Global trends of measured surface air temperature. *J. Geophys. Res.*, 92, 13345-13372.

Harvey, L.D. and Schneider, S.H., 1985. Transient climate response to external forcing on 10^0-10^4 year time scales Part 1: Experiments with globally-averaged, coupled, atmospheric and ocean energy balance models. *J. Geophys. Res.* 90, 2191-2205.

Hasselmann, K., Sausen, R., Maier-Reimer, E., Reinhard, V., 1993. On the cold start problem in transient simulations with coupled atmosphere - ocean models. *Climate Dynamics*, 9, 53-61.

Hoffert, M.I., Callegari, A.J. and Hsieh, C.-T., 1980. The role of deep sea heat storage in the secular response to climate forcing. *J. Geophys. Res.*, 85, 6667-6679.

Hoffert, M.I. and Flannery, B.P., 1985. Model projections of the time-dependent response to increasing carbon dioxide. In: M.C. MacCracken and F.M. Luther (eds.), Projecting the Climatic Effects of Increasing Carbon Dioxide. US Department of Energy, Carbon Dioxide Research Division, pp.149-190.

Houghton, J.T., Callander, B.A. and Varney, S.K., (eds.) 1992. *Climate Change 1992. The Supplementary Report to the IPCC Scientific Assessment.* Cambridge Univ. Press, 200pp.

Houghton, J.T., Jenkins, G.J. and Ephraums, J.J., (Eds.) 1990. *Climate Change. The IPCC Scientific Assessment.* Cambridge University Press, 364pp.

Jones, P.D., 1990. Antarctic Temperatures over the Present Century - A Study of the Early Expedition Record. *J. Climate*, 3, 1193-1203.

Jones, P.D., 1994. Hemispheric surface air temperature variations: A reanalysis and an update to 1993. *J. Climate*, 7, 1794-1802.

Jones, P.D. and Briffa, K.R., 1992. Global surface air temperature variations over the twentieth century. Part 1: Spatial, temporal and seasonal details. *Holocene*, 2, 174-188.

Jones, P.D., Kelly, P.M., Goodess, C.M. and Karl, T.R., 1989. The effect of urban warming on the northern hemisphere temperature average. *J. Climate*, 2, 285-290.

Jones, P.D., Groisman, P.Ya., Coughlan, M., Plummer, N., Wang, W-C. and Karl, T.R., 1990. Assessment of urbanisation effects in the time series of surface air temperature over land. *Nature*, 347, 169-172.

Jones, P.D. and Wigley, T.M.L., 1990. Global warming trends. *Scientific American*, 263, 84-91.

Jones, P.D., Wigley, T.M.L. and Farmer, G., 1991. Marine and land temperature data sets: a comparison and a look at recent trends. In: M.E. Schlesinger (ed.). *Greenhouse-gas-induced Climatic Change: A Critical Appraisal of Simulations and Observations.* Elsevier Science Publ., Amsterdam (in press).

Kelly, P.M. and Wigley, T.M.L., 1990. The influence of solar forcing trends on global mean temperature since 1861. *Nature*, 347, 460-462.

Kelly, P.M. and Wigley, T.M.L., 1992. Solar cycle length, greenhouse forcing and global climate. *Nature*, 360, 328-330.

Klige, R.K., 1982. Oceanic level fluctuations in the history of the earth in sea and oceanic level fluctuations for 15,000 years. *Acad. Sc. USSR, Inst. Geogr., Moscow, Nauka* (in Russian) 11-22.

Komar, P.D. and Enfield, D.B., 1987. Short-term sea-level changes and shoreline erosion. In: D. Nummedal, O.H. Pikey and J.D. Howard (eds.), *Sea-Level Fluctuation and Coastal Evolution*. Soc. of Econ. Paleont. Mineral. Spec. Publ. 41, 17-27.

Leggett, J.A., Pepper, W.J. and Swart, R.J., 1992. Emission Scenarios for IPCC: An update. In: J.T. Houghton, B.A. Callander and S.K. Varney (eds.), *Climate Change 1992. The Supplementary Report to the IPCC Scientific Assessment*. Cambridge Univ. Press, 69-95.

Leuenberger, M. and Siegenthaler, U., 1992. Ice-age atmospheric concentration of nitrous oxide from an Antarctic ice core. *Nature*, 360, 449-451.

Levitus, S., 1982. Climatological atlas of the world oceans. NOAA Professional Paper, 13. U.S. Gov. Printing Office, Washington, D.C.

Lisitzin, E., 1974. *Sea Level Changes*. Elsevier Publ., New York, 286pp.

Manabe, S. and Stouffer, R.J., 1993. Century-scale effects of increased atmospheric CO_2 on the ocean-atmosphere system. *Nature*, 364, 215-217.

Manabe, S. and Stouffer, R.J., 1994. Multiple-century response of a coupled ocean-atmosphere model to an increase of atmospheric carbon dioxide. *J. Climate*, 7, 5-23.

Maul, G.A. and Hanson, K., 1991. Interannual coherence between North Atlantic atmospheric surface pressure and composite southern U.S.A. sea level. *Geophys. Res. Letters*, 18, 653-656.

Meier, M.F., 1984. Contributions of small glaciers to global sea level. *Science*, 226, 1418-1421.

Mitchell, J.F.B., Johns, T.C., Gregory, J.M. and Tett, S.F.B., 1995. Climate response to increasing levels of greenhouse gases and sulphate aerosols. *Nature*, 376, 501-504.

Mitchell, J.F.B., Manabe, S., Meleshko, V. and Tokioka. T., 1990. Equilibrium Climate Change - and its Implications for the Future. In: J.T. Houghton, G.J. Jenkins and J.J. Ephraums (eds.), *Climate Change. The IPCC Scientific Assessment*. Cambridge Univ. Press, 131-172.

Mitchell, J.F.B. and Gregory, J.M., 1992. Climatic consequences of emissions and a comparison of IS92a and SA90. In: J.T. Houghton, B.A. Callander and S.K. Varney (eds.), *Climate Change 1992: The Supplementary Report to the IPCC Scientific Assessment.* Cambridge Univ. Press, 171-175.

Murphy, J.M., 1994. Transient response of the Hadley Centre coupled ocean-atmosphere model to increasing carbon dioxide Part III. Analysis of global mean response using simple models. *J. Climate* (in press).

Oerlemans, J., 1989. A projection of future sea level. *Climatic Change*, 15, 151-174.

Oerlemans, J. and Fortuin J.P., 1992. Sensitivity of glaciers and small ice caps to greenhouse warming. *Science*, 258, 115-116.

Osborn, T.J. and Wigley, T.M.L., 1994. A simple model for estimating methane concentration and lifetime variations. *Climate Dynamics*, 9, 181-193.

Parker, D.E., Jones, P.D., Folland, C.K., Bevan, A., 1994. Interdecadal changes of surface temperature since the late nineteenth century. *J. Geophys. Res.*, 99, 14,373-14,399.

Peltier, W.R. and Tushingham, A.M., 1989. Global sea level rise and the greenhouse effect: might they be connected? *Science*, 244, 806-810.

Peltier, W.R. and Tushingham, A.M., 1991. The influence of glacial isostatic adjustment on tide gauge measurements of secular sea level. *J. Geophys. Res.*, 96, 6779-6796.

Pirazzoli, P.A., Grant, D.R. and Woodworth, P., 1987. Trends of relative sea level change: past, present and future. XII International INQUA Congress, Ottawa, July 31 - August 9, Special Session 18: Global Change.

Pugh, D.T., Spencer, N.E. and Woodworth, P.L., 1987. Data holdings of the Permanent Service for Mean Sea Level, Bidston, Birkenhead, Permanent Service for Mean Sea Level, 156pp.

Raper, S.C.B., Briffa, K.R., Wigley, T.M.L., 1995. Glacier changes in northern Sweden from A.D. 500: A simple geometric model of Storglaciaren. *J. Glaciology* (submitted).

Raper, S.C.B. and Cubasch, U., 1995. Emulation of the results from a coupled general circulation model using a simple climate model. *Geophys. Res. Letts.* (submitted).

Röthlisberger, F., 1986: 10 000 Jahre Gletchergeschichte der Erde. Aarau-Verlag Säuerlander.

Santer, B.D., Mikolajewicz, Brüggemann, W., Cubasch, U., Hasselmann, K., Hock, H., Maier-Reimer, E. and Wigley, T.M.L., 1995. Ocean variability and its influence on the detectability of greenhouse warming signals. *J. Geophys. Res. (Oceans).* submitted.

Schimel, D., Enting, I., Heimann, M., Wigley, T.M.L., Joos, F., Raynaud, D., Alves, D. and Siegenthaler, U., 1995. CO_2 and the carbon cycle. In: J.T. Houghton, L.G. Meira Filho, J.P. Bruce, Hoesung Lee, B.A. Callander, E.F. Haites, N. Harris and K. Maskell (eds.), Radiative Forcing of Climate and an Evaluation of the IPCC IS92 Emissions Scenarios. Cambridge Univ. Press (in press).

Schlesinger, M.E. and Ramankutty, N., 1992. Implications for global warming of intercycle solar irradiance variations. *Nature*, 360, 330-333.

Sear, C.B., Kelly, P.M., Jones, P.D., Goodess, C.M., 1987. Global surface-temperature responses to major volcanic eruptions. *Nature*, 330, 365-367.

Shine, K.P., Derwent, R.G., Wuebbles, D.J., Moncrette, J-J., 1990. Radiative forcing of climate. In: J.T. Houghton, G.J. Jenkins and J.J. Ephraums (eds.), *Climate Change. The IPCC Scientific Assessment.* Cambridge Univ. Press,41-68.

Shine, K.P., Fouquart, Y., Ramaswamy, V., Solomon, S. and Srinivasan, J., 1995. Radiative forcing of climate. In: J.T. Houghton, L.G. Meira Filho, J.P. Bruce, Hoesung Lee, B.A. Callander, E.F. Haites, N. Harris and K. Maskell (eds.), *Radiative Forcing of Climate and an Evaluation of the IPCC IS92 Emissions Scenarios.* Cambridge Univ. Press (in press).

Siegenthaler, U. and Joos, F., 1992. Use of a simple model for studying oceanic tracer distributions and the global carbon cycle. *Tellus*, 44B, 186-207.

Trupin, A. and Wahr, J., 1990. Spectroscopic analysis of global tide gauge sea level data. *Geophys. J. Int.*, 100, 441-454.

Warrick, R.A. and Oerlemans, J., 1990. Sea Level Rise. In: J.T. Houghton, G.J. Jenkins and J.J. Ephraums (eds.), *Climate Change. The IPCC Scientific Assessment*. Cambridge Univ. Press, 257-281.

Wigley, T.M.L., 1989. Possible climate change due to SO_2-derived cloud condensation nuclei. *Nature*, 339, 365-367.

Wigley, T.M.L., 1993. Balancing the carbon budget: Implications for projections of future carbon dioxide changes. *Tellus*, 45B, 409-425.

Wigley, T.M.L., 1994. The contribution from emissions of different gases to the enhanced greenhouse effect. In: T. Hanisch (ed.), *Climate Change and the Agenda for Research*. Westview Press, Boulder, 193-222.

Wigley, T.M.L., 1995. Global-mean temperature and sea level consequences of greenhouse gas concentration stabilization. *Geophys. Res. Letters*, 22, 45-48.

Wigley, T.M.L. and Barnett, T., 1990. Detection of the Greenhouse Effect in the Observations. In: J.T.Houghton G.J.Jenkins and J.J.Ephraums (eds.), *Climate Change. The IPCC Scientific Assessment*. Cambridge Univ. Press, 239-255.

Wigley, T.M.L., Hall, A., Raper, S.C.B., Warrick, R.A., 1995. Past sea level contribution from glacier melt using a regional degree-day model. *J. Geophys. Res.*, (submitted).

Wigley, T.M.L. and Raper, S.C.B., 1987. Thermal expansion of sea water associated with global warming. *Nature*, 330, 127-131.

Wigley, T.M.L. and Raper, S.C.B., 1990. Natural variability of the climate system and detection of the greenhouse effect. *Nature*, 344, 324-327.

Wigley, T.M.L. and Raper, S.C.B., 1991a. Internally Generated Natural Variability of Global-Mean Temperatures. In: M.E. Schlesinger (ed.), *Greenhouse-gas-induced Climatic Change: A Critical Appraisal of Simulations and Observations*. Elsevier Science Publ., Amsterdam.

Wigley, T.M.L. and Raper, S.C.B., 1991b. Detection of the Enhanced Greenhouse Effect on Climate. In: J. Jager and H. Ferguson (eds.), *Climate Change: Science, Impacts and Policy*. Cambridge Univ. Press, 231-242.

Wigley, T.M.L. and Raper, S.C.B., 1992. Implications for climate and sea level of revised IPCC emissions scenarios. *Nature*, 357, 293-300.

Wigley, T.M.L. and Raper, S.C.B., 1993. Future changes in global mean temperature and thermal expansion-related sea level rise. In: R.A. Warrick, E.M. Barrow and T.M.L. Wigley (eds.), *Climate and Sea Level Change: Observations, Projections and Implications*. Cambridge Univ. Press, 111-133.

Wigley, T.M.L. and Raper, S.C.B., 1995. An Heuristic Model for sea level rise due to the melting of small glaciers. *Geophys. Res. Letters* 22, 2749–2752.

Woodworth, P.L., 1990. A search for accelerations in records of European mean sea level. *Intern. J. Climatology*, 10, 129-143.

Appendix: Radiative Forcing

The radiative forcing values used here are based on those being used by the Intergovernmental Panel on Climate Change (IPCC). For background, the reader is referred to the relevant IPCC reports, and to Wigley and Raper (1992) and Wigley (1994). At the time of writing (April 1995), a set of procedures and values has been agreed by the IPCC Working Group I scientists involved in the production of the 1995/6 second assessment of the climate change issue (a group that includes the present authors). This assessment will not be published until 1996 and minor changes may still be made.

For CO_2, the past concentration history is the standard IPCC data set, an update of that derived for an IPCC inter-model comparison exercise by Wigley and Enting (Enting et al., 1994). The revision only affects the most recent values, changed to give a 1980s-mean atmospheric mass increase of 3.3 GtC/year, the value given in the 1995/6 IPCC report (an update from Schimel et al., 1995). The precise mean is 3.28 GtC/year.

For the future, we use the model of Wigley (1993). IPCC uses the model of Siegenthaler and Joos (1994), but the differences between these are negligible. Both models use the initialization procedure devised for the IPCC inter-model comparison exercise described by Enting et al. (1994), updated to use a different 1980s-mean carbon budget. The current budget has a 1980s-mean net deforestation flux of 1.1 GtC/year and the above atmospheric mass increase. The CO_2-fertilisation factor is adjusted to give a balanced 1980s-mean budget, as described in Wigley (1993b) and Enting et al. (1994).

For CH_4, the concentration history is that used by Osborn and Wigley (1994). Future projections are made using the Osborn–Wigley model with parameter values adjusted to match closely the concentration projections being used by IPCC. In terms of 1990–2100 radiative forcing, the differences are negligible, ranging from 0.00 W/m^2 to 0.03 W/m^2 depending on emissions scenario. In calculating future concentrations we follow IPCC in assuming the emissions of CO, NO_x and VOCs remain constant at 1990 levels. Although IPCC92 gives scenarios for the future emissions of these gases, these are not used because their influence on methane lifetime depends on the spatial details of the emissions, details that are not specified. In addition to direct CH_4 forcing, we also include the indirect effects of CH_4-related increases in stratospheric water vapour and tropospheric ozone. The former follows Shine et al. (1990), but with a scaling factor reduced from 0.30 to 0.05 following Shine et al. (1995). The latter assumes a global-mean ozone increase of 4 ppbv since pre-industrial times due to CH_4 increases and a radiative forcing sensitivity ($\delta Q/\delta C$) of 0.02 W/m^2 per ppbv.

A "direct" tropospheric ozone term due to emissions other than CH_4 is also included. This is assumed to parallel fossil fuel emissions to 1990 and remain constant thereafter. The forcing for this component is assumed to occur only in the Northern Hemisphere and to be equivalent to a global-mean forcing to 1990 of 0.32 W/m^2 (giving a total tropospheric ozone forcing of 0.40 W/m^2). This forcing component is split 90/10 between the land and ocean areas, cf. aerosol forcing — see below.

For N_2O, the history is based on ice-core data and, more recently, observed data. The initial (pre-industrial) value is taken as 270 ppbv based on Leuenberger and Siegenthaler (1992) and more recent unpublished results from D. Raynaud. Future concentration projections assume a lifetime of 120 years.

For halocarbons, the suite of gases used is that used by Wigley and Raper (1992). To account for stratospheric ozone depletion effects, we use the chlorine loading method devised by these authors, but with a non-linear exponent (the ozone term is proportional to $(nCl)^{1.7}$ rather than $(nCl)^{1.0}$) and a higher bromine weighting (Wigley, 1994). This gives a 1990 direct halocarbon forcing of 0.27 W/m^2, offset by 0.16 W/m^2 due to stratospheric ozone depletion; values in accord with IPCC (Shine et al., 1995).

The same stratospheric ozone model is used for past and future changes. Only one emissions scenario is used, that corresponding to IS92d,e. This scenario is the one that corresponds most closely to the latest (Copenhagen) amendment of the Montreal Protocol. Emissions reductions for CFC11 and CFC12 in this scenario are probably slower than should occur under the Protocol, but this has a negligible effect on the forcing results.

The most complex and uncertain part of the radiative forcing is that due to aerosols; aerosols from biomass burning and from fossil fuel combustion. For the former, we assume a total forcing of -0.2 W/m^2 in 1990, with the history following that of gross deforestation (see Wigley, 1993, for more on the differences between gross and net deforestation). After 1990, this forcing value is kept constant. We also assume spatially uniform forcing for this component, although there is evidence to suggest that the forcing should be greater over land than ocean. For fossil-fuel-derived aerosols we assume a 1990 direct (clear-sky) effect of -0.3 W/m^2, a value that accounts for the offsetting effect of soot. This term is distributed between the hemispheres in the ratio NH/SH = 4/1, and between land and ocean in the ratio 9/1, based on a range of three-dimensional modelling results.

The indirect forcing (through aerosol-induced changes in cloud albedo) is much more uncertain. Shine et al. (1995) give a range of possible global-mean values of 0.0 to -1.5 W/m^2. For the present calculations, three values have been used (0.0, -0.4 and -0.8 W/m^2) to give an indication of the range of uncertainty. The main results presented here are based on a forcing of -0.8 W/m^2, to accord with the IPCC's main calculations. Indirect forcing is split NH/SH = 2/1 and land/ocean = 9/1 based on three-dimensional model results.

A choice of -0.4 W/m^2 for the 1990 indirect aerosol forcing gives a global warning over 1880–1990 more in accord with observations, see Table A1. This Table shows results for global-mean changes for four different values of the climate sensitivity and four different aerosol possibilities: zero aerosol forcing, and biomass and direct forcing with indirect forcing in 1990 set at 0.0 ("D+B" in Table), -0.4 ("Low I") and -0.8 W/m^2 ("Mid I"). Temperature changes are split into results for four "boxes" (NH land, NH ocean, SH land and SH ocean). While the "low I" case gives a global-mean value in the range 0.5 ± 0.15°C, the split between the boxes is more even and more consistent with observations for the "Mid I" aerosol case. In all cases, however, the "Mid I" values are less than 0.40°C for the global-mean temperature change. In our judgement, this does not make the "Mid I" forcing case less likely, because of the uncertain magnitude of any natural warming component. IPCC uses the "Mid I" case, and we follow them.

These simple calculations do not preclude higher values of the indirect (or direct) aerosol forcing — to do this more convincingly would require using a more sophisticated climate model. The only such study carried out to date (Mitchell et al., 1995) supports the above analysis, but the conclusions are still subject to doubt because these authors did not consider tropospheric ozone (which has a similar, but opposite sign, pattern to aerosol forcing, and so has an offsetting effect).

TABLE A1
Model estimates of 1880–1990 changes in temperature and sea level (temperatures in °C, sea-level changes in cm).

ΔT_{2x}	Aerosol[1]	ΔT-NHL	ΔT-NHO	ΔT-SHL	ΔT-SHO	ΔT	Δz(L)[2]	Δz(M)	Δz(H)
1.5	Zero	1.04	0.51	0.59	0.41	0.59	2.5	10.7	20.3
	D+B	0.71	0.40	0.45	0.35	0.45	1.4	9.0	17.6
	Low I	0.42	0.32	0.29	0.29	0.33	0.2	7.1	14.7
	Mid I	0.15	0.23	0.13	0.24	0.21	-0.9	5.3	11.9
2.5	Zero	1.53	0.75	0.84	0.59	0.86	5.7	15.1	26.5
	D+B	1.06	0.58	0.65	0.49	0.65	4.0	12.5	22.5
	Low I	0.65	0.44	0.43	0.40	0.47	2.1	9.7	18.2
	Mid I	0.25	0.31	0.22	0.31	0.29	0.3	6.8	13.9
3.5	Zero	1.91	0.93	1.02	0.72	1.06	8.1	18.6	31.3
	D+B	1.33	0.72	0.80	0.59	0.81	6.0	15.3	26.4
	Low I	0.83	0.54	0.54	0.48	0.58	3.6	11.7	20.9
	Mid I	0.33	0.36	0.29	0.37	0.35	1.3	8.0	15.4
4.5	Zero	2.21	1.08	1.17	0.82	1.23	10.1	21.3	35.0
	D+B	1.55	0.83	0.92	0.68	0.93	7.6	17.5	29.5
	Low I	0.97	0.62	0.63	0.54	0.66	4.8	13.2	23.1
	Mid I	0.40	0.41	0.34	0.41	0.40	2.0	9.0	16.6

[1] D+B denotes that only forcing due to direct sulphate aerosols and biomass-burning-derived aerosols is included. Low I indicates that, in addition to D+B, an indirect aerosol forcing of -0.4 W/m^2 (1990 value is included. Mid I is the same, but with -0.8 W/m^2 (1990) for the indirect aerosol forcing component.

[2] L, M, H denote Low, Medium and High ice-melt parameters.

TABLE A2

Forcing values for 1765 to 1990 (W/m^2); present values compared with IPCC90 (Shine et al., 1990).

ITEM	IPCC90	This study
CO_2	1.50	1.52
CH_4 + Strat. H_2O	0.54	0.44
N_2O	0.10	0.15
Halocarbons	0.29	0.27
Strat. O_3	—	-0.16
Trop. O_3	—	0.40
SO_4 aerosol	—	-1.10
biomass aerosol	—	-0.20
TOTAL	2.45	1.32

TABLE A3

Radiative forcing changes (W/m^2) 1990–2100. WR92 indicates results from Wigley and Raper (1992, Table 1). Bracketted items after "present ΔQ" indicate the aerosol assumption, keyed to Table A1.

Scenario	a	b	c	d	e	f
WR92	5.44	5.26	2.98	3.84	6.74	6.26
present (Zero)	5.76	5.55	2.55	3.45	7.90	7.07
present (D + B)	5.47	5.29	2.63	3.50	7.28	6.65
present (Low I)	5.29	5.11	2.71	3.54	6.95	6.40
present (Mid I)	5.10	4.94	2.79	3.57	6.62	6.15

TABLE A4

1990–2100 global-mean temperature changes (°C) for IPCC95 emissions scenarios, for different aerosol assumptions and different climate sensitivities. Aerosols keyed to Table A1.

		Scenario					
ΔT_{2x}	**Aerosol**	**a**	**b**	**c**	**d**	**e**	**f**
1.5	Zero	1.65	1.60	0.83	1.06	2.20	1.99
	D+B	1.53	1.46	0.84	1.05	1.96	1.82
	Low I	1.46	1.42	0.85	1.05	1.84	1.73
	Mid I	1.39	1.35	0.87	1.06	1.72	1.63
2.5	Zero	2.46	2.39	1.30	1.61	3.24	2.94
	D+B	2.27	2.20	1.29	1.59	2.87	2.67
	Low I	2.15	2.09	1.30	1.58	2.68	2.52
	Mid I	2.03	1.98	1.31	1.58	2.48	2.37
4.5	Zero	3.62	3.51	2.02	2.44	4.69	4.28
	D+B	3.31	3.22	1.98	2.38	4.12	3.86
	Low I	3.12	3.04	1.97	2.36	3.82	3.62
	Mid I	2.93	2.85	1.96	2.33	3.53	3.38

TABLE A5

1990–2100 global-mean sea-level changes (cm) for IPCC95 emissions scenarios, for different aerosol assumptions and different ice-melt parameter values (L,M,H) and different climate sensitivities. Aerosols keyed to Table A1.

		Scenario					
ΔT_{2x}	**Aerosol**	**a**	**b**	**c**	**d**	**e**	**f**
1.5 (L)	Zero	27	26	18	20	33	31
	D+B	24	24	16	18	29	28
	Low I	22	21	14	17	27	25
	Mid I	20	19	13	15	24	23
2.5 (M)	Zero	64	63	48	52	74	70
	D+B	58	57	44	48	66	63
	Low I	54	53	41	45	61	59
	Mid I	49	48	38	43	55	54
4.5 (H)	Zero	107	105	85	90	121	116
	D+B	98	97	79	85	109	106
	Low I	92	91	76	81	102	99
	Mid I	86	85	72	78	94	93

For both past and future sulphate aerosol forcing, we use the forcing-emissions relationships employed by Wigley and Raper (1992).

The total radiative forcing change over 1765–1990 based on the above is 1.32 W/m^2 which compares with a value of 2.45 W/m^2 used in IPCC90 (see Table A2). Using our best guess model parameters, these forcings lead to estimates of climate sensitivity which produce a best-fit between model and observed temperature changes over 1861–1994 of ~6°C and ~1.3°C, respectively. The high value of 6°C may indicate that our aerosol forcing choice is too high; or that part of the observed warming is natural.

Finally, in Table A3 we show the total radiative forcing changes over 1990–2100 for the six IPCC95 scenarios, compared with values given previously in Wigley and Raper (1992), while in Tables A4 and A5 we show 1990–2100 temperature and sea-level changes for the four aerosol cases given in Table A1.

Note that, following IPCC, we now refer to the IS92 scenarios as IS95a-f to indicate that changes have been made to the emissions of CO, NO_x and VOCs, and halocarbons. In addition, CH_4 and N_2O emissions have been adjusted by constant amounts determined by the difference between the IS92 1990 values and the emissions required to give balanced budgets given the latest concentration and lifetime data.

CHAPTER 2

Land Subsidence in Coastal Lowlands

Saskia Jelgersma

ABSTRACT: Causes and ranges of coastal lowland subsidence are varied, but in general natural subsidence amounts to centimeters to a few meters per century, while human-induced subsidence can amount to meters in a few decades. As a result, coastal subsidence must be taken into account when predicting relative sea-level rise over the next century.

Introduction

An increasing body of evidence suggests that within the near future, global warming due to the greenhouse effect could lead to a substantial rise in sea level. Most estimates of this rise range from 0.2 to 1.4 m for the next 100 years (Villach conference, 1985). These contrast with Warrick and Oerleman's estimates (1990), given in the report of the Intergovernmental Panel of Climate Change (IPCC). For their best estimate for the year 2030, the Business-as-usual scenario, 10 cm comes from thermal expansion, 7 cm from the partial melting of mountain glaciers, 1.8 cm by melting of Greenland ice, and −0.6 cm from the buildup of Antarctic ice due to increased precipitation. Their best estimate for 2090 is +66 cm, which is higher that the +48 cm estimate of Wigley and Raper (1992). This wide range in estimates is due to the many uncertainties of the input to the predictive models.

A large portion of the world population lives in low lying areas near the sea and, consequently, a rise in sea level will have an important impact on society. Most low lying coastal areas, especially deltas, are subject to marked subsidence. In densely populated deltaic areas with heavy agriculture and industrial development, accelerated subsidence due to overdrawing of ground water is already a serious problem. Further human induced subsidence is caused by draining of the land to make polders in these areas. Because local subsidence can be much greater than the maximum expected sea-level rise due to global warming, these must be considered when estimating the expected sea-level rise in any given area.

J.D. Milliman and B.U. Haq (eds.), Sea-Level Rise and Coastal Subsidence, 47-62.

Geological Setting

Many coastal lowlands and deltaic plains are underlain by subsiding basins. Such basins are the sites of a considerable accumulation of Cenozoic sediments, sometimes several thousand meters thick. The position of the shoreline in these basins depends on the rate of subsidence and the rate of influx of sediments. In a geological time frame, these two factors have varied and together with eustatic sea-level shifts have caused alternating periods of shoreline regression and transgression. In general, sea level during glacial periods dropped 100 m or more, while during interglacials sea level remained near or slightly above the present level. Because of these factors there are often alternating marine and fluvial deposits buried below the surface of many coastal areas.

These deposits from different environments show various lithologies. In general marine deposits are fine-grained, in contrast to the alluvial deposits, which are much coarser grained. It is also clear that the pore water of marine deposits was originally saline or brackish and that the alluvium originally contained fresh water.

The environment of deposition and their lithologies can be summarized as follows: In an *alluvial valley*, meandering river systems the point bars of the river deposits consist largely of coarse grained sands; levees are silty clays; and the backswamp contains clays and peats. The braided river system shows a different sedimentation pattern due to the irregular and abundant sediment supply. With tectonic subsidence, river deposits can contain alternating layers of fine and coarse grained sediments (Fig. 1). Changes in tectonic movement and or precipitation in the drainage basin can give rise to successive cycles of meandering and braided river deposits in the subsurface of the alluvial valley.

River *deltas* consist of deltaic and tidal plains. In the deltaic plain, distributaries with sandy deposits are bordered by marshes, swamps and muds. The tidal plain shows tidal channels with sand bordered by tidal deposits like mangrove swamps and muds. In general, deltaic sedimentation on tidal plains is influenced by waves and tides as well as the rate of sediment influx. Therefore sediments can range from sand (beach ridges) to mud (tidal flats, lagoonal clays and peats).

The *offshore zone* can show an accumulation of mud deposits, in some places alternating with sandy shoals and banks. These environments shift landward and seaward with time due to tectonic subsidence and sea-level change.

Due to the postglacial rise in sea level, recently formed wedge-shaped layers of sediment often are found in low-lying coastal areas. In The Netherlands and the Mexican Nayarit coast, sandy coastal barriers and ridges border lagoonal and peaty deposits (Fig. 2). In the Louisiana coastal segment west of the Mississippi Delta, a thick layer of bay mud sediment is present with a series of sandy cheniers imbedded in this bay mud (Fig. 3). The same can be found in the coastal plain north of the Yangtze Delta (China) and along the northern coastline of Malaysia. In other areas, such as Bangkok Bay, a thick layer of clay represents the sediment deposited during the post-glacial rise in sea level.

Between 3000 and 8000 m of Cenozoic sediments have been deposited in the sedimentary basins of the Nile, the Mississippi and the North Sea (Figs. 4, 5 and 6,

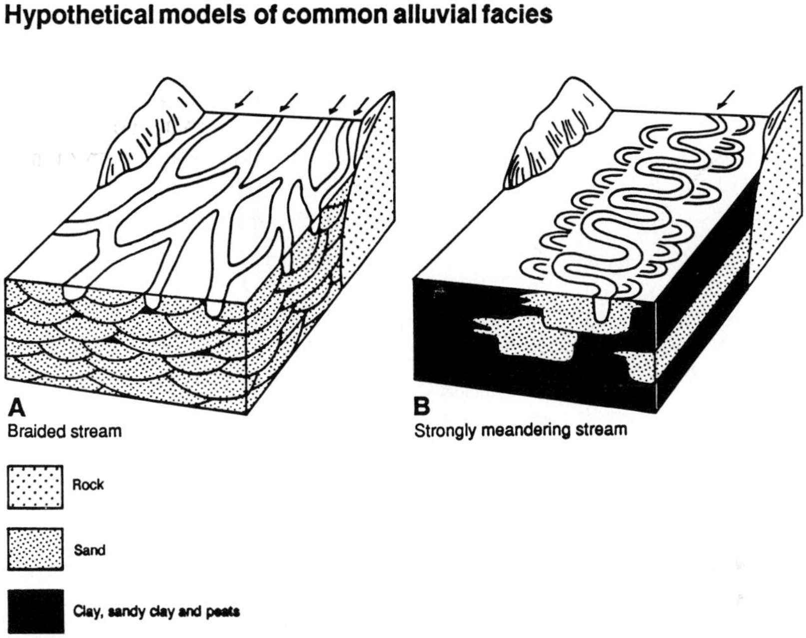

Fig. 1. Hypothetical models illustrating textural and geometrical characteristics of common alluvial facies (redrawn after Allen, 1970).

respectively). The sediment sequence shows alternating marine, littoral and fluvial deposits; the lithology an alternation of semi-consolidated sands and clays.

In the coastal zone the uppermost stratum of the sedimentary succession may consist of a wedge of unconsolidated sediments, whose thickness can exceed several tens of meters, deposited during the post-glacial rise in sea level.

Subsidence in Coastal Lowlands

Natural Subsidence

Natural subsidence can result from tectonic action, compaction by sediment during loading, oxidation of peats, vibration (earthquakes) and dewatering of sediments (dry periods and or drop of sea level).

Most coastal lowlands are situated on top of huge sedimentary basins that undergo slow tectonic subsidence, which consists of basement subsidence enhanced by subsidence due to isostatic loading of sediments in the basin. Not much is known about the amount of tectonic subsidence in sedimentary basins, but it generally is not more than a few cm/century. In The Netherlands, for example, the depth and the age of the marine deposits from the last interglacial (Eemian) are 8 m below MSL, in Belgium they are at MSL and in southern England (Selsey, Sussex) and northwestern France (Normandy) they are 8 m above MSL. The latter areas are thought to be rather stable. Accordingly, a difference in level of 16 m took

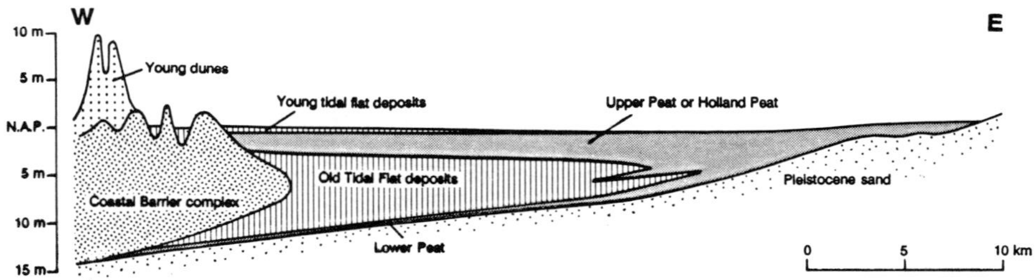

Fig. 2. Cross-section through the Holocene wedge-shaped layer of sediments deposited in the western part of The Netherlands during the post-glacial sea-level rise (after Jelgersma, 1961).

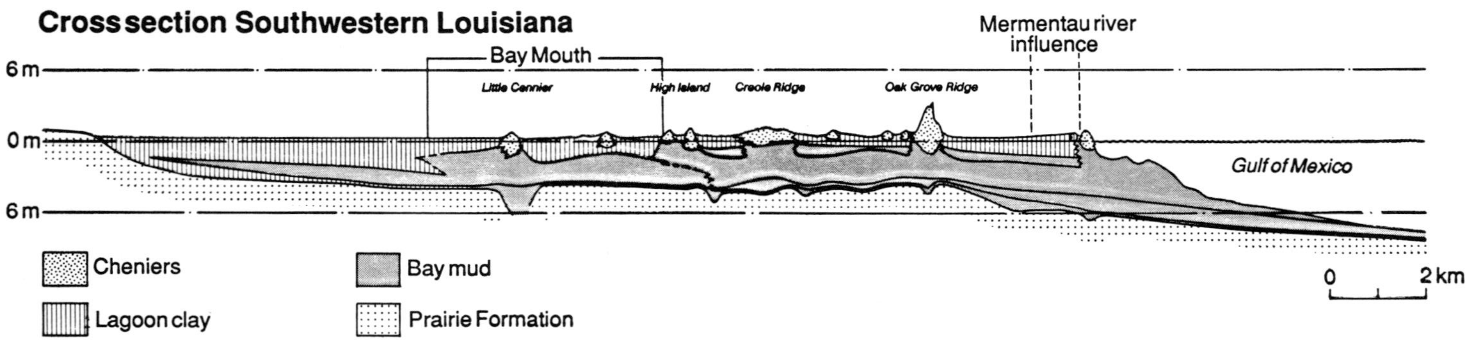

Fig. 3. Cross-section through the Holocene wedge-shaped layer of sediments deposited in coastal Louisiana during the post-glacial sea-level rise (redrawn after Gould and McFarlan, 1959).

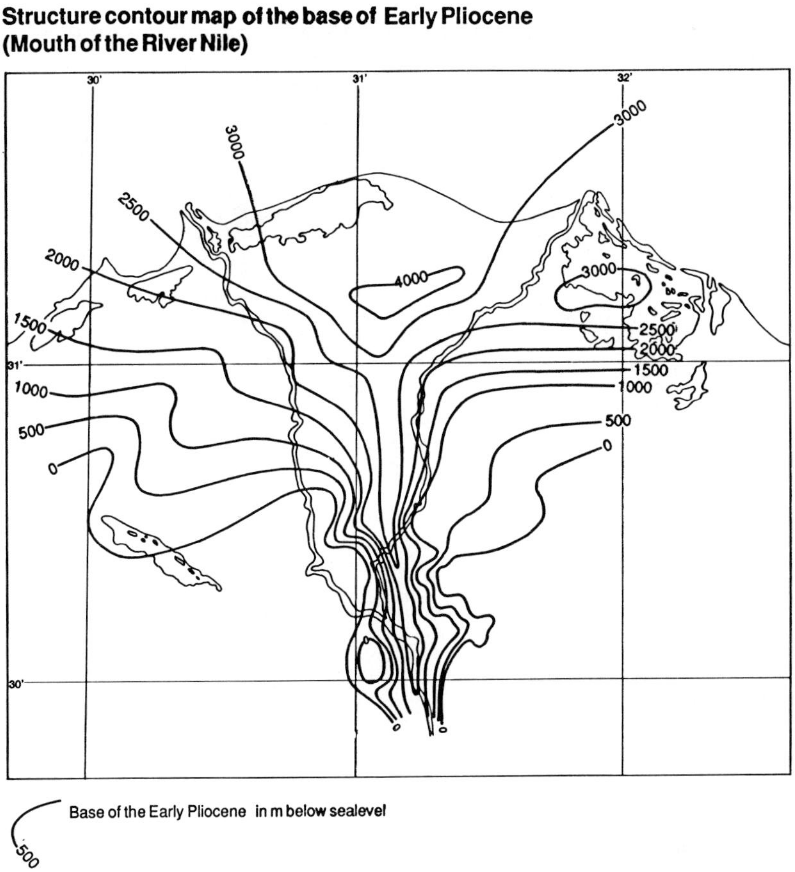

Fig. 4. Depth contour of the base of the Early Pliocene in the Nile Delta (redrawn after Said, 1981).

place during the last 120,000 years (age of the Eemian) between The Netherlands and France-England. This indicates a tectonic subsidence of 1.4 cm/century in the Dutch part of the North-Sea basin (Zagwijn, 1983).

This assumes, of course, that tectonic subsidence is a continuous process, which may be not true. Evidence from the sedimentary basin of The Netherlands indicates shifts in zones of maximum subsidence, the so-called depocenters (Figs. 6a and 6b). Some sedimentary basins show greater tectonic activity. An example is the Ganges-Brahmaputra basin, where there are active faults and subsequent earthquakes. The latter caused catastrophic river changes in the delta in the last century (Fergusson, 1863).

In general, slow tectonic subsidence in deltaic areas is compensated by fluvial sediment input. In recent decades, however, sediment deposition in many coastal areas has decreased (or locally, such as the Nile Delta, ceased) due to river diversion and damming.

Dewatering occurs when fine-grained sediments, particularly clays, can show the effects of natural loading. Recent muds with clay contents greater than 40%

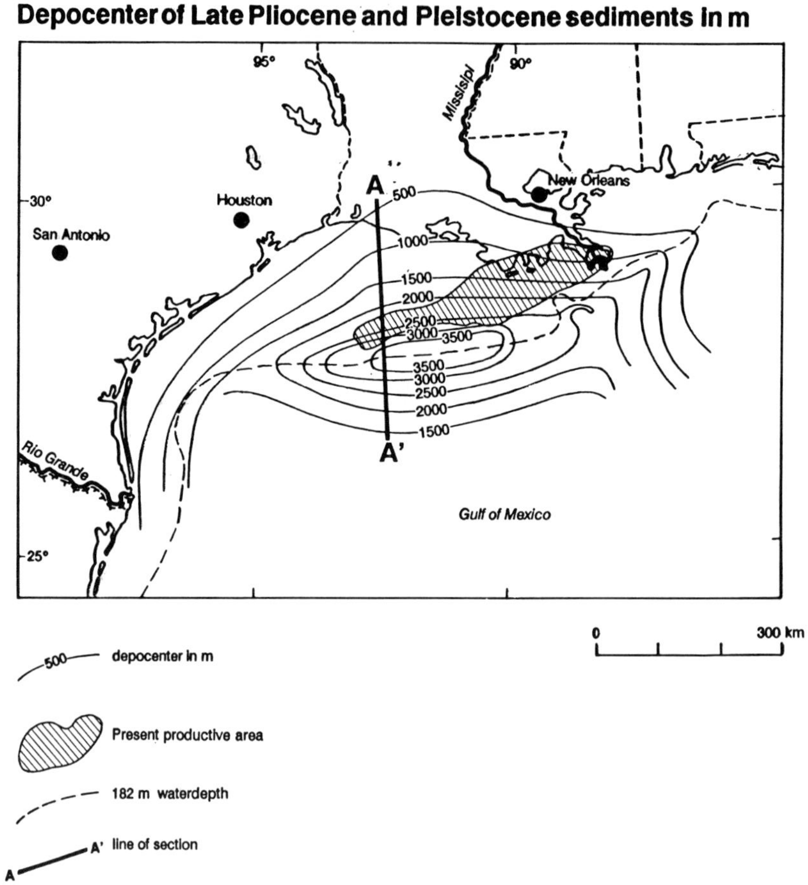

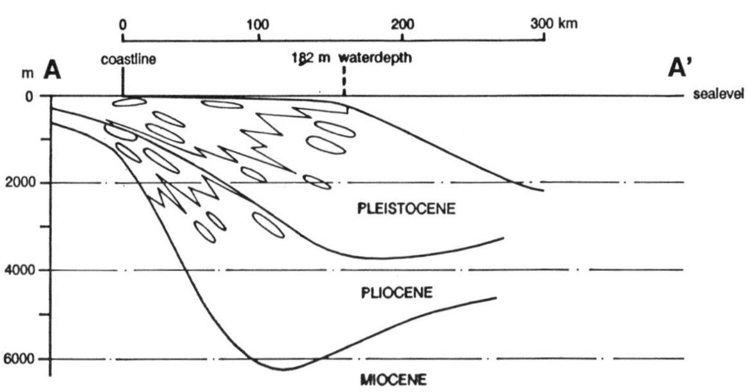

Fig. 5. Depth contour of the base of the Pleistocene and a cross-section of the depocenter of the Mississippi (redrawn after Woodbury et al., 1973).

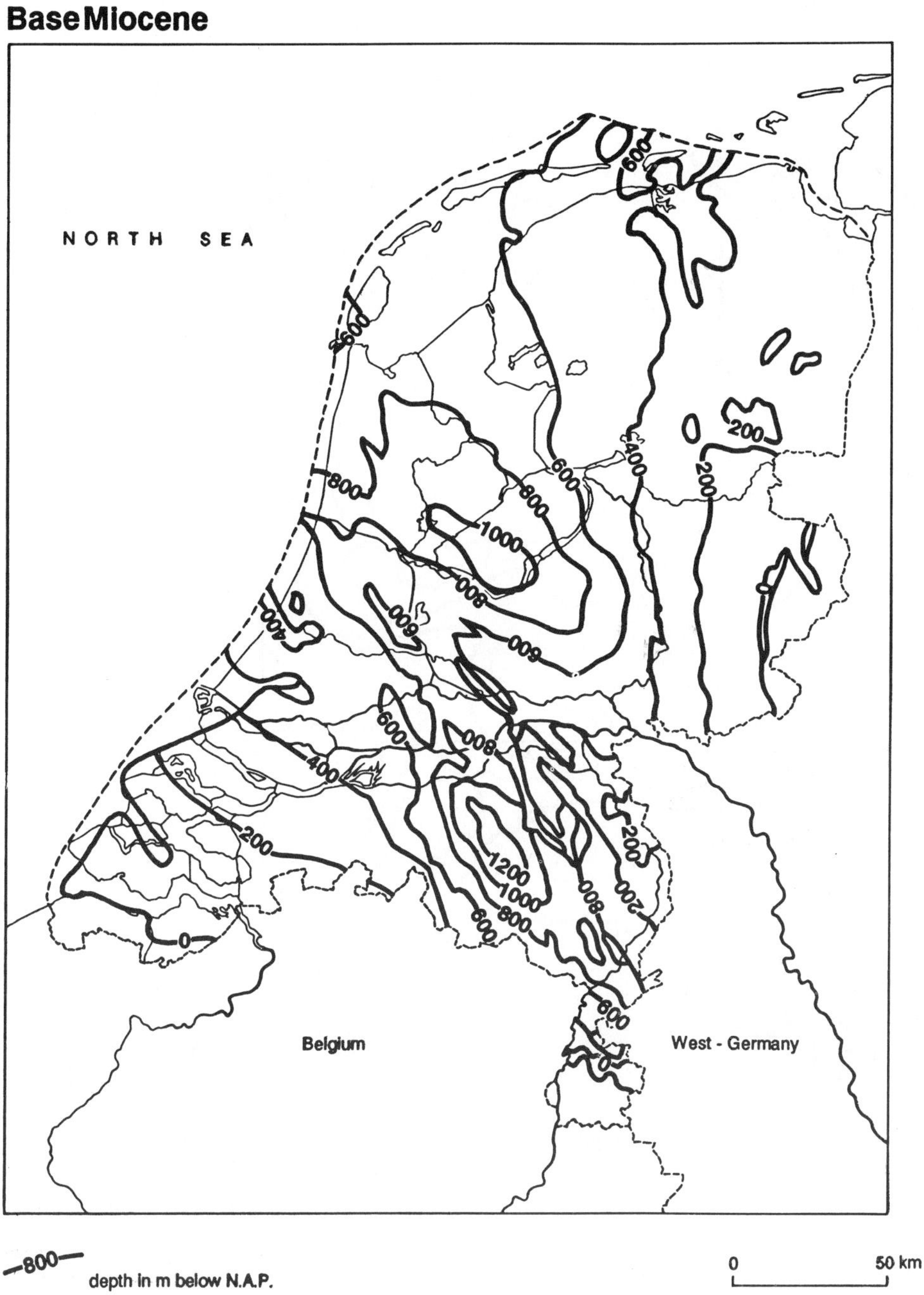

Fig. 6A. Depth contour of the base of the Miocene in The Netherlands (redrawn after Zagwijn, 1989).

Base Quaternary

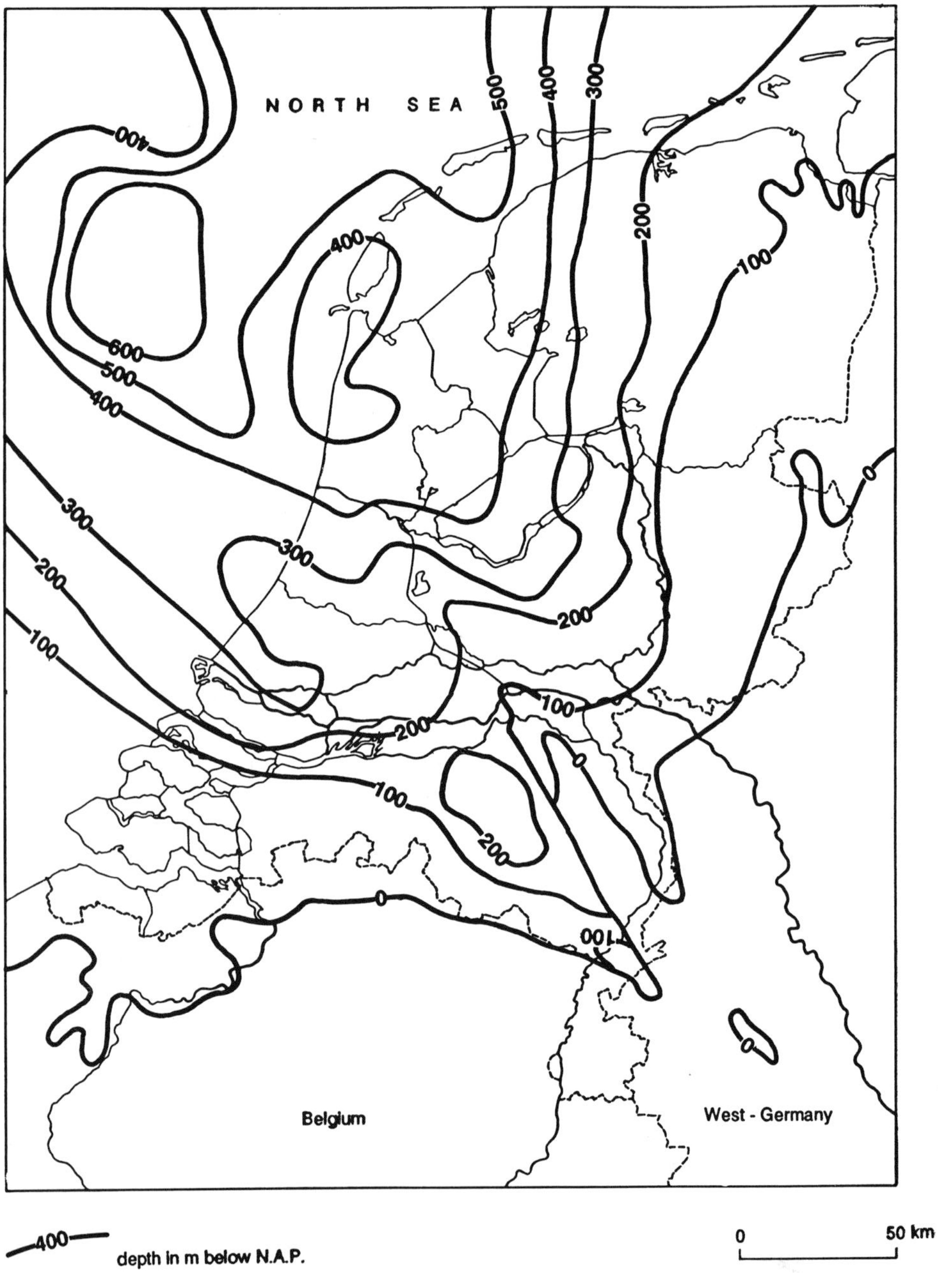

Fig. 6B. Depth contour of the base of the Quaternary in The Netherlands (redrawn after Zagwijn, 1989).

have porosities of about 85%. The effect of loading reduces the porosity to 40–50%, and accordingly, the density increases while the volume decreases. The water table also can be lowered during dry periods or lowering of sea level by reducing the porosity of the drained sediment and subsequently causing sediment compaction. An example of compaction due to loading is the site of New York's La Guardia Airport: the natural compaction of an 18 m-thick water-saturated organic silt and clay deposit was accelerated by placing 7.5 m of fill material atop the soft sediment. After 25 years the land subsided 2.5 m.

Oxidation of superficial peat layers occurs during dry periods, when the swamp dries out. This oxidation results in surface lowering.

Vibration of soft fine grained sediment during earthquakes can cause significant compaction due to the decreased porosity and corresponding increased density.

HUMAN INDUCED SUBSIDENCE

Human induced subsidence can result from extraction of gas, oil and water, irrigation and land reclamation.

Land reclamation

When water-saturated fine grained sediments are to be irrigated and/or reclaimed, the lowering of the water table stimulates compaction of the sediments with accompanying subsidence of the surface.

In The Netherlands, Bennema et al. (1954) did extensive research on compaction of reclaimed land. Clays have porosities greater than 80%, and thus they can be compressed to about half of their original thickness after reclamation over 100 years. Peats are compressed much more: to about 1/9 of their original thickness. Holocene sands, because of their lower porosity, are not subject to much compaction.

Striking compaction differences can be seen in land reclaimed centuries ago, where peat layers have been eroded by creeks. The latter have been filled up by sand and silt and now emerge up to 2 m above the adjacent part of the landscape due to compaction of the peat and clay. Significant compaction has been found in some land reclaimed in the late 19th century. In an old creek filled with peat mud a compaction of 2.5 m occurred over seventy-five years. Outside this creek, the surface lowering in these polders amounts to 1–1.5 m over the same period of time.

After World War II, new polders were reclaimed in The Netherlands in the former Zuiderzee area. To forecast subsidence of the surface after reclamation, consolidation constants were estimated from the pore space and the thickness of the sediment (de Glopper, 1969, 1973 and 1984) before reclamation. Before the polder was pumped, permanent measuring sites were installed to observe the regional pattern of subsequent subsidence. In 25 years the observed subsidence varied between 10 and 100 cm, and is related to the thickness of the Holocene layer and its composition. Due to poor drainage conditions, compaction during the first 10 years is much lower than 15 years that followed.

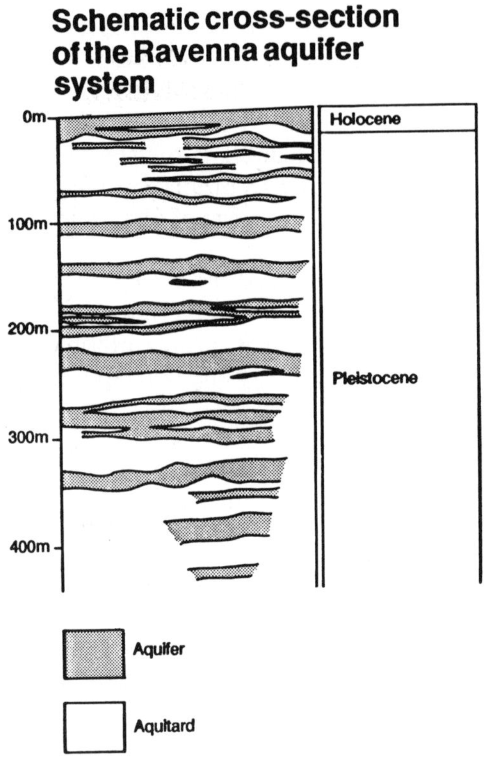

Fig. 7. Schematic cross-section of an aquifer system in a sedimentary basin (redrawn from Guidebook, studies of land subsidence 1984).

Ground water, Oil and Gas Withdrawal

A serious anthropogenic geological hazard is land subsidence due to the excessive extraction of ground water in areas underlain by Cenozoic sediments. Detailed information about this problem can be derived from the proceedings of international symposia dealing with land subsidence (particularly the International Association of Hydrological Sciences, 1969, 1976, 1984, 1991) and the UNESCO Guidebook to studies of land subsidence due to ground water withdrawal.

Surface and subsurface layers of subsiding basins in coastal areas consist of young, unconsolidated material deposited in shallow, alluvial, and lacustrine marine environments. Depending on their respective lithologies, permeabilities can vary considerably. Accordingly, subsurface layers generally consist of confined or semi-confined aquifers of gravel and sand with high permeability and low compressibility alternating with aquitards of low vertical permeability and high compressibility (Fig. 7). Ground water withdrawal will cause artesian head decline, resulting in compaction due to increased effective stress. The compaction of these aquifers is immediate, but can be mostly recoverable if fluid pressure is restored.

This compaction, however, is small compared to that which occurs in the aquitards, due to time-dependent pore-pressure reduction in the fine-grained sed-

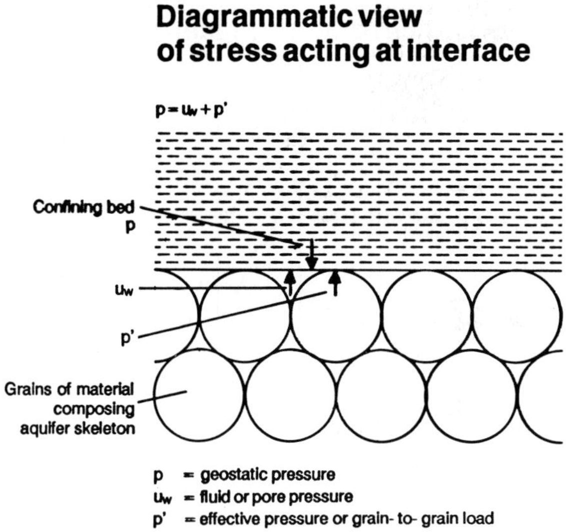

Fig. 8. Artesian aquifer and confining bed (redrawn after Guidebook, studies of land subsidence, 1984).

iments. The decrease in head in the aquifer creates a hydraulic gradient from the clays of the aquitards to the aquifers. Because these fine-grained beds have low hydraulic conductivities and high compressibilities, the vertical escape of water and the consequent decrease in pore pressure and increase in effective stress are slow and time-dependent. The resulting compaction can be large and irreversible (Fig. 8).

There are various problems associated with subsidence due to ground water withdrawal. In coastal areas, tidal encroachment can cause salt water intrusion in the aquifer and flooding of populated areas. Other problems include differential changes in elevation and gradient of stream channels, drains, canals and failure of pipelines. In water wells, casings may fail due to compressive stresses caused by compaction of aquifers. Moreover, because of increased subsidence, some areas have become subject to inundation by normal tides, and even a larger area may be subject to catastrophic flooding by hurricanes.

These problems must be taken into account if heavy withdrawal is allowed to occur in deltaic areas with heavy population and intense industrial development. Resulting accelerated subsidence may then necessitate protection from both river and marine flooding by construction of an extensive system of dikes, flood walls, locks and pumping stations. The costs of such construction are obviously extremely high.

In the USA, the Houston-Galveston region has experienced subsidence as much as 2.8 m since 1915 in areas of highest ground water withdrawal. As demonstrated in Figure 9, there is a close relation between the observed decline of the water table in the wells and the observed subsidence.

In Japan, 59 areas, most of them located on the coast, are subject to serious land subsidence due to ground water withdrawal. One of these areas, the Nobi fluvial

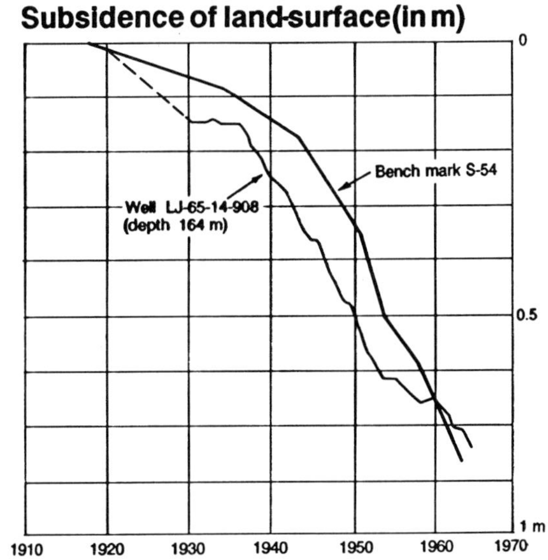

Fig. 9. The relation between water level decline and land surface decline in the Houston area, Texas (redrawn after Guidebook, studies of land subsidence, 1984).

and deltaic plain, is underlain by more than 2000 m of young, clastic sediments. Ground water withdrawal caused a surface lowering of the land between 20 cm and 140 m in 15 years. Due to this subsidence, 248 km^2 of the area is now below mean sea level. To deter this disastrous situation, withdrawal of ground water had to be severely regulated. In the northeastern part of Tokyo the overpumping of the aquifers resulted in a maximum subsidence of 4.57 m between 1920 and 1975 and a ground water lowering of 60 m. Dikes had to be build to protect the area because the subsidence had placed the area at high risk of flooding during typhoon-induced floods.

In Taiwan, land subsidence in the Taipei River basin has amounted to 2 m over 22 years of heavy ground water withdrawal. This situation has hampered the normal drainage of heavy rainfall that can lead to flooding (Fig. 10).

Finally, the example of increased land subsidence in Bangkok is the subject of several papers later in this volume.

In coastal areas with heavy ground water withdrawal, it is advisable to monitor ground movement through regular geodetic surveys and to install a network of piezometers to check pressure heads in the utilized aquifers. Measurements of compaction in the aquifer itself can be carried out by extensometers in boreholes. Investigations of permeability, grain size, clay chemistry and compressibility of aquifers and aquitards should be made to identify the physical properties of the subsurface formations. These data can be used to construct a numerical model, that predicts the compaction of the ground water reservoir caused by ground water withdrawal. Based on the results of these investigations withdrawal of ground

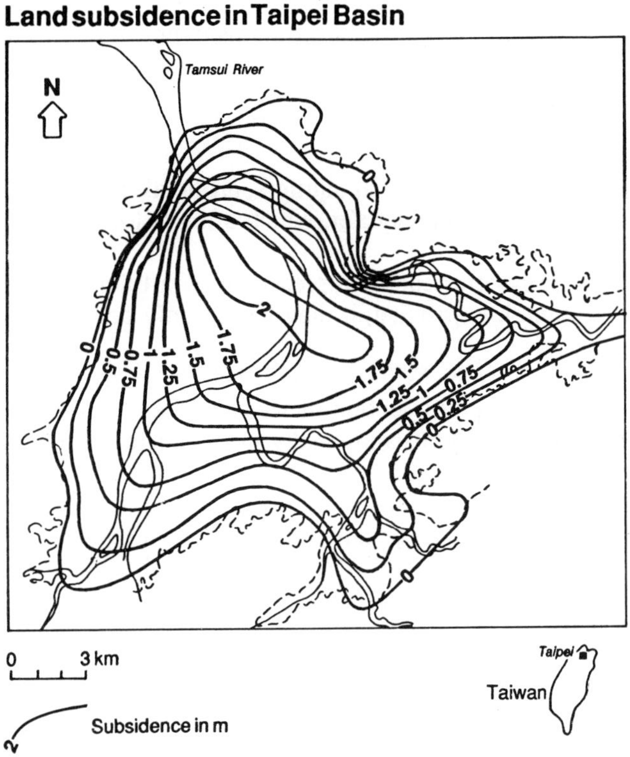

Fig. 10. Land subsidence in the Taipei basin due to ground water extraction between 1955 and 1977 (redrawn after Poland, 1984).

water in coastal plain aquifers should be managed carefully by local or national authorities.

Methods to Control or Inhibit Human Induced Subsidence

Methods to control or inhibit subsidence generally are costly and require the development of organizational and legal infrastructures. Whether any of the methods mentioned below is economically justified depends on the costs compared with the cost of continued or even further accelerated land subsidence.

Reduction of Pumping

Ground water withdrawal can be reduced by regulations, while increased water demands met by importing surface water. Examples are the major aqueduct that imports Colorado River water to south-central Arizona and the aqueduct in the Venice area that supplies the industrial area with river water. Reduction of pumping also can be facilitated by using surface water for irrigation and reusing treated water in industrial plants. In many places, ground water withdrawal is concentrated in certain areas. It is advisable to spread the pumping over a much larger area in order to reduce land subsidence.

Artificial Recharging of Aquifers

To restore a falling hydraulic head in aquifers, artificial recharging can be undertaken in the outcrop area of the aquifer, which often is located at the edge of the sedimentary basin.

Confined aquifer systems can be repressured by artificial recharging directly through wells. Although expensive, it may be the only practical way to slow or stop land subsidence.

A classical example is the Wilmington oil field in southern California, where subsidence was controlled by injection of water through wells. This reduced the subsiding area from 58 to 8 km^2, and locally the land surface rebounded as much as 30 cm (Mayuga and Allen, 1969). In Shanghai, China, land subsidence due to ground water withdrawal was stopped by injection of river water through wells. Within a few years bench marks began to rise, and from 1966 to 1976 as much as 3.4 cm of rebound took place locally. However, the recharged water must be free of particulates or the wells and aquifers may become clogged.

Conclusions

In the last decades, increased global population has been responsible for extensive urban, agriculture and industrial development in the coastal lowlands. Interference with the natural environment is one of the negative side effects of these activities. These include interference in the hydro-dynamic systems of shorelines and river valleys and in land subsidence. The latter in particular has been triggered or enhanced by human activities.

The negative effects usually become noticeable after subsidence has accelerated, and the lack of understanding of the geological and hydrological processes may prevent sound counter measures. Moreover, most effects caused by man-induced processes are irreversible.

In the coming decades, human population and activities in coastal lowlands will continue to increase. Therefore, it is increasingly important that proven methods of controlling or inhibiting subsidence become part of integrated coastal lowland management programs. This requires development of organizational and legal infrastructures. Sea-level changes should be recorded by tide-gauge measurements. Assistance in construction of tide gauges and training in measurements and analysis techniques could be provided by the Global Sea-Level Observing System coordinated by the Intergovernmental Oceanographic Commission. A regular geodetic survey can be used to monitor surface level movements caused either by natural or anthropogenic processes. The latter process should be carefully observed by annual geodetic surveys and by observing compaction in the subsurface layers.

References

Allen, J.R.L., 1970. *Physical processes of sedimentation.* Allen & Unwin, London. 248 pp

Bennema, J., Geuse, E.C.W.A., Smits, H. and Wiggers, A.J., 1954. Soild compaction in relation to Quaternary movements of sea level and subsidence of the land, especially in The Netherlands. Geo. Mijn., 16, 173-178.

Fergusson, J., 1863. Delta of the Ganges. Geol. Soc. London Quart. Jour., 29, 321-354.

Ferris, J.G., D.B. Knowles, R.H. Brown and R.W. Stallsman, 1962. Theory of aquifer tests. U.S. Geol. Survey Water Supply Paper 1536-E, 69-174.

Glopper, R.J. de, 1973. Subsidence after drainage of the deposits in the former Zuyder Zee and in the brackish and marine forelands in the Netherlands: Van Zee tot Land n0. 50, Staatsuitgeverij, Den Haag, The Netherlands, 205 p.

Glopper, R.J. de, 1984. Shrinkage of subaqueous sediments of Lake IJssel (The Netherlands) after reclamation. Proceedings of the First International Symposium on Land Subsidence, Tokyo, September 1969.

Glopper, R.J. de, 1973. Subsidence after drainage of the deposits in the former Zuyder Zee and in the brackish and marine forelands in the Netherlands: Van Zee tot Land no. 50. Staatsuitgeverij, 's-Gravenhage, The Netherlands, 205 pp.

Glopper, R.J. de, 1984. Subsidence in the recently reclaimed IJsselmeerpolder "Flevoland". Proceedings of the International Symposium on Land Subsidence. I.A.H.S., Venice 1984.

Gould, H.R. and McFarlan, E., 1959. Geological history of the chenier plain, southwestern Louisiana. Trans. Gulf Coast Assoc. Geol. Soc., 9, 261-270.

Guidebook to studies of land subsidence due to ground water withdrawal. 1984. Poland, J. (ed.), Prepared for the I.H.P. working group 8.4, UNESCO

International Association of Hydrological Sciences, 1969. Proceedings of the First International Symposium on Land Subsidence, Tokyo, September 1969, Publ. 88 and 89.

International Association of Hydrological Sciences, 1976. Proceedings of the Second International Symposium on Land Subsidence, Anaheim, California, December 1976, Publ. 121.

International Association of Hydrological Sciences, 1984, Proceedings of the Third International Symposium on Land Subsidence, Venice, March 1984, Publ. 151

International Association of Hydrological Sciences, 1991. Proceedings of the Fourth International Symposium of Land Subsidence, Houston, May 1991, Publ. 200.

Journal of Irrigation and Drainage Division, ASCE, vol. 107, 1981. Legal, Socioeconomic, and Environmental Significance of Land Subsidence in the United States.

Jelgersma, S. 1961. Holocene sea-level changes in The Netherlands. Ned. Geol. Stichting, Series C, VI, 100.

Mayuga, M.N. and Allen, D.R., 1969. Subsidence in the Wilmington Oil Field, Long Beach, California, USA. Int. Assoc. Hydrol. Sci., 88 and 89.

Said, R. 1981: The Geological Evolution of the River Nile. Springer-Verlag, Berlin.

Vlam, A.W., 1952. Historisch morfologisch onderzoek van eenige Zeeuwsche eilanden. Tijd. Kon. Ned. Aard. Gen., 6, 1-20.

Woodbury, H.O., Murray, I.B., Pickford, P.J. and Akers, W.H., 1973. Pliocene and Pleistocene depocenters, outer continental shelf Louisiana and Texas. Am. Assoc. Pet. Geol. Bull. 49, 1893-1907.

Yen, C.L. et al., 1979. A study of the influence of land subsidence on the capacity of drainage system in Taipei. Research Report no. HY 6805, dept. of civil engineering, National Taiwan University, Taipei (in Chinese).

Zagwijn, W.H., 1983. Sea level changes in The Netherlands during the Eemian. Geol. Mijnb., 62, 437-450.

Zagwijn, W.H., 1989. The Netherlands during the Tertiary and Quaternary. A case history of Coastal lowland evolution. Geol. Mijnb. 68, 107-120.

Ziegler, P.A. and C.J. Louwerens, 1977. Tectonics of the North Sea. In: E. Oele, R.T.E. Schüttenhelm and A.J. Wiggers (eds.), *The Quaternary history of the North Sea*. Acta Univ. Ups. Symp. Uppsala.

CHAPTER 3

River-Sediment Inputs to Major Deltas

Robert H. Meade

ABSTRACT: River sediments are distributed unevenly in space and time, and they are markedly susceptible to human influences. Half the world's river sediment is derived from the Himalayan region and its environs. Most of the remainder is derived from other tectonically active regions such as the western Pacific islands, the Andes, and southern Alaska. River-sediment loads are variable at many time scales: seasonal, annual, decadal, and longer. The storage of sediment in river systems confounds our ability to predict the delivery of sediment to coastal zones. Natural river-sediment loads are increased by deforestation and crop farming, and decreased by dams and reservoirs.

Introduction

Many of the subsiding coastal areas of the world are located on or near major river deltas where the balance between sea-level rise and the submergence of coastal lands is affected by inputs of river sediment. The total discharge of river sediment to the coastal zones of the world under present-day conditions averages 15–20 $\times 10^9$ tonnes per year (Milliman and Meade, 1983; Walling and Webb, 1983; Milliman and Syvitski, 1992). To assess the degree and extent to which river sediment may augment or offset the effects of coastal submergence, one needs some understanding of the variability of sediment inputs in space and time, as well as some appreciation of the marked susceptibility of river-sediment loads to human influences.

The ultimate sink for river sediments, on a millennial to multi-millennial time scale, is the coastal zone. The degree to which river sediments have influenced the coastal areas of the world is a reflection of the balance between river-sediment discharges and the Holocene rise of sea level (i.e., the Flandrian transgression). Where river-sediment inputs during the last 10,000 years have been small, the rising sea has inundated the lower river valleys and formed large embayments. Where the inputs of river sediment have been just sufficient to fill the lower valleys as their water levels were being raised by the rising sea, the result has been a coastline free of major indentations or protrusions. Where the inputs of river sediment were

J.D. Milliman and B.U. Haq (eds.), Sea-Level Rise and Coastal Subsidence, 63-85.

more than sufficient to compensate for the rising sea level, large deltas have been constructed out onto continental shelves. For purposes of this volume, we probably can minimize any discussions of large estuarine embayments (the St. Lawrence, for example) where the quantity of river sediment reaching the coastal zone is small, or of areas (like the southern California borderland or Peru-Chile Trench) where river sediments move directly from the continents to the floor of the deep sea. The principal areas of our concern are deltas and the coastal lowlands adjacent to them. And the important deltas of the world are associated, by and large, with the mouths of the important sediment-carrying rivers.

All modern deltas of major rivers appear to be less than 10,000 years old. Stanley and Warne (1994) compiled radiocarbon dates on basal sediments from deltas around the world (including those of the Ganges, Yangtze, Mississippi, Mackenzie, and Nile Rivers) to show that the deposition of most modern deltas began between 8500 and 6500 years ago. They concluded "that deceleration in sea-level rise was the key to initiation of deltas between ~8500 and 6500 years B.P., and that Holocene deltaic sequences began to accumulate as the rate of fluvial sediment input overtook the declining rate of sea-level rise along coasts. As determined by radiocarbon data, this threshold occurred on a worldwide basis within a timespan of ~2000 years". Some remarks on terminology are appropriate in the introduction to this chapter. "Sediment load" is used here in a qualitative sense to include all the particulate matter being transported by a river. "Sediment discharge" is used in more quantitative contexts, and it is expressed as mass per unit time (tonnes per day, tonnes per year). "Suspended sediment" describes the material that a river transports in suspension, which includes suspended sand grains as well as silt and clay particles. "Bedload" refers to the sediment that is moved along the river bed, by rolling, sliding, or slipping, within a few grain diameters of the bed. Bedload is not to be confused with "bed material", which refers to the sediment on the river bed regardless of whether it is in motion or at rest.

Because it is difficult to measure accurately, bedload is often ignored in estimates of the sediment discharge of large rivers. In a few large rivers where its quantity has been measured or calculated, bedload amounts to 10 percent or less of the total sediment discharge (Burrows and Harrold, 1983; Emmett, 1981; Jordan, 1965; Mertes, 1985). Ten percent is less than the range of probable error in most measurements (and especially in most estimates) of the discharge of suspended sediment. Therefore, the inclusion of bedload does not improve the accuracy of the estimate of sediment discharge of most major rivers. In some exceptional deltas, however — those of the Brahmaputra and Zaire Rivers (Coleman, 1969; Peters, 1978) — bedload may account for substantially more than 10 percent of the total input of river sediment.

Tabulations of Discharge Data for Major Rivers

Discharges of suspended sediment and water by the world's major rivers are listed in Tables 1 and 2. The 25 largest discharges of river sediment are listed in descending order in Table 1. For comparison, the 25 largest discharges of water are listed in descending order in Table 2. Rivers that do not rank among the

TABLE 1

Average suspended-sediment discharge (in 10^6 t y^{-1}) of major rivers to the coastal zone. Mostly from compilation by Milliman and Meade (1983). Where newer estimates of present-day or earlier sediment discharges are available, their sources are listed in the right column.

	Year 1990	(Year 1890)	Presumed natural level	Source of newer estimate
A. Twenty-five largest sediment discharges:				
1. Amazon	1000–1300	(1000–1300)	1000–1300	Meade et al. (1985)
2. Yellow (Huanghe)	1100	(1200)	100	Milliman et al. (1987)
3. Ganges/Brahmaputra	900–1200			Abbas/Subramanian (1984) Goswami (1985)
4. Yangtze (Changjiang)	480			
5. Irrawaddy	260	(260)		
6. Magdalena	220			
7. Mississippi	210	(500)	400	
8. Godavari	170			Biksham/Subramanian
9. Red (Hunghe)	160			
10. Mekong	160			
11. Orinoco	150	(150)	150	Meade et al. (1990)
12. Purari/Fly	110			
13. MacKenzie	100	(100)	100	
14. Parana/Urugua	100			Drago/Amsler (1988)
15. Salween	~100			Order-of-magnitude guess
16. Pearl (Zhu Jiang)	80			Ren Mei-e, p.c., 1988
17. Copper	70	(70)	70	
18. Choshui (Taiwan)	66			
19. Yukon	60	(60)	60	
20. Amur	52			
21. Indus	50	(250)		Milliman et al. (1984)
22. Zaire	43			
23. Liaohe	41			
24. Niger	40			
25. Danube	40	(70)		Varga et al. (1989); Petschinov (1968)
B. Rivers that formerly discharged large sediment loads:				
Nile	0	(125)		
Colorado	<1	(125)		
C. Rivers that discharge large volumes of water (but have low sediment loads):				
Zambesi	20			
Ob	16			
Lena	11			Shiklomanov/Skakalsky
Columbia	8	(15)		
Yenisey	5	(13)		N.N. Bobrovitskaya, p.c.
St. Lawrence	3			

TABLE 2
Average water discharges (in 10^9 m^3/y) of major rivers to the oceans, circa 1980.

A. Twenty-five largest water discharges:	
1. Amazon	6300
2. Zaire	1250
3. Orinoco	1200 (1)
4. Ganges/Brahmaputra	970
5. Yangtze (Changjiang)	900
6. Yenisey	630 (2)
7. Mississippi	530
8. Lena	510 (2)
9. Mekong	470
10. Parana/Uruguay	470
11. St. Lawrence	450
12. Irrawaddy	430
13. Ob	400 (2)
14. Amur	325
15. MacKenzie	310
16. Pearl (Zhu Jiang)	300
17. Salween	300
18. Columbia	250
19. Indus	240
20. Magdalena	240
21. Zambesi	220
22. Danube	210
23. Yukon	195
24. Niger	190
25. Purari/Fly	150
B. Other rivers that discharge(d) large quantities of water:	
Red (Hunghe)	120
Godavari	92
Yellow (Huanghe)	49
Copper	39
Liaohe	6
Choshui	6
Colorado	1 (was 20)
Nile	0 (was 39)

(1) New estimates provided by D. Pérez Hernandez, Venezuelan Minstry of the Environment and Renewable Resources, Caracas.

(2) New estimates provided by Vladimir Babkin, Russian State Hydrological Institute, St. Petersburg.

first 25 in one table but do fall into the first 25 in the other table, such as the St. Lawrence (large water discharge but small sediment discharge) and Yellow (large sediment discharge but small water discharge), are listed at the bottom of each table. Where two large rivers share a delta, such as the Ganges/Brahmaputra and the Parana/Uruguay, they are listed as a single entity in the tables. Also included in Table 1 are incomplete estimates of the sediment discharges that prevailed a century ago and those that prevailed under natural conditions free of any significant human influence. Most of the entries in these two columns are for rivers in the less populated areas of North and South America whose natural sediment loads can still be measured or inferred. The numbers in these columns represent a first try at such a compilation; where further information or insights are available, other investigators are invited to fill in the blank spaces.

Most of the estimates of suspended sediment listed in Table 1 are subject to considerable error. Of the first three numbers listed, only the estimate for the Yellow River is based on several decades of high-quality data. The estimate for the Amazon is based on only 10 measurements (Meade, 1985). The data from which the Brahmaputra estimate is made are said to be extensive (Goswami, 1985), but they have not been made available to the world hydrologic community in a form that allows for evaluation. Because of the range of error in the estimates of the suspended-sediment discharges of the Amazon and Ganges/Brahmaputra Rivers, the rankings that have been assigned to the first three rivers listed in Table 1 are quite arbitrary. What can be stated with a reasonable degree of certainty is that three of the world's largest river deltas are receiving suspended sediment at average rates of about a billion (10^9) tonnes per year. Furthermore, these three rivers account for 15–20 percent of the global discharge of river sediment to the oceans.

Considerable error is involved in many of the other estimates as well. The estimate of the sediment discharge of the Irrawaddy River is based on samples collected during the 1870s, by methods not necessarily comparable to those used today. The new estimate of 150×10^6 tonnes per year for the Orinoco River probably carries an error of X 50×10^6 tonnes per year. The new estimate of the sediment discharge of the Salween River is an unvarnished guess, based on the measured discharges of the adjacent Irrawaddy and Mekong Rivers.

Spatial Distributions of Sediment Sources and Sinks

With the notable exception of the loess region of China, the greatest sediment discharges on a global scale are associated with the areas of most intense tectonic activity. In their compilation of the global distribution of river-sediment discharges to the coastal zones, Milliman and Meade (1983) estimated that nearly half the river sediment that reached the world's coastline was delivered by the large rivers of southern and eastern Asia whose headwaters rise in the tectonic collision zone between the Asian continent and the Indian subcontinent. They estimated further that another 20 percent was delivered by smaller rivers that drain the tectonically active islands of the western Pacific region; this percentage has been virtually doubled in the more recent compilation by Milliman and Syvitski (1992). Other large sediment loads are delivered by rivers that drain the tectonically and glacially active areas of southern Alaska. Even the large sediments loads carried to the Atlantic Ocean by the major rivers of South America are derived largely from the tectonic regions of the Andes.

Spatial variability of sediment yield is also evident within large river basins. In the Amazon River basin, for example, the distribution of sediment sources differs markedly from that of water sources. Gibbs (1967, p. 1218) estimated that 82 percent of the total suspended solids discharged by the Amazon River is supplied from the 12 percent of the total drainage basin that lies in the Andes. The more recent data portrayed in Figure 1 (upper row) suggest that more than 90 percent of the suspended sediment in the Amazon system is derived from the tectonically active area of the Andes — even though most of the water in the river is derived from the more slowly eroding areas of the basin such as the Guayana and Brazilian shields. Likewise, in the Orinoco River basin, the water is derived in roughly equal amounts from the areas underlain by the Andes and the Guayana Shield, whereas at least 90 percent of the suspended sediment is derived from the Andes (Fig. 1, lower row). In the Yellow River basin of China, more than 90 percent of the sediment in the river comes from the loess region that occupies no more than 40 percent of the total basin area (Fig. 4 of Long and Xiong, 1981).

Although the sediment discharges of the world's large rivers are commonly reported in terms of delivery "to the oceans" (e.g., Milliman and Meade, 1983), this practice can be misleading. A river usually delivers different quantities of sediment to its delta, to the coastline, to the continental shelf, and to the deep sea. Figure 2 shows the different proportions of sediment delivered to these destinations by the world's three major conveyors of river sediment. Of the approximately 10×10^8 tonnes that the Amazon River delivers annually to its delta, about 2×10^8 tonnes are deposited in the delta, 6×10^8 tonnes are deposited on the continental shelf, and the remaining 2×10^8 tonnes are transported northwestward along the coasts of the Guyanas; virtually no Amazon sediment reaches the deep sea under present conditions. Of the 11×10^8 tonnes that the combined Ganges and Brahmaputra Rivers deliver to Bangladesh during an average year, about 6×10^8 tonnes are deposited in the delta, about 3×10^8 tonnes are deposited on the continental shelf, about 1×10^8 tonnes may be transported across the shelf to be deposited in the deep sea. Of the 11×10^8 delivered by the Yellow River during an average year, 9×10^8 tonnes are deposited within the delta, and the remaining 2×10^8 tonnes are transported alongshore or deposited on the continental shelf, virtually all within the Gulf of Bohai.

Temporal Variations in River-Sediment Discharge

Interseasonal and interannual variations in the suspended-sediment discharges of two large rivers, the Yangtze and the Mississippi, are shown in Figure 3. These two rivers are the largest in the world (in terms of water discharge) for which daily suspended-sediment data are available for continuous periods of several decades. Both rivers are influenced by human activities. The Yangtze River basin is home to nearly 10 percent of the world's population and it has been heavily cultivated for millennia. Although some of the forests in the basin still are being converted to

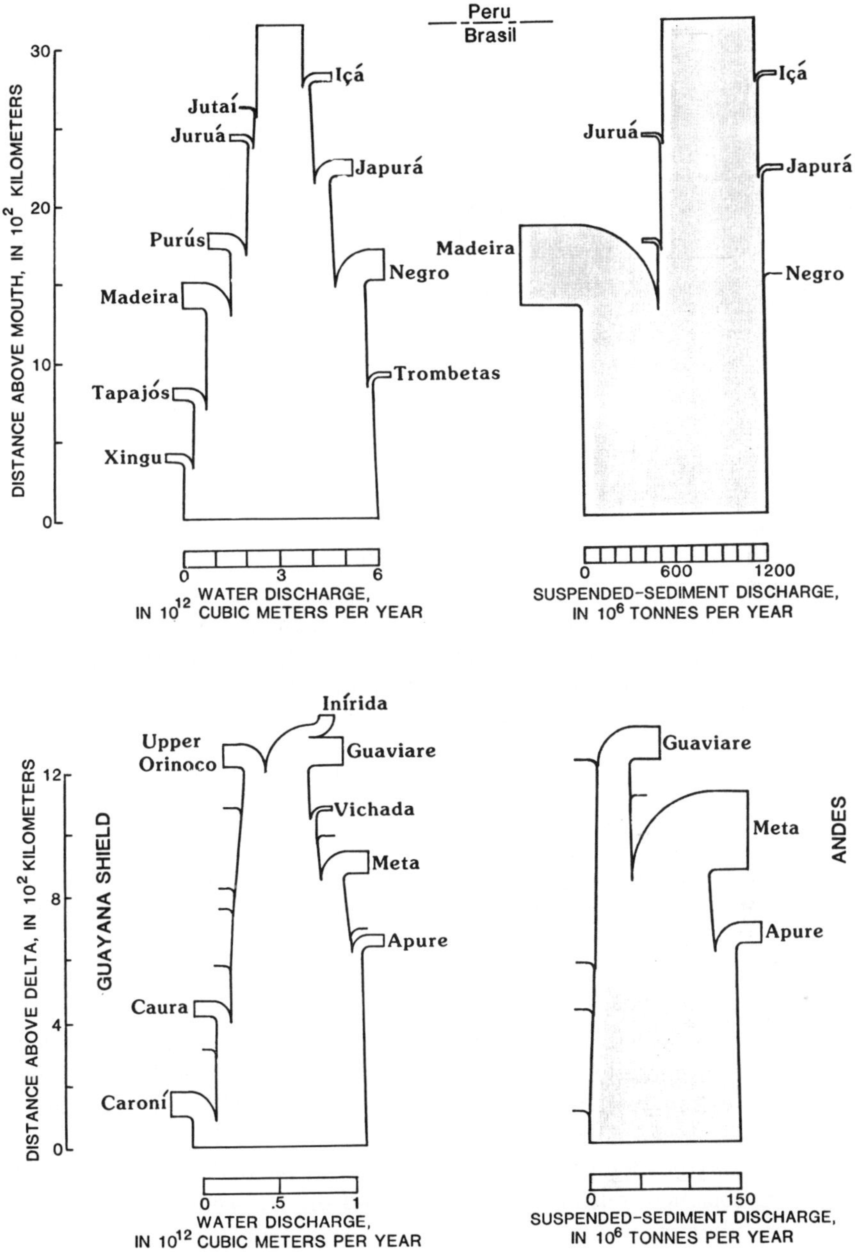

Fig. 1. Spatial distributions of long-term average discharges of water and suspended sediment in two large river basins (Meade, 1994). Upper row, Amazon River of Brazil. Water-discharge data from unpublished records of Companhia de Pesquisas de Recursos Minerais, Manaus and Belèm. Sediment data from Meade (1985). Presumptive losses of suspended sediment by deposition in the delta or on the floodplains of the lowermost river are not shown here (see Fig. 2, left). Lower row, Orinoco River of Venezuela and Colombia. Water-discharge data from Avellan Vegas et al. (1969). Sediment data from Meade et al. (1990b).

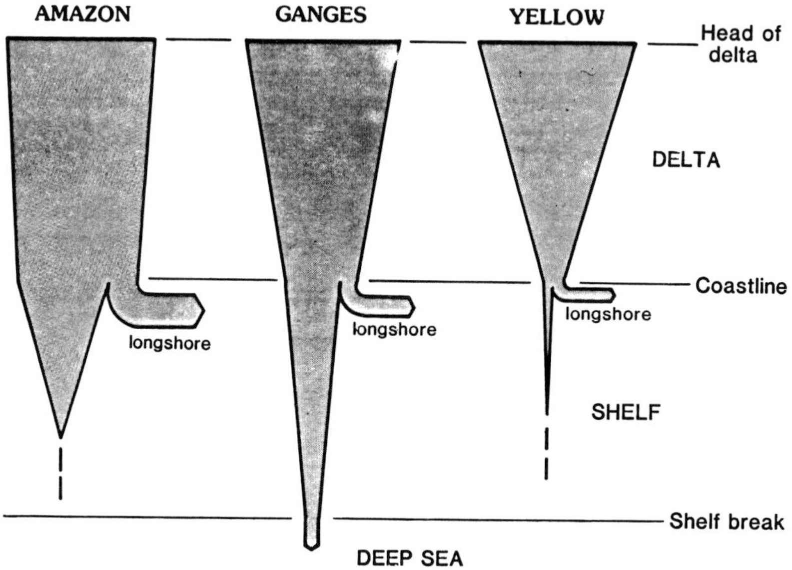

Fig. 2. Different proportions of average suspended-sediment discharges delivered to deltas, coastlines, continental shelves, and the deep sea by the world's three major sediment-discharging rivers (Awosika et al., 1994). Sediment discharge at the head of each delta is portrayed here as the same quantity (100%) regardless of its absolute value (probable range, 900 to 1300 $\times 10^6$ tonnes per year) and regardless of changes in sediment discharges upriver of the delta. *Left*: Amazon River of Brazil (Eisma et al., 1991; Kuehl et al., 1986; Meade, 1994). *Middle*: Ganges (including Brahmaputra) River of Bangladesh (Kuehl et al., 1989; Milliman, unpublished data). *Right*: Yellow River of China (Bornhold et al., 1986; Milliman et al., 1987; Ding, 1989).

croplands (Shi et al., 1985), no increases or decreases in the sediment loads of the Yangtze River have been discerned during the last 30 years (Zhang and Liu, 1989). The Mississippi River's principal contributor of sediment, the Missouri River, has been controlled since 1953 by a series of large dams (see Fig. 27 of Meade and Parker, 1985, or Fig. 13 of Meade, Yuzyk and Day, 1990). For the periods of record used in Figure 3, therefore, we can assume at least quasi-stationary conditions — i.e. conditions that have remained sufficiently stable to allow the use of these periods of record to illustrate the nature of the interseasonal and interannual differences in the sediment discharges of large rivers.

The most obvious temporal variations in the suspended-sediment discharges of rivers are seasonal, reflecting the wetter and drier seasons of the years that follow each other in cyclic patterns. Because sediment discharges usually are power functions (simple or, more often, complex) of water discharges, the seasonal variations of sediment discharge usually are exaggerated versions of the seasonal variations of water discharge (see, for example, Fig. 2 of Yang et al., 1983). In two of the world's largest rivers, the median sediment discharges during seasons of highest runoff (July–September in the Yangtze River, March–June in the Mississippi River) are 5 to 20 times the median sediment discharges during the low-runoff months (Figs. 3A and 3B).

Interannual differences in the sediment discharges of large rivers are significantly smaller than the interseasonal differences. Figures 3C and 3D show differences of only 2 to 3 times between sediment discharges during wetter and drier years in the Yangtze and Mississippi Rivers. Interannual variations in sediment discharge are much greater in smaller rivers (see Fig. 33 of Meade and Parker, 1985, or Fig. 3 of Meade, Yuzyk and Day, 1990), but, in general, the larger the river, the smaller the difference in sediment discharge from one year to the next.

On a decadal time scale, large floods can cause marked increases in the sediment discharges of rivers of moderate size. The Eel River of northern California, for example, transported as much sediment during 10 days of record flooding in December 1984 as it normally transported during 10 average years (Brown and Ritter, 1971). In the largest river basins, however, the effects of such locally extreme events apparently are damped by more normal flows from less affected parts of the basin. Notice in Figures 3C and 3D that the total sediment discharges during the exceptional flood years of 1954 and 1981 in the Yangtze and 1973 in the Mississippi were not extraordinarily large.

On a multi-decadal time scale, some natural variations in river-sediment loads seem to be related to fluctuations in climate. A marked decrease in suspended-sediment discharge, for example, was recorded in the middle 1940s in the Colorado River basin of the southwestern United States. Although it was not related to the construction of dams and reservoirs, this decrease was first thought to be due to artificial causes: a change in sediment-sampling equipment, or changes in livestock grazing practices (Hadley, 1974). More recent field studies have delineated a pattern of increased aggradation and accumulation of sediment since the 1940s in the tributary basins of the Colorado River (Hereford, 1984). A consensus among field investigators favors the conclusion that the present episode of aggradation followed a subtle shift in climate, which perhaps was the obverse of the conditions that triggered the episode of intense degradation and gullying that characterized the region during the late 1800s (Cooke and Reeves, 1976). In any event, the causes of this century-scale cycle of degradation and aggradation are presently considered to have been natural rather than man-induced.

Storage of Sediment in Large River Systems

Sediment is stored in large rivers systems at many different time scales. This storage confounds our understanding of temporal variation in sediment loads because very little direct linkage is visible between the erosion of uplands and the discharges of sediment at the mouths of large rivers. In the conterminous United States, for example, upland erosion rates exceed the rates of river-sediment discharge into the coastal zones by a factor of ten (Curtis et al., 1973; Holeman, 1981). Said another way, ninety percent of the sediment presently being eroded off the land surface of the United States is being stored somewhere in the river systems between the uplands and the sea, most likely on hillslopes and floodplains. Furthermore, the time lags between erosion and sediment transport are such that the sediments carried by large rivers today may represent episodes of erosion that occurred decades, centuries, or even millennia ago. Excellent summaries of the problems involved in

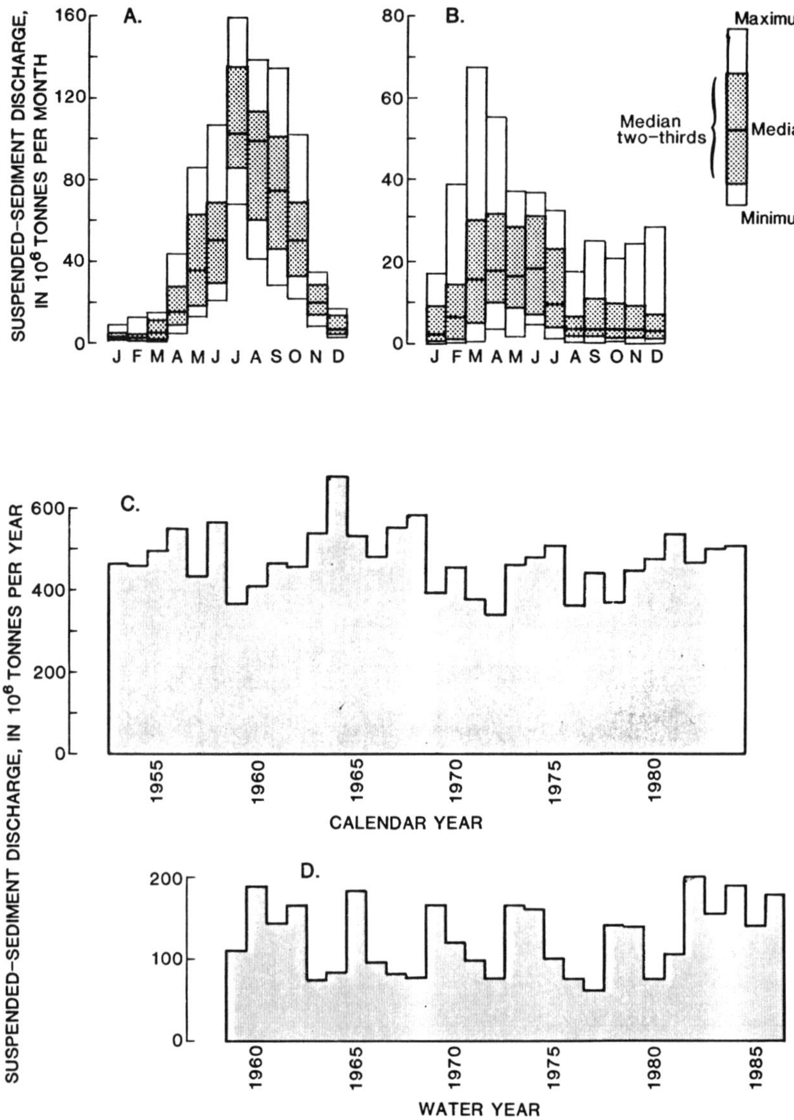

Fig. 3. Monthly and yearly variations of suspended-sediment discharges in two large rivers, based on long-term records of daily measurements. (A) Monthly discharges of suspended sediment of Yangtze River at Datong, showing maximum, median and minimum values recorded during 32-year period, 1953–1984 (data of Yangtze Valley Planning Office). (B) Monthly discharges of suspended sediment of Mississippi River at St. Louis during 28-year period, 1959–1986 (data of U.S. Geological Survey). (C) Yearly discharges of suspended sediment, Yangtze River at Datong, 1953–1984. Average for 32-year period, 470×10^6 tonnes per year. (D) Yearly discharges of suspended sediment, Mississippi River at St. Louis, 1959–1986. Average for 28-year period, 128×10^6 tonnes per year. Water year extends from 1 October through 30 September.

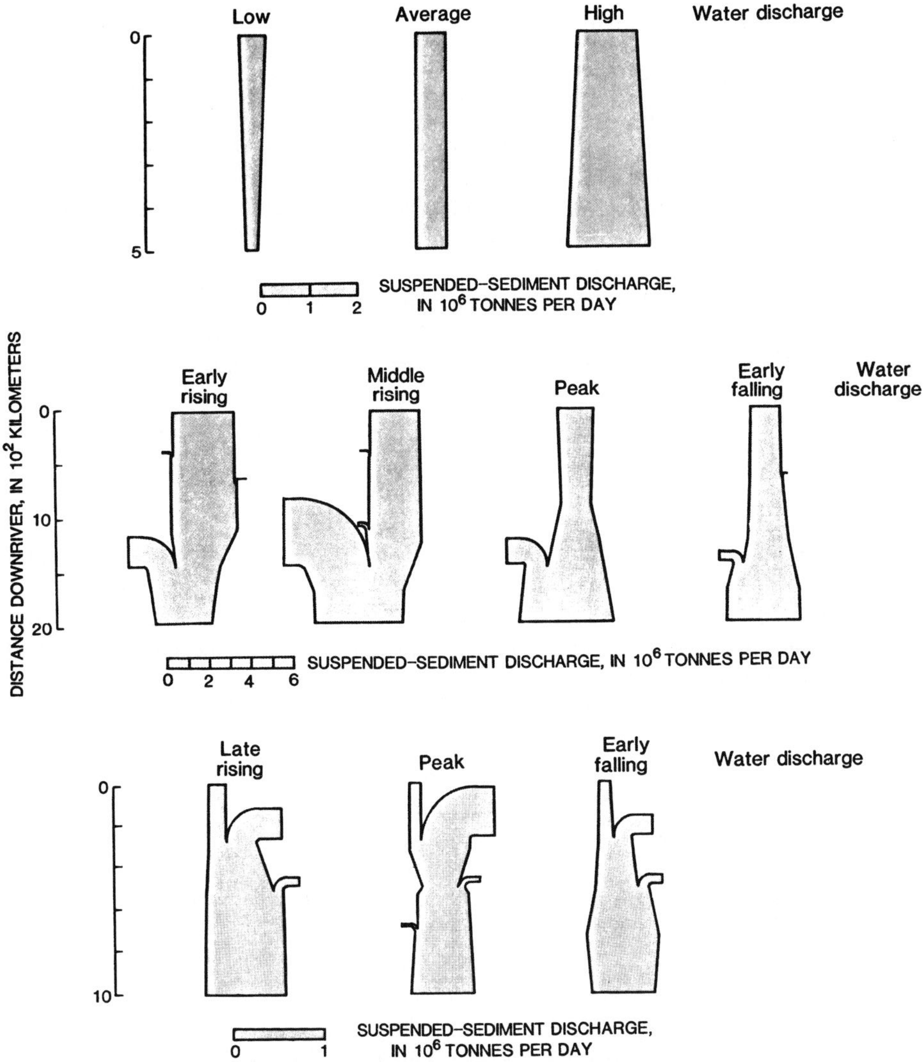

Fig. 4. Patterns of seasonal storage of river sediment, shown by measurements of suspended-sediment discharge made under different water-discharge conditions. *Upper row*: Mississippi River in the 500-km reach nearest the mouth. Compiled from data of Everett (1971, p. 14) and Wells (1980, p. 13). *Middle row*: Amazon River in the 2000-km reach between Santo Antonio do Ica and Obidos, Brazil (Meade et al., 1985). *Lower row*: Orinoco River in the 1000-km reach upriver of Ciudad Guayana, Venezuela (Meade et al., 1983; Meade, 1994).

understanding the linkage (or lack of it) between soil erosion and river-sediment transport have been prepared by Schumm (1977) and Walling (1983).

Storage on Seasonal Time Scales

On a seasonal time scale, rivers generally tend to deposit sediments on their beds and along their banks at low or falling discharges, and to resuspend these sediments at high or rising discharges. But there are some exceptions to these general tendencies, as we shall see below in examples taken from the three biggest rivers of North and South America. Whether the storage and resuspension of sediments on seasonal time scales have any residual effects on the longer-term storage of river sediment is not clear.

In the lowermost 500 km of the Mississippi River of the United States, suspended sediment is stored and remobilized in response to seasonal changes in water discharge and river slope (Fig. 4, upper row). At less than average water discharges and water-surface slopes, suspended sediment is dropped by the river and stored on the river bed. At average water discharges, sediment is neither stored nor resuspended, on a net basis, in this reach. At greater than average water discharges and water-surface slopes, at least part of the previously stored sediment is resuspended from the river bed. The pattern of storage and resuspension in this reach exaggerates the direct correlation between water discharges and sediment loads: at lower discharges, the already small sediment loads being transported into the upper end of the reach are further reduced by deposition and storage on the river bed; at higher discharges, the already large sediment loads are increased further by resuspension from the river bed.

In a 750-km reach of the lower Amazon River of Brazil (downriver distances 1250–2000 km in Fig. 4, center row), the seasonal pattern of storage and remobilization of suspended sediment is more closely correlated with river slope than with water discharge. Because of the timing of inputs from several large tributaries, the slope of the river surface in this reach is, contrary to the usual pattern, smaller on rising stages than on falling stages. Consequently, sediment is stored during rising river stages and resuspended during peak and falling river stages, in quantities on the order of 10^6 tons per day (Fig. 4, center row). This pattern of sediment storage and resuspension, in contrast to the pattern observed in the lower Mississippi, regulates the discharge of sediment to the sea. During rising river stages, when sediment loads from the upper Amazon are large, the lower Amazon stores part of the sediment before it reaches the sea. During falling stages, when sediment loads from the upper river are smaller, the lower river augments the loads being transported seaward by resuspending sediment from the river bed.

A 200-km reach of the middle Orinoco River of Venezuela shows still another pattern of seasonal storage and resuspension. Sediment discharges through this reach are shown at downriver distances between 300 and 500 km on the ordinate scale in the lower row of Figure 4. When the river stage is near its annual peak, this reach of the Orinoco forms a large area of backwater that includes the lower reaches of a number of tributaries as well as large tracts of intervening low-lying flood plains. During peak stage, nearly half the inflowing sediment is stored in this reach. During subsequent falling stages and during the early part of the next rising stage, this reach of the Orinoco yields sediment from storage to be transported farther downriver. Sediment loads in the lower Orinoco, therefore, are smaller at peak water discharge than on rising or early falling stages.

Storage on Decade to Century Time Scales

Perhaps our perceptions of sediment storage in river systems are biased toward the time scale of 10 to 100 years which is, after all, the secular scale, or the scale of a human life span. Any direct sensory perceptions we may have of changes in rivers are confined to processes that operate over periods of 100 years or less.

On the basis of suspended-sediment discharges measured in the Brahmaputra River and its tributaries in the Assam province of northeastern India, Goswami (1985) calculated a sediment budget for a 607-km reach of the river during the period 1971–1979. Some 2100×10^6 tonnes of sediment, or 70 percent of the total sediment brought into the reach during the period 1971–1979, remained stored in the reach at the end of the period. This quantity of stored sediment must have aggraded the river bed by an average of 10–30 cm during the period 1971–1979. Goswami (1985, p. 977) speculates that the river channel "is currently experiencing a secular period of rapid aggradation", perhaps triggered by the great earthquake of 1950 in the Himalayas, which should be followed in due time by a period of relatively slower removal. Such an episode of remobilization and degradation of the river bed would send a large pulse of sediment out of Assam and into Bangladesh and the delta at the head of the Bay of Bengal. Because only 8–9 years of sediment data were available here, we can only speculate on the period of such a cycle of aggradation and degradation, but it is perhaps most easily visualized at a time scale somewhere between several decades and a century.

The classic case study of the movement and storage of sediment in a river system on time scales of 10 to 100 years is that of the hydraulic-mining debris in the Sacramento River valley of California (Gilbert, 1917; Kelley, 1959). Between 1855 and 1885, enormous quantities of coarse sediment were washed into some of the tributaries of the Sacramento River during hydraulic mining for gold. The resulting problems that developed downstream (flooding, filling of navigation channels, destruction of flood-plain farms) became so serious that hydraulic mining was curtailed by a court decision in 1884. By that time, however, the large mass of sediment, characterized as a "wave" by Gilbert (1917), was already in the stream channels and was moving slowly down the tributaries and in the Sacramento River. As the mass of sediment advanced, it raised the levels of the channel beds, much as an ocean swell raises the level of the sea as it passes through. Bed levels rose 5 m in the tributary Yuba River at Marysville and nearly 3 m in the Sacramento River at Sacramento. The river beds at these towns reached their highest elevations 10 to 20 years after the mining was stopped, and then they declined steadily during the next 30 to 40 years to their previous elevations. All in all, the great wave of hydraulic-mining debris took nearly a century to pass through the channels of the Sacramento River system and finally to reach San Francisco Bay (Meade, 1982, p. 243–244; Smith, 1965).

Storage on Century to Millennium Time Scales

The pattern of storage and remobilization described in the preceding paragraph, however, applies only to the sediment in and near the river channels. It does not apply to the debris that overflowed onto the flood plains. The hydraulic-mining debris that was carried out of the river channel during floods and deposited on the flood plains was sufficient in many places to cover entire houses and orchards (Kelley, 1959, p. 134–135, 203–204). Most of that debris still remains where it was deposited a century ago: In some tributary basins of the Sacramento River, about 90 percent of the hydraulic-mining debris still remains stored on flood plains (Adler, 1980; James, 1989). The time required to remove sediment from storage on the flood plain is much longer than the century that was required to remove the debris from the main river channels. Flood-plain deposits are removed mainly by erosion of channel banks as streams slowly migrate laterally, a process that proceeds at a substantially slower pace than the vertical removal of material stored in the bottom of the river channel. The complete remobilization of flood-plain deposits may require time periods of an order of a millennium or more (Leopold et al., 1964, p. 328).

Many of the problems associated with the prediction of storage and remobilization of sediment on time scales of 100 to 1000 years were demonstrated in a study carried out on Coon Creek, a small tributary of the Mississippi River that drains 360 km^2 of southwestern Wisconsin. Originally covered by forests, Coon Creek basin was settled by European immigrants and cleared for farming about 1850. As the forests were cleared and the land was plowed, a cycle of erosion and sedimentation began, the consequences of which are still strongly in effect today. Sediments were eroded at a greatly accelerated rate from upland and tributary areas, and were transferred to the lower hillslopes and valleys of the creek basin. Much less than 10 percent (5×10^6 tons) of the sediment eroded from the uplands during the years since 1850 was exported out of the basin by the creek. More than 90 percent (80×10^6 tons) of the sediment was deposited along the way, on hillslopes and flood plains, where most of it still remains in storage. Upland erosion rates, therefore, are not reflected in the sediment yields at the mouth of Coon Creek (Trimble, 1983).

In tectonically active regions of the world, cycles of erosion, storage, and remobilization of sediment in river valleys may be greatly accelerated. Grant (1985) presents evidence for 8 major episodes of erosion and alluvial sedimentation in New Zealand during the last 1,800 years. He believes that the major episodes of erosion and sedimentation were related to fluctuating magnitudes of major rainstorms and floods. During periods of more intense rainfall, sediment was eroded from steep hillslopes and deposited as alluvium in the stream valleys. During intervening periods of less intense rainfall, the alluvial deposits were degraded, leaving behind terraced valleys. These valleys were then refilled with sediments during the next episode of heavier rainfall. In tectonically less active regions of the world, such episodes of erosion and sedimentation proceed more slowly (Leopold and Miller, 1954; Judson, 1963).

Human Influences on River-Sediment Loads

Man's activities have both increased and decreased the natural quantities of river sediment that are delivered to the coastal zones. Much has been written on this subject, particularly on those human activities that increase river-sediment loads (see Douglas, 1967; Jansson, 1982, p. 38–41; Gregory and Walling, 1973, p. 342–358; Judson, 1968). The "presumed natural level" column of Table 1 represents a preliminary attempt at estimating the natural sediment loads of major rivers so that they can be compared with those of the present day.

Human Activities that Increase River-Sediment Loads

On a global scale, the most massive anthropogenic increases in river sediment loads have come about as a result of crop farming, especially in areas where forests have been cut down to make way for croplands. The settlement of the eastern parts of Canada and the United States during the 17th, 18th, and 19th centuries was marked by the replacement of native forests by croplands. During the 19th century, especially after the general adoption of the moldboard plow, prairie grasslands were converted to deeply plowed fields. These conversions caused order-of-magnitude increases in soil erosion and corresponding increases in the sediment yields of rivers (Meade, 1969, 1982). Similarly, the extensive agricultural development of the loess plateau of northern China that began about 200 BC also is believed to have caused an order-of-magnitude increase in the suspended-sediment discharge of the Yellow River (Milliman et al., 1987). Although the evidence of such changes in other large river basins are obscured in antiquity and by other long-term effects such as climatic shifts, we can surmise that the anthropogenic additions to the natural sediment loads of rivers such as the Indus and Ganges must have contributed considerably to the accelerated growth of their deltas. Because of the storage of sediment in river systems and the time lags between upland erosion and the delivery of river sediment to the coastal zone, the agriculturally-induced increases in the growth rates of large river deltas are most appropriately considered in temporal frameworks of millennial scale. More recent conversions of virgin forest lands to cattle-grazing lands in the Amazon region of Brazil, for example, so far have not caused any discernible increases in the sediment loads of the major trunk streams or in the delivery of sediment to the Amazon Delta. Nor is there any sound hydrologic evidence for any increase in the flooding of the Amazon that can be related to deforestation (see the discussions by Nordin and Meade, 1982, Richey et al., 1989, and Sternberg, 1987).

Human Activities that Decrease River-Sediment Loads

One of the most pervasive influences on sediment loads delivered to the coastal zones of the world is exerted by the dams and reservoirs that have been built in large numbers across large rivers. Dams are built to impound water for various purposes, and the reservoirs they form interrupt the downriver flow of sediment. Although the river water that enters a reservoir is released eventually (through a power plant, into a diversion canal, or over a spillway), much of the sediment is trapped permanently in the reservoir. Nearly all reservoirs on major rivers trap at least one-half of the river sediment that flows into them. Some of the largest reservoirs in the world, like Lake Mead on the Colorado River and Lake Nasser on the Nile River, trap virtually all the sediment that flows into them (Brune, 1953).

Examples are described below from rivers of North America where sufficient historical data are available. The effects shown in the North American examples can be presumed to follow the construction of dams and reservoirs on other major rivers of the world, judging from recent assessments of the Indus and Yenisey Rivers and rivers of northern Turkey (Milliman et al.,1984; Bobrovitskaya, 1994; Hay, 1994).

The classic example in North America of the interruption of a large discharge of river sediment to the coastline is that of the Colorado River. Before about 1930, the Colorado River delivered an average of 125 to 150 $\times$ 10^6 tonnes of suspended sediment per year to its delta at the head of the Gulf of California. Since the closure of Hoover Dam, which began in 1935, this rate of sediment delivery has declined, first precipitously and then more gradually, to an average annual amount today of about 100,000 tonnes. Aside from a period between 1934 and 1938, when 30×10^9 m^3 of the river water was appropriated for the initial filling of Lake Mead behind Hoover Dam, the quantity of water carried by the Colorado River past Yuma, Arizona, has declined more or less progressively (Fig. 5). This decline has been in response to the increasing diversion of water from the Colorado River for irrigation of croplands and for municipal water supplies. The more abrupt decline in sediment discharge at Yuma (Fig. 5) clearly was related to a single event, the closing of Hoover Dam. This sequence of events is strongly analogous to that in the lower Nile River, wherein the annual suspended-sediment discharges were decreased from about 125×10^6 tonnes to less than 3×10^6 tonnes following the closure of the High Dam at Aswan (Shahin, 1985, p. 460). Since the closure, the annual accumulations of new sediment in Aswan High Dam Reservoir have averaged about 100×10^6 m^3 (El Dardir, 1994), while the Mediterranean shoreline of the outer Nile River Delta has receded significantly (see the chapter in this volume by El-Sayed).

Another large river system whose sediment loads are strongly influenced by reservoirs is the Mississippi. Previous to extensive European settlement of the Mississippi Valley, and up to about 1950, the major sources of sediment in the river were the western tributaries, the Missouri River and, to a lesser extent, the Arkansas River. Following World War II, a series of five large dams was completed on the Missouri River for irrigation, hydroelectric power, and navigation control (Meade and Parker, 1985). During the same time, a series of locks and dams was constructed on the Arkansas River (Madden, 1965). Beginning in the early 1950s, downstream sediment loads were diminished, and the effect could be observed all the way down to the mouth of the Mississippi River. The diagrams in Figure 6 compare the estimated sediment discharges as of circa 1700 with those measured circa 1980. Even though the decrease in sediment loads from the western tributaries is partly offset by an increase in the sediment inputs from the more humid Ohio River Valley (related mainly to the conversion of the original forests to croplands, mostly during the late 19th and early 20th centuries), sediment discharges to the Gulf of Mexico by the Mississippi River are now less than one-half of what they were before 1950. The reduction of river-sediment discharges during the last four decades certainly has affected the rates of retreat of coastal wetlands in parts of the Mississippi River Delta (see the chapter in this volume by J.T. Wells).

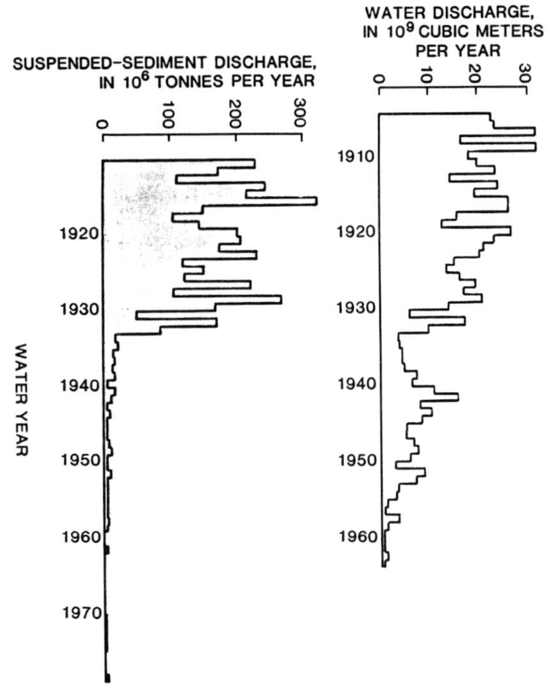

Fig. 5. Annual discharges of water (1905–64) and suspended sediment (1911–79) in the Colorado River at Yuma, Arizona. Compiled from data of U.S. Geological Survey and U.S. Bureau of Reclamation by Curtis et al. (1973) and Meade and Parker (1985). The abrupt decrease in suspended-sediment discharge in the middle 1930s coincided with the closure of Hoover Dam, 500 km upriver of Yuma.

Amelioration of Human Influences on River-Sediment Loads

Considering how river-sediment loads have been changed by human activities, it seems reasonable to suggest that these changes might be reversed by other, compensatory, activities. For example, if sediment loads have been increased by the conversion of forests to croplands, might not such loads be decreased by introducing soil-conservation measures or even by converting croplands back to woodlands? Where such measures have been introduced, however, their effects have not been discernible in rivers of moderate size (Meade and Trimble, 1974; Trimble, 1977; Meade, 1982), and their effects are likely to be even less detectable in large rivers. The importance of the storage effects discussed above probably increases with increasing river size. The remobilization of stored materials from the bed and banks of a large river can overwhelm the effects of any change in the sediment supply from eroded uplands for time periods measurable in centuries or even millennia.

More immediate effects are possible in the design and operation of dams and reservoirs. Sediment that was formerly trapped in a reservoir can be flushed downriver by changing the design of the dam or by altering the operating schedule of the reservoir. For example, the Sanmenxia Dam on the Yellow River was

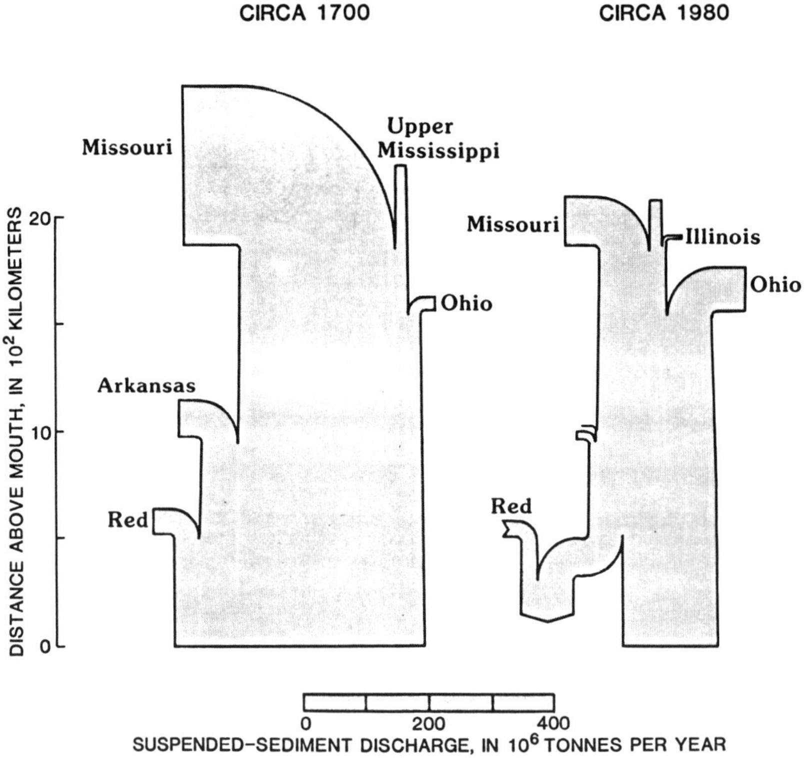

Fig. 6. Long-term average discharges of suspended sediment in the lower Mississippi River circa 1700 and circa 1980. Estimate of discharge circa 1700 from Meade et al. (1990a). Discharge circa 1980 from compilation by Keown et al. (1986).

originally completed in 1960 with no provision for bypassing large sediment loads. After only four years of operation, the deposition of sediment had decreased the water-storage capacity of the reservoir by more than 60 percent. A major reconstruction project (1965-1973) installed new sluicing outlets through which the sediment is now flushed downriver toward the Yellow River Delta. Based on the experience at Sanmenxia, similar sluicing works were incorporated in the Gezhouba Dam on the Yangtze River, and similar structures are being included in the design for the proposed Three Gorges Project on the same river. Consequently, the effects of these large dams on the sediment discharge to the Yangtze Delta are expected to be slight (Qian et al., 1987; Tang and Lin, 1987).

What of the future of large reservoirs? In the absence of widespread adoption of bypassing measures such as those at Sanmenxia and Gezhouba, many of the world's major reservoirs may well be completely filled with deposited sediment by the end of the 21st Century. Many of the major reservoirs built during the middle years of the 20th Century were designed to accommodate only 100 years worth of sediment accumulation. More recent changes — alterations of land-use patterns, increased soil conservation, construction of additional reservoirs upstream — have extended the projected life of many of these major reservoirs. However, at some point in the second half of the 21st Century, the capacities of many major reservoirs for trapping sediment will be approaching zero.

What scenarios can we project for the years 2050–2100? Are the sediment loads of major river systems, which were increased by man-induced acceleration of soil erosion during the last few millennia, and which have been decreased by dams and reservoirs during the last few decades, likely to be increased again when the reservoirs become too full to trap any more sediment? Are the time frames of anticipated sea-level rise or deltaic subsidence such that they may be significantly offset by resumed inputs of large loads of river sediment?

Acknowlegements

New data on river sediment discharges were provided by Ren Mei-e of Nanjing University, V. Subramanian of Jawaharlal Nehru University, New Delhi, and Nelly N. Bobrovitskaya of the Russian State Hydrological Institute, St. Petersburg. New estimates of river water discharges were provided by David Pérez Hernandez of the Venezuelan Ministry of the Environment and Renewable Natural Resources, Caracas, and Vladimir Babkin of the Russian State Hydrological Institute. Permission to use the Yangtze River data in Figure 3 were obtained by Zhao Chuanshao and Yang Dingyuan of the Ministry of Water Resources, Beijing. The manuscript was reviewed by J.D. Milliman and Christopher Shaw.

References

Abbas, N. and Subramanian, V., 1984. Erosion and sediment transport in the Ganges River basin (India). *Jour. Hydrol.*, 69, 173-182.

Adler, L. L., 1980. Adjustment of the Yuba River, California, to the Influx of Hydraulic Mining Debris, 1849-1979. University of California, Los Angeles, M. A. thesis, 180 pages.

Avellan Vegas, F., Porras Gonzalez, P., Rivero Pantoja, G. and Chacon Mendoza, F., 1969. Inventario Nacional de Aguas Superficiales (Venezuela) Comision del Plan Nacional de Aprovechamiento de los Recursos Hidraulicos; v. 1, 127 pages; v. 2, 14 plates.

Awosika, L.F., Meade, R.H., Milliman, J.D., Wells, J.T. and Windom, H.L., 1994). Anthropogenic influences on sediment discharge to the coastal zone and environmental consequences. United Nations Joint Group of Experts on the Scientific Aspects of Marine Environmental Protection (GESAMP) Rep. and Studies, 51, 67 pages.

Biksham, G. and Subramanian, V., 1988. Sediment transport of the Godavari River basin and its controlling factors. *Jour. Hydrol.*, 101, 275-290.

Bobrovitskaya, N.N., 1994. Assessment of trends to sediment discharge variations in the rivers of the Former Soviet Union (FSU), in Proceedings, International Symposium, East-West, North-South Encounter on the State-of-the-art in River Engineering Methods and Design Philosophies. State Hydrological Institute, St. Petersburg, 2, 32-39.

Bornhold, B.D., Yang, Z.-s., Keller, G.H., Prior, D.B., Wiseman, W.J., Jr., Wang, Q., Wright, L.D., Xu, W.-d. and Zhuang, Z.-y., 1986. Sedimentary framework of the modern Huanghe (Yellow River) Delta. *Geo-Marine Letters*, 6, 77-83.

Brown, W. M., III and Ritter, J. R., 1971. Sediment transport and turbidity in the Eel River basin, California. U. S. Geol. Surv. Water-Supply Paper 1986, 70 pp.

Brune, G. M., 1953. Trap efficiency of reservoirs. *Am. Geophys. Union Trans.*, 34, 407-418.

Burrows, R. L. and Harrold, P. E., 1983. Sediment transport in the Tanana River near Fairbanks, Alaska. U. S. Geol. Surv. Water-Res. Invest. Rep. 83-4064, 116 pp.

Cooke, R. U. and Reeves, R. W., 1976. *Arroyos and Environmental Change in the American South-West.* Clarendon Press, Oxford: 213 pages.

Coleman, J. M. (1969). Brahmaputra River: channel processes and sedimentation. Sed. Geol. 3, 131-239.

Curtis, W. F., Culbertson, J. K. and Chase, E. B., 1973. Fluvial-sediment discharge to the oceans from the conterminous United States. U.S. Geol. Surv. Circ. 670, 17 pp.

Ding, L.-y., 1989. Improvement of the mouth of the Yellow River and sediment disposal. In: Brush, L.M., Wolman, M.G. and Huang, B.-W.(eds.), *Taming the Yellow River: Silt and Floods*. Kluwer Academic Publishers, Dordrecht, 637-655.

Douglas, I., 1967. Man, vegetation, and the sediment yields of rivers. *Nature*, 215, 925-928.

Drago, E. E. and Amsler, M. L., 1988. Suspended sediment at a cross section of the Middle Parana River: concentration, granulometry and influence of the main tributaries. In: Bordas, M. P. and Walling, D.E. (eds.), Sediment Budgets. *Internat. Assoc. Hydrol. Sci. Pub.* , 174, 381-396.

El Dardir, M., 1994. Sedimentation in Nile High Dam Reservoir, 1987-1992, and sedimentary futurologic aspects. *Sedimentology of Egypt*, 2, 23-39.

Eisma, D., Augustinus, P.G.E.F. and Alexander, C., 1991. Recent and subrecent changes in the dispersal of Amazon mud. *Netherlands Jour. Sea Res.*, 28, 181-192.

Emmett, W. W., 1981. Measurement of bed load in rivers, in Erosion and Sediment Transport Measurement. *Internat. Assoc. Hydrol. Sci. Pub.* 133, 3-15.

Everett, D. E., 1971. Hydrologic and quality characteristics of the lower Mississippi River. Louisiana Dept. Public Works Tech. Rep. 5, 48 pp.

Gibbs, R. J., 1967. The geochemistry of the Amazon River system: Part I. The factors that control the salinity and the composition and concentration of the suspended solids. *Geol. Soc. America Bull.*, 78, 1203-1232.

Gilbert, G. K., 1917. Hydraulic-mining debris in the Sierra Nevada. U. S. Geol. Surv. Prof. Paper 105, 154 pp.

Goswami, D. C., 1985. Brahmaputra River, Assam, India: physiography, basin denudation, and channel aggradation. *Water Resources Res.*, 21, 959-978.

Grant, P. J., 1985. Major periods of erosion and alluvial sedimentation in New Zealand during the Late Holocene. *Royal Soc. New Zealand Jour.*, 15, 67-121.

Gregory, K. J. and Walling, D. E., 1973. *Drainage Basin Form and Process.* John Wiley & Sons, New York: 456 pp.

Hadley, R. F., 1974. Sediment yield and land use in southwest United States. In: Effects of Man on the Interface of the Hydrological Cycle with the Physical Environment. *Internat. Assoc. Hydrol. Sci. Pub.* 113: 96-98.

Hay, B.J., 1994. Sediment and water discharge rates of Turkish Black Sea rivers before and after hydropower dam construction. *Environm. Geol.*, 23, 276-283.

Hereford, R., 1984) Climate and ephemeral-stream processes: twentieth-century geomorphology and alluvial stratigraphy of the Little Colorado River, Arizona. *Geol. Soc. America Bull.* 95: 654-668.

Holeman, J. N., 1981. The national erosion inventory of the Soil Conservation Service, U. S. Department of Agriculture, 1977-79, in Erosion and sediment transport measurement. *Internat. Assoc. Hydrol. Sci. Pub.* 133, 315-319.

James, L. A., 1989. Sustained storage and transport of hydraulic gold mining sediment in the Bear River, California. *Assoc. Am. Geogr. Annals* 79, 570-592.

Jansson, M. B., 1982. Land erosion by water in different climates. Uppsala Univ. Naturgeografiska Inst. Rapp., 57, 151 pp.

Jordan, P. R., 1965. Fluvial sediment of the Mississippi River at St. Louis, Missouri. U. S. Geol. Survey Water-Supply Paper 1802: 89 pp.
Judson, S., 1963. Erosion and deposition of Italian stream valleys during historic time. *Science*, 140, 898-899.
Judson, S., 1968. Erosion of the land, or what's happening to our continents? *Am. Scientist*, 56, 356-374.
Kelley, R. L., 1959. *Gold vs. Grain - The Hydraulic Mining Controversy in California's Sacramento Valley*. Arthur H. Clark, Glendale, California: 327 pp.
Keown, M. P., Dardeau, E. A., Jr. and Causey, E. M., 1986. Historic trends in the sediment flow regime of the Mississippi River. *Water Resources Res.*, 22, 1555-1564.
Kuehl, S.A., DeMaster, D.J. and Nittrouer, C.A., 1986. Nature of sediment accumulation on the Amazon continental shelf. *Continental Shelf Res.*, 6, 209-225.
Kuehl, S.A., Hariu, T.M., and Moore, W.S., 1989. Shelf Sedimentation off the Ganges-Brahmaputra river system: evidence for sediment bypassing to the Bengal fan. *Geology*, 17, 1132-1135
Leopold, L. B. and Miller, J. P., 1954. A postglacial chronology for some alluvial valleys in Wyoming. U. S. Geol. Surv. Water-Supply Paper1261, 90 pages.
Leopold, L. B., Wolman, M. G. and Miller, J. P., 1964. *Fluvial Processes in Geomorphology*. W. H. Freeman & Co., San Francisco: 522 pp.
Long, Y-q. and Xiong, G.-s., 1981. Sediment measurement in the Yellow River, in Erosion and sediment transport measurement. *Internat. Assoc. Hydrol. Sci. Pub.* 133, 275-285.
Madden, E. B., 1965. Channel design for modified sediment regime conditions on the Arkansas River, in Proceedings, Federal Inter-Agency Sedimentation Conference, 1963. U. S. Dept. Agr. Misc. Pub. 970, 335-352.
Meade, R. H., 1969. Errors in using modern stream-load data to estimate natural rates of denudation. *Geol. Soc. America Bull.*, 80, 1265-1274.
Meade, R. H., 1982. Sources, sinks, and storage of river sediment in the Atlantic drainage of the United States. *Jour. Geol.*, 90, 235-252.
Meade, R. H., 1985. Suspended sediment in the Amazon River and its tributaries in Brazil during 1982-84. U. S. Geol. Surv. Open-File Rep. 85-492, 39 pp.
Meade, R.H., 1994. Suspended sediments of the modern Amazon and Orinoco Rivers. *Quaternary Internat.*, 21, 29-39.
Meade, R. H., Nordin, C. F., Jr., Pérez Hernandez, D., Mejia B., A. and Pérez Godoy, J. M., 1983. Sediment and water discharge in Rio Orinoco, Venezuela and Colombia, in *Proceedings, Second International Symposium on River Sedimentation, Nanjing*. Water Resources and Electric Power Press, Beijing: 1134-1144.
Meade, R. H., Dunne, T., Richey, J. E., Santos, U. de M. and Salati, E., 1985. Storage and remobilization of suspended sediment in the lower Amazon River of Brazil. *Science*, 228, 488-490.
Meade, R. H. and Parker, R. S., 1985. Sediment in rivers of the United States. In: National Water Summary 1984. U. S. Geol. Surv. Water-Supply Paper 2275, 49-60.
Meade, R. H. and Trimble, S. W., 1974. Changes in sediment loads in rivers of the Atlantic drainage of the United States since 1900. In: Effects of Man on the Interface of the Hydrological Cycle with the Physical Environment. *Internat. Assoc. Hydrol. Sci. Pub.* 113, 99-104.
Meade, R. H., Yuzyk, T. R. and Day, T. J., 1990a. Movement and storage of sediment in rivers of the United States and Canada. In: Wolman, M. G. Colorado, The Geology of North America, O-1.
Meade, R. H., Weibezahn, F. H., Lewis, W. M., Jr. and Pérez Hernandez, D., 1990b. Suspended-sediment budget for the Orinoco River, in Weibezahn, F. H., Alvarez, H. and

Lewis, W. M., Jr. (eds.),*The Orinoco River as an Ecosystem*. Impresos Rubel, Caracas: 55-79.

Mertes, L. A. K., 1985. Floodplain Development and Sediment Transport in: The Solimoes-Amazon River, Brazil. University of Washington, Seattle, M. S. thesis, 108 pages.

Milliman, J. D. and Meade, R. H., 1983. World-wide delivery of river sediment to the oceans. *Jour. Geol.*, 91, 1-21.

Milliman, J. D., Quraishee, G. S. and Beg, M. A. A., 1984. Sediment discharge from the Indus River to the ocean: past, present and future, in Haq, B. U., and Milliman, J. D. (eds.), *Marine Geology and Oceanography of Arabian Sea and Coastal Pakistan*. Van Nostrand Reinhold Co., New York: 65-70.

Milliman, J. D., Qin, Y.-s., Ren, M.-e. and Saito, Y., 1987. Man's influence on the erosion and transport of sediment by Asian rivers: the Yellow River (Huanghe) example. *Jour. Geol.*, 95, 751-762.

Milliman, J.D. and Syvitski, J.P.M., 1992. Geomorphic/tectonic control of sediment discharge to the ocean: the importance of small mountainous rivers. *Jour. Geol.*, 100, 525-544.

Nordin, C. F. and Meade, R. H., 1982. Deforestation and increased flooding of the upper Amazon. *Science*, 215, 426-427.

Peters, J. J., 1978. Discharge and sand transport in the braided zone of the Zaire estuary. *Netherlands Jour. Sea Res.*, 12, 273-292.

Petschinov, D., 1968. Schwebstoffe der Donau (Unterlauf), in *Limnologische Berichte der X. Jubilaumstagung der Arbeitsgemeinschaft Donauforschung, Bulgarien 10-20 Oktober 1966*. Verlag der Bulgarischen Akademie der Wissenschaften, Sofia: 69-81.

Qian, N., Zhang, R. and Chen, Z.-c., 1987. On some sedimentation problems of the Three Gorges Project. *Internat. Jour. Sed. Res.* (Beijing), 1, 5-38.

Richey, J.E., Nobre, C. and Deser, C., 1989. Amazon River discharge and climate variability: 1903 to 1985. *Science*, 246, 101-103.

Schumm, S. A., 1977. *The Fluvial System*. John Wiley & Sons, New York: 338 pages.

Shahin, M. ,1985. *Hydrology of the Nile Basin*. Elsevier, Amsterdam, 575 p.

Shi, Y.-l., Yang, W. and Ren, M.-e., 1985. Hydrological characteristics of the Changjiang and its relation to sediment transport to the sea. *Continent. Shelf Res.*, 4, 5-15.

Shiklomanov, I.A., and Skakalsky, B.G., 1994. Studying water, sediment and contaminant runoff of Siberian rivers: Modern status and prospects, in Workshop on Arctic Contamination, May 2-7, 1993, Anchorage, *Alaska Arctic Research of the United States*, 8, 295-306.

Smith, B. J., 1965. Sedimentation in the San Francisco Bay system, in Proceedings, Federal Inter-Agency Sedimentation Conference, 1963. U. S. Dept. Agr. Misc. Pub. 970, 675-708.

Stanley, D.J. and Warne, A.G., 1994. Worldwide initiation of Holocene marine deltas by deceleration of sea-level rise. *Science*, 265, 228-231.

Sternberg, H. O., 1987. Aggravation of floods in the Amazon River as a consequence of deforestation? *Geogr. Ann.*, 69A, 201-219.

Tang, R.-c. and Lin, W.-q., 1987. A study on sedimentation problems of the Gezhouba Project. *Internat. Jour. Sed. Research* (Beijing), 1, 69-101.

Trimble, S. W., 1977. The fallacy of stream equilibrium in contemporary denudation studies. *Am. Jour. Sci.*, 277, 876-887.

Trimble, S. W., 1983. A sediment budget for Coon Creek basin in the Driftless Area, Wisconsin, 1853-1977. *Am. Jour. Sci.*, 283, 454-474.

Varga, S., Bruk, S. and Babic-Mladenovic, M., 1989. Sedimentation in the Danube and tributaries upstream from the Iron Gates (Djerdap) Dam, in Proceedings, *Fourth In-*

ternational Symposium on River Sedimentation, Beijing. China Ocean Press, Beijing: 1111-1118.

Walling, D. E., 1983. The sediment delivery problem, in Rodriguez-Iturbe, I. and Gupta, V. K. (eds.), Scale problems in hydrology. *Jour. Hydrol.*, 65, 209-237.

Walling, D. E. and Webb, B. W., 1983. Patterns of sediment yield, in Gregory, K. J. (ed.), *Background to Palaeohydrology*. John Wiley & Sons, New York: 69-100.

Wells, F. C., 1980. Hydrology and water quality of the lower Mississippi River. Louisiana Office Public Works Tech. Rep. 21, 83 pages, 5 plates.

Yang, Z.-s., Milliman, J. D. and Fitzgerald, M. G., 1983. Transfer of water and sediment from the Yangtze River to the East China Sea, June 1980. *Can. Jour. Fisheries Aquat. Sci.*, 40 (supp. 1), 72-82.

Zhang, P. and Liu, Z.-s., 1989. Trend of sediment load over the years in the upper region of the Three-Gorges in the Yangtze River, in Proceedings, *Fourth International Symposium on River Sedimentation, Beijing*. China Ocean Press, Beijing: 286-293.

CHAPTER 4

Coastal Erosion and Rising Sea-Level

Eric C.F. Bird

ABSTRACT: Coastal erosion is facilitated by rising sea level, but it can also occur on coastlines where the sea level is stable or even falling. The modern prevalence of beach erosion on the world's coastlines is not, therefore, an indication of global sea-level rise. However, if such a sea-level rise develops, perhaps as a consequence of the greenhouse effect, there will be an acceleration of existing beach erosion, and erosion will begin on many beaches that are now stable or growing. Accelerated erosion also will be seen on cliffs, deltas, coastal swamps, and developed coastlines. The changes predicted globally are already in evidence on sectors of the world's coastline where land subsidence has produced a relative sea-level rise.

Introduction

Coastal erosion is facilitated by a rising sea level, which brings wave action to progressively higher levels and permits larger waves to reach the coast through deepening near-shore waters. Throughout geological time erosion has been widespread during marine transgressions, which have been due to actual global (eustatic) sea-level rise, coastal land subsidence, or some combination of the two.

The world's coastlines retreated rapidly during the world-wide late Quaternary marine transgression which took place between 18,000 and 6,000 years ago when climatic warming caused extensive melting of glaciers, thereby increasing the volume of the oceans and submerging continental shelves. Elevated areas were cut back as cliffs, while low-lying areas were encroached by the sea to form embayments, inlets and estuaries. About 6,000 years ago this marine transgression came to an end, except on subsiding coasts, where a relative rise of sea level has continued. As the late Quaternary marine transgression abated, coastal deposition became extensive. On gently-shelving coasts, sand washed in from the sea floor built beaches, spits and barriers; in the vicinity of river mouths, fluvial sediments accumulated as deltas; and in sheltered inlets and estuaries muddy sediments accreted, sustaining salt marshes and mangrove swamps (Bird, 1984).

Coastal erosion also occurs where the sea level is stable or falling, especially where weak or unconsolidated geological materials face high wave energy, and

J.D. Milliman and B.U. Haq (eds.), Sea-Level Rise and Coastal Subsidence, 87-103.

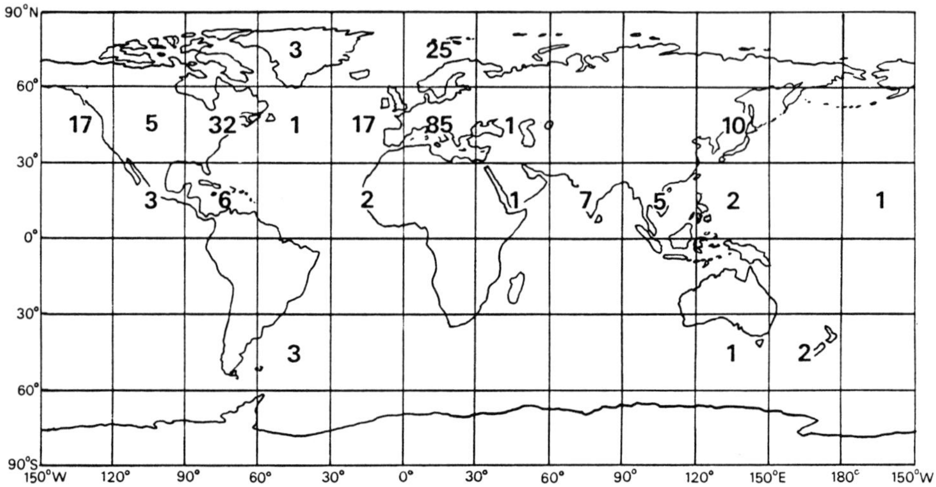

Fig. 1. Global distribution of 229 tide gauge stations with records for more than 30 years. 65 of these (28.5%) show a mean sea-level rise of more than 2 mm/year; 52 (22.5%) show a rise of 1 to 2 mm/year; 47 (20.5%) show a rise of less than 1 mm/year; and 65 (28.5%) show a mean sea-level fall. After Pirazzoli (1986).

where a deficit has developed in the availability of deposits to maintain or enlarge the coastline. Thus, there are retreating cliffs and eroding beaches on coasts where tide gauge records indicate little, if any, change in mean sea level over the past century. Emerging coastlines in northern Canada and Scandinavia include some sectors of receding cliff cut into glacial or periglacial drift deposits; beach erosion is in progress on the shores of the Indal Delta, in northern Sweden, due to the damming of the river and consequent interception of the sediment supply that would have maintained the shoreline; relative sea level here is *falling* by about one meter per century (Bird, 1987).

Coastal erosion is therefore not a reliable indication of a rising sea level; but it is certainly more extensive, and often more rapid, in areas where tide gauge records show that the sea is rising. In these areas, coastlines have been maintained, or have even grown, where deposition has continued in areas that are relatively sheltered from strong wave action or current scour, and where there is still an abundant sediment supply. Undoubtedly, the predicted sea-level rise produced by the greenhouse effect will initiate erosion on coasts that are presently stable or growing, and intensify existing coastal erosion.

Submerging Coastlines

Most tide gauge records around the world's coastlines show a relative rise of sea level, but over the past few decades, the scale of this rise has varied. It has been suggested that global sea level has risen 10–15 cm over the past century, and is now rising about 1–1.5 mm per year (Barnett, 1984; Gornitz and Lebedeff, 1987). However, when Pirazzoli (1986) analyzed 1178 tide gauge records obtained from published literature and the Permanent Service for Mean Sea Level, he found that only 229 were reliable over as period of more than thirty years, and that these were unevenly distributed around the world's coastline — 195 (85%) being in Western Europe and North America, and only 6 (2.6%) in the Southern Hemisphere (Fig. 1). Of the 229 long-term tide gauge records, only 30 (13%) suggested the 1–1.5 mm rise per year was "average", but 164 (71.6%) showed a sea-level rise of at least 0.1 mm per year. The remaining 65 (28.4%), showing a fall in mean sea level, were mainly on the coasts of Scandinavia and northern North America, which are rising isostatically as a result of deglaciation. Similar conclusions were reached in an independent survey by Emery and Aubrey (1991). A geographical pattern of rising and falling relative sea level has been outlined (Pirazzoli, 1991), but it remains difficult to separate actual changes of sea level from upward and downward movements of land borders, and some (if not all) of the records showing a sea-level rise are on coasts where land subsidence has been taking place.

A representative network of at least 300 tide gauge stations around the world is being set up under the Global Sea Level Observing System (GLOSS) by UNESCO's Intergovernmental Oceanographic Commission, and in due course a more reliable statement of global sea level trends will be possible (Pugh, 1987; Woodworth, 1991). Meanwhile, the prediction is that global warming — due to enhancement of the greenhouse effect in the Earth's atmosphere by accumulation of carbon dioxide and other gases generated by urban, industrial and agricultural activities - will lead to a world-wide sea-level rise (Barth and Titus, 1984). In 1990, The Intergovernmental Panel on Climatic Change (IPCC) estimated that global mean sea level would rise 20 cm by the year 2030 and 30–100 cm by the end of the 21st century (Houghton et al., 1990). This is shown in Fig. 2, with high (one meter rise by 2090) and low (one by 2140) estimates.

Coastal land subsidence is generally due either to tectonic activity or the extraction of underground water or minerals. Parts of the world's coastline where such subsidence has occurred are indicated in Figure 3, but it is likely that further research will extend these. They include the Gulf and Atlantic coasts of the United States, southeastern of England, the Netherlands, northern Germany, southwestern Denmark, the northern Adriatic, and parts of the Japanese coast, all of which are subsiding tectonically, while groundwater or mineral extraction has contributed to subsidence in and around such places as Long Beach (California), Venice, Ravenna, Bangkok, Taipei, Tokyo, and Niigata (Fig. 4) (Bird, 1993).

It is in these areas that coastal erosion and associated ecological changes accompanying sea-level rise can be studied, together with human responses (including engineering structures) to ongoing submergence of the kind that will affect the entire world's coastlines if the predictions of a sea-level rise due to global warming prove correct.

Coastline Changes

Between 1972 and 1984 the International Geographical Union's Commission on the Coastal Environment carried out a global survey of changing coastlines. It was found that erosion had become prevalent on the world's sandy coastlines, and that

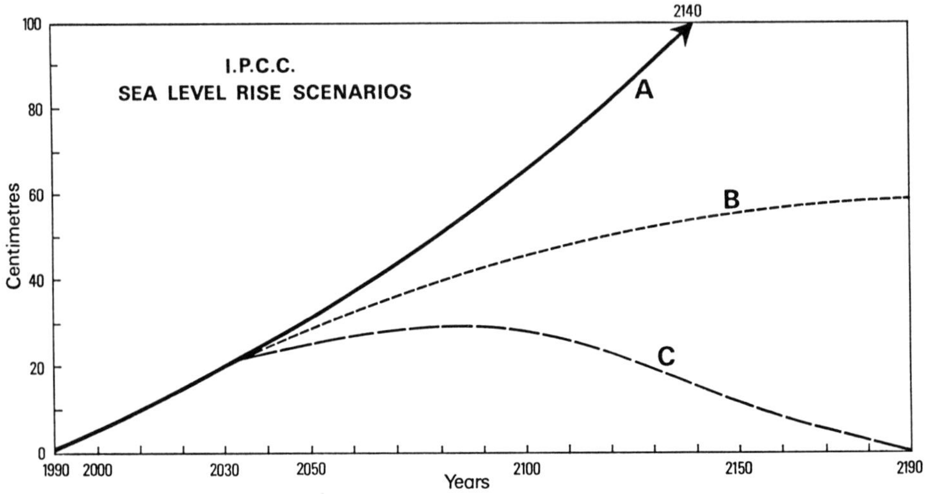

Fig. 2. Sea level rise scenarios (Intergovernmental Panel on Climatic Change) After Houghton et al. (1990).

in addition to continued regression of cliffed coastlines, many deltaic and swampy coasts were in retreat (Bird, 1985).

Measurements of rates of coastal erosion and deposition showed that on the world scale most coastlines have advanced or retreated less than 1 m per year; only a very few have changed by more than 100 m per year. A gain or loss of more than 10 per year is regarded as exceptionally rapid.

Cliff Erosion

Rates of regression of cliffs have varied with their height, rock resistance, geological structure, and exposure to wave action. Erosion's effects have been minor on cliffs of hard, massive rocks even when exposed to violent seas, while cliffs made of soft materials have receded quickly. The most rapid changes were measured where volcanic activity had deposited ash (pyroclastic sand and gravel), which is readily eroded in a high wave energy environment. Cliffs cut in such materials after the volcanic construction of Isla San Benedicto off the Pacific coast of Mexico in 1952 retreated initially at up to one meter per day (Richards, 1960). The explosive eruption of the Indonesian volcanic island of Krakatau in 1883 left residual islands thickly carpeted with ash, and on the outer coasts, exposed to ocean waves, these have since receded by up to 1 km. Within the central crater, the younger volcanic island of Anak Krakatau has grown since 1927, and measurements of cliff regression on volcanic ash on the south coast of this island ranged up to 6.6 m per year (Bird and Rosengren, 1984).

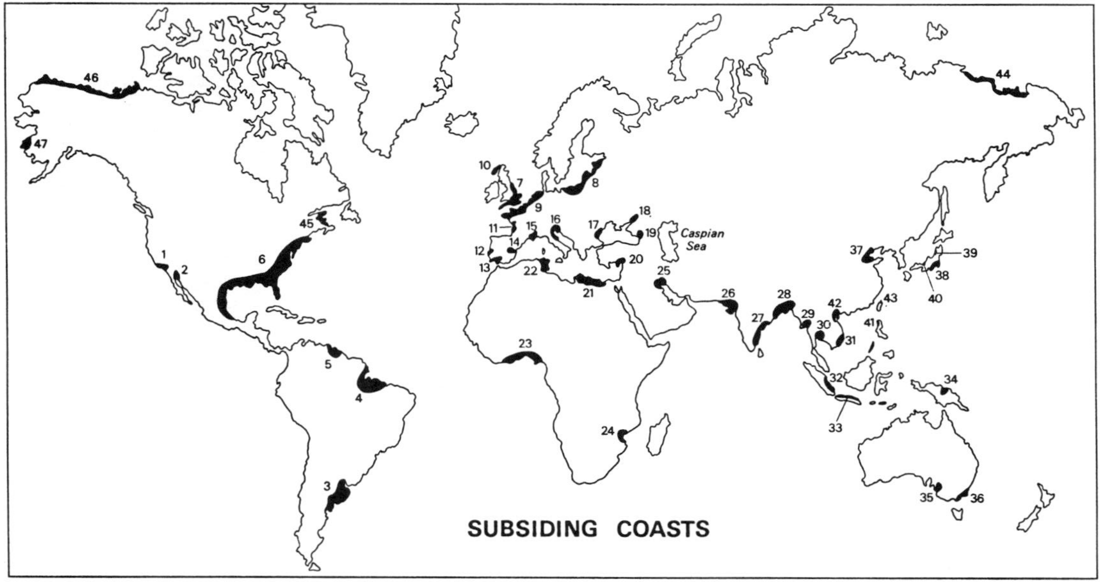

Fig. 3. Sectors of the world's coastline that have been subsiding in recent decades, as indicated by evidence of tectonic movements, increased marine flooding, geomorphological and ecological indications, geodetic surveys, and groups of tide gauges recording a rise of mean sea level greater than 2 mm per year over the past three decades. Compiled from published information and advice from R.W.G. Carter, V. Gornitz, Al Guilcher, P.A. Pirazzoli, P. Paskoff, V.P. Zenkovich, and M. Zunica. *Key to Map: 1. Long Beach area, southern California; 2. Colorado River Delta; 3. Gulf of La Plata; 4. Amazon Delta; 5. Orinoco Delta; 6. Gulf and Atlantic coast, Mexico and United States; 7. Southern and eastern England; 8. Southern Baltic from Estonia to Poland; 9. North Germany, The Netherlands, Belgium and northern France; 10. Loire estuary; 11. Vendé, western France; 12. Lisbon region; 13. Guadalquavir Delta; 14. Ebro Delta; 15. Rhone Delta; 16. Northern Adriatic from Rimmi to Venice and Grado; 17. Danube Delta; 18. Eastern Sea of Azov; 19. Poti Swamp; 20. Southeast Turkey; 21. Nile Delta to Libya; 22. Northeast Tunisia; 23. Nigerian coast, especially the Niger Delta; 24. Zambesi Delta; 25. Tigris-Euphrates Delta; 26. Rann of Kutch; 27. Southeastern India; 28. Ganges-Brahmaputra Delta; 29. Irrawaddy Delta; 30. Bangkok coastal region; 31. Mekong Delta; 32. Eastern Sumatra; 33. Northern Java deltaic coast; 34. Sepik Delta; 35. Port Adelaide region; 36. Corner Inlet region; 37. Huang-he Delta; 38. Head of Tokyo Bay; 39. Niigata, Japan; 43. Northern Taiwan; 44. Eastern Siberian coastal lowlands; 45. Maritime Provinces, Canada; 46. Northern Alaska and the MacKenzie Delta; 47. Yukon Delta.*

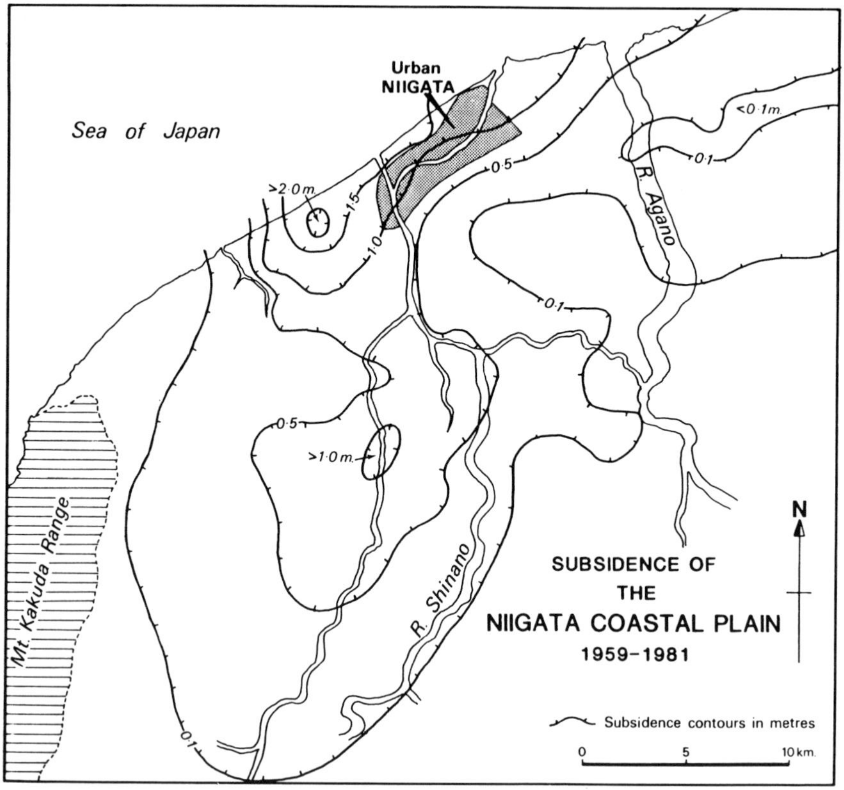

Fig. 4. Land subsidence in the Niigata region, Japan. Based on a map published by Nakano and Takehisa (1983).

Rapid recession also has also taken place on cliffs cut into glacial and periglacial deposits subject to seasonal wave attack on Arctic coasts, notably in Siberia, where Zenkovich (1967) quoted rates of up to 100 m per year for tundra bluffs that collapse as the result of thawing of buried ice, and are washed away by wave action. The regression of cliffs of soft materials tends to be intermittent, with phases of rapid erosion during stormy periods. Zenkovich (1967) found that cliffs cut in soft glacial drift on the Polish coast retreated at an average rate of a meter a year, but up to 5 m were lost during a severe storm. May and Heeps (1985) documented regression rates of up to 1.3 m per year on English Channel chalk cliffs, noting that most change occurred during winter storms, or when a winter freeze was followed by a thaw.

Cliff regression is generally more rapid where nearshore waters are deep, allowing large waves to reach the coast, rather than behind gentle gradient sea floor areas or reef structures. Rates of cliff regression also vary in relation to the degree of protection offered by beaches. On the Holderness coast, in eastern England, lobes of beach material migrating southward along the shore temporarily protect adjacent sectors of cliff cut in glacial drift, but erosion restarts as the lobe

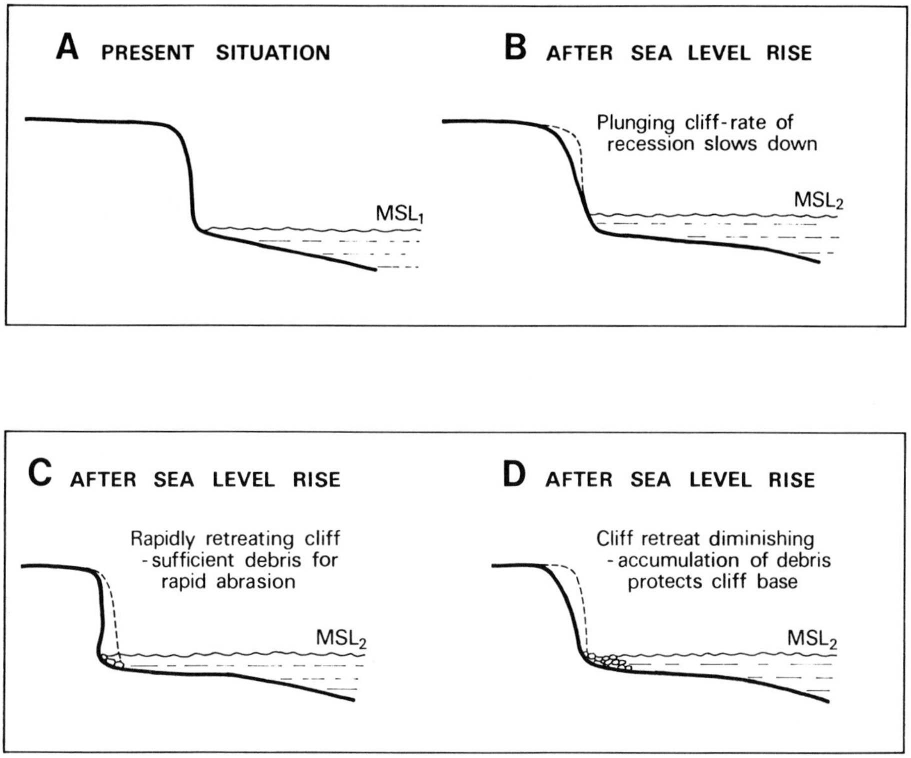

Fig. 5. Effects of a mean sea level (MSL) rise on a cliffed coast.

moves on (Pringle, 1981). Some cliffs cut in soft materials are kept steep or vertical by basal erosion, but subaerial weathering, runoff and slumping usually contribute to the intermittent regression of the coastline. On the Dorset and Norfolk coasts of England, slumping forms lobes that temporarily advance the coastline, but are soon washed away (Hutchinson, 1976). Brunsden and Jones (1980) suggested that cliffs on the Dorset coast showed cycles of slumping and steepening (restoring the transverse profile at intervals of a century or more) accompanying a periodic sea-level rise of about two mm/year.

Cliff regression is likely to accelerate on coasts where sea level is rising (Fig. 5), but the accuracy of monitoring rates of cliff retreat has rarely been sufficient for such an acceleration to be detected. The rising sea will submerge shore platforms wherever they develop, and cliff retreat will accelerate as stronger waves cross them to attack the cliff base. The latter will occur even at low tide. On the Japanese island of Niijima, Sunamura (1992) calculated that cliffs cut in poorly consolidate volcanic gravel and ash, now retreating at 1.2 m per year, will be cut back by up to 2.3–2.9 m per year by the time sea level has risen 1 m. On some coasts sediment may be deposited and colonies of marine organisms become established, on submerging shore platforms, at least partly offsetting this augmentation of wave attack and cliff regression.

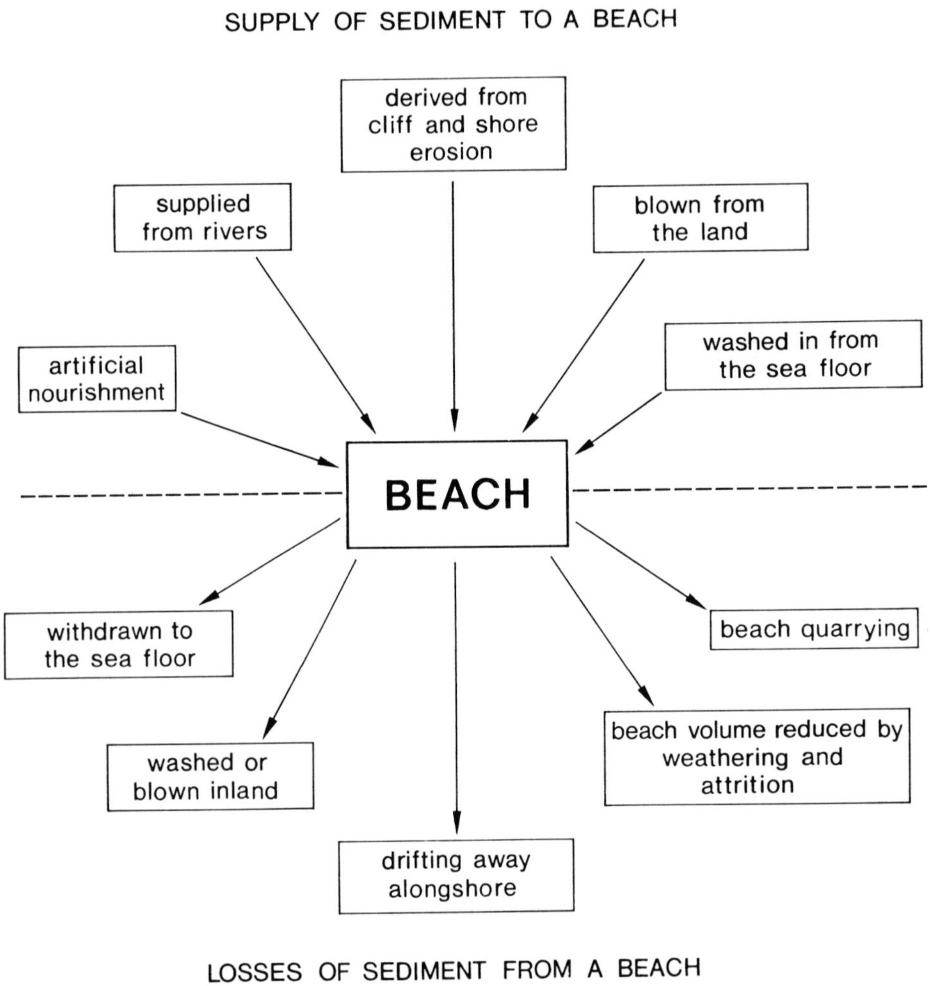

Fig. 6. How sand is gained and lost from a beach.

Beach Erosion

Changes to beaches are the outcome of variations in the rate of sediment supply to, and losses from, a beach system (Fig. 6). Beaches grow where the rate of sediment gain exceeds the rate of loss, and shrink where the reverse applies.

Beach erosion is already widespread. Studies by the International Geological Union's Commission on the Coastal Environment found that at least 70% of the world's sandy coastline has retreated during the past few decades, less than 10% having advanced steadily, and the remaining 20–30% having been stable or shown no net change (Bird, 1985). Bruun (1962) demonstrated that a rising sea level would lead to regression of a sandy coastline (Fig. 7), and beach erosion along much of the Gulf and Atlantic coasts of the United States is believed to be partly

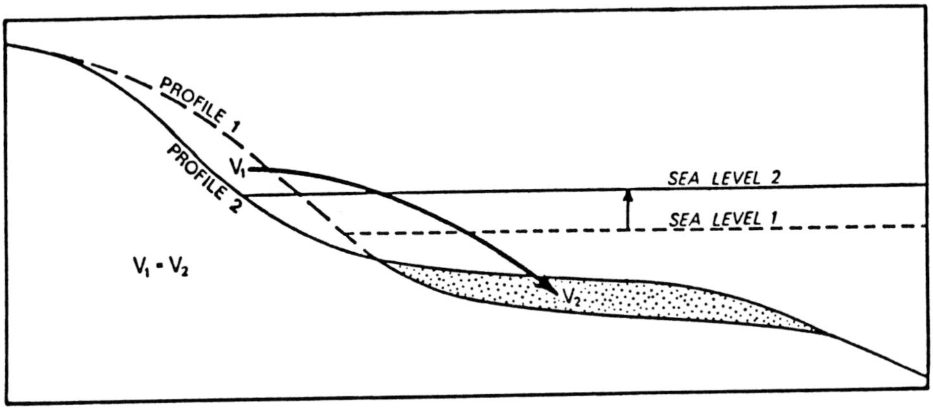

Fig. 7. The Bruun Rule: response of the beach profile to a sea-level rise.

due to sea-level rise resulting largely from coastal land subsidence (Fisher, 1980). However, coastal submergence is only one of many factors that have contributed to the modern prevalence of beach erosion (Table 1); so beach erosion is not necessarily an indication that sea level is rising. As has been noted, beach erosion has occurred even on parts of the Swedish coastline where sea level is falling.

Nevertheless, there is no doubt that beach erosion will become more extensive, and more rapid, as a consequence of the global sea-level rise expected as a consequence of the greenhouse effect (Leatherman, 1990). Prediction of the rate and extent of erosion using the model proposed by Bruun (1962) has proved difficult, partly because this model presupposes initial and terminating equilibrium (Scientific Committee on Ocean Research Working Group, 1991), while the scenarios presented (Fig. 2) offer only an accelerated sea-level rise, with no indication of eventual stabilization. Nevertheless, the prospect is of increasingly severe beach erosion except on a few coasts where there is sufficient natural replenishment of sand and gravel (e.g., from rivers with augmented sediment loads) to maintain them. The armoring of beach-fringed coasts with sea walls, boulders, or other structures may hold the coastline; but the beaches are unlikely to survive in front of them (Zunica, 1990). Dune systems that have developed behind beaches are likely to be modified by coastline retreat and a rising water table (Carter, 1991), and beach ridge plains, like the permatange of the Malaysian coast, will also be trimmed back (Teh Tiong Sa, 1993). It is of interest that the rapid rise in sea level in the Caspian Sea over the past decade has resulted in the formation of nearshore sand barriers and lagoons in front of the submerging coastline (Leontiev and Vekliev, 1990).

The possibility that low-lying islands, such as sandy cays on coral reefs, will be swept away by the rising sea has alarmed the inhabitants of such islands, as in the Maldives (Pernetta and Sestini, 1989; Hulsbergen and Schröder, 1989), but the revival of coral growth on the surrounding reef flats may protect them from erosion, and even supply additional sediment to enlarge them (Fig. 8) (Neumann and Macintyre, 1985; Hopley and Kinsey, 1988).

TABLE 1 : CAUSES OF BEACH EROSION

1) Submergence and increased wave attack as the result of a rise in sea level relative to the land (including subsiding coasts).
2) Decrease of fluvial sand supply to the coast as a result of reduced runoff or sediment yield from a river catchment (e.g. because of a lower rainfall, or dam construction leading to sand entrapment in reservoirs or successful soil conservation works).
3) Reduction in sand supply from eroding cliffs or shore outcrops (e.g. because of diminished runoff, a decline in the strength and frequency of wave attack, or the building of sea walls to halt cliff recession).
4) Reduction of sand supply to the shore where dunes that had been moving from inland are stabilized, either by natural vegetation colonization or by conservation works, or where the sand supply from this source has run out.
5) Decrease of sand supply washed in by waves and currents from the adjacent sea floor, either because the sand supply has run out or because the transverse profile has attained a concave form which no longer permits such shoreward drifting.
6) Reduction in sand supply from the sea floor because of increased growth of seagrasses or other marine vegetation, which impedes shoreward drifting.
7) Diminished production of sand from sea floor biological sources because of ecological changes restricting the production of shelly material.
8) Reduction in sand supply from coastal sources as the result of interception (e.g. by a constructed breakwater).
9) Increased loss of sand from the beach to the backshore and hinterland areas by landward drifting of dunes, notably where backshore dunes have lost their retaining vegetation cover and drifted inland, lowering the terrain immediately behind the beach and thus reducing the volume of sand to be removed via coastline regression.
10) Removal of sand from the beach by quarrying
11) Loss of sand from prolonged recreational usage.
12) Increased wave energy reaching the shore because of the deepening of near-shore water (e.g. where a shoal has drifted away, where seagrass vegetation has disappeared, or where dredging has taken place).
13) Increased wave attack due to a climatic change that has produced a higher frequency, duration, or severity of storms in coastal waters.
14) Diminishing beach caliber and nearshore material as a result of attrition of beach sand grains, leading to winnowing and loss of increasingly fine sediment from the shore.
15) Decrease in the volume of beach and nearshore material as the result of weathering, solution, or attrition, resulting in the lowering of the beach face and a consequent increase in penetration of wave attack to the backshore.
16) A rise in the water table within the beach, due to increased rainfall or changes in local drainage, rendering the beach sand wet and thus more readily eroded.
17) Increased loss of sand alongshore as a result of a change in the angle of incidence of waves (e.g. as the result of the growth or removal of a shoal or reef, or breakwater construction).
18) Intensification of wave attack as the result of lowering of the beach face on an adjacent area (e.g. as the result of reflection scour induced by sea wall construction).
19) Removal of beach lobes or promonotories as the result of longshore drifting.
20) Removal of a protective sea ice fringe by melting, so that waves reach the beach (e.g. during a longer-than-usual summer).

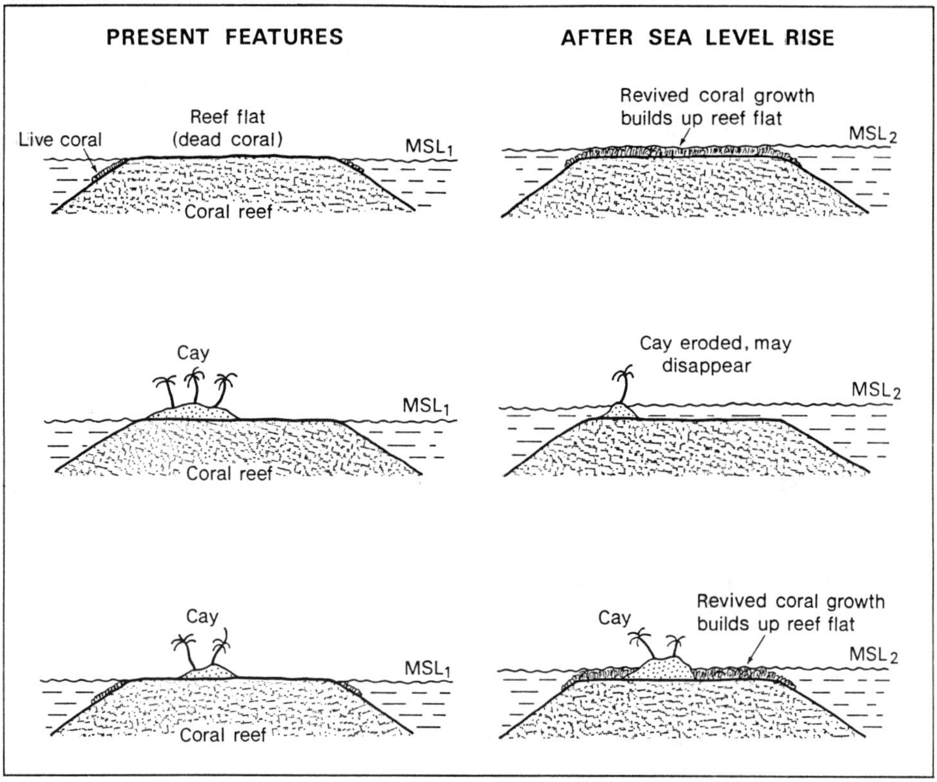

Fig. 8. Possible response of coral on a reef flat during a mean sea level (MSL) rise.

Erosion on Deltaic Coasts

Deltas have experienced rapid expansion, at rates of up to 180 m per year on the tropical deltas in Indonesia (Verstappen, 1966). The advancement of deltaic land above high tide level results from deposition from river floodwaters, sometimes accompanied by the formation of wave-built beaches and spits along the seaward borders. Some deltas have advanced by the seaward spread of sediment-trapping salt marshes or mangrove swamps, others by the deposit on their shores of sandy sediment carried down to the mouths of rivers.

Erosion of parts of deltaic coastline has often followed the natural or artificial diversion of a river mouth, thereby cutting off the sediment supply that previously maintained or expanded the shore. Well-known examples of this include the Huanghe in China, where the river mouth changed in 1852 (Chen et al., 1985), the Medjerda in Tunisia (Paskoff, 1981), and the Cimanuk in Java (Bird and Ongkosongo, 1980). In each case the abandoned delta has been cut back by erosion at the same time a new delta has grown around the mouth of the diverted river. Diversion of a river mouth by canalization, as on the Brazos Delta in Texas (Wadsworth, 1966), has had the same effect.

Erosion also has occurred on the shores of deltas where fluvial sediment supply has been reduced by the building of dams and the consequent entrapment of sediment in reservoirs. For example, there has been rapid erosion on sectors of the Nile Delta close to distributary mouths following the completion of the Aswan High Dam in 1964 (Orlova and Zenkovich, 1974).

Most large deltas are subsiding because of isostatic adjustment to accumulating sedimentary loads. This is why the successively abandoned Holocene sub-deltas of the Mississippi have become submerged and eroded once the river moved to a new outlet (Kolb and Van Lopik, 1966). Erosion is thus prevalent on sections of delta coastlines that are no longer being maintained or enlarged by fluvial deposition (Day and Templett, 1990).

If this deltaic subsidence is augmented by a global sea-level rise, submergence and erosion of delta coastlines will become increasingly extensive, and some of the smaller deltas may be replaced by estuarine inlets.

Erosion of Coastal Swamps

Salt marshes and mangrove swamps have spread rapidly where there is an abundant sediment supply, especially in areas sheltered from strong wave or current action. Ranwell (1974) found that the seaward edge of salt marshes in the Fal estuary, southwestern England, advanced by up to 900 m between 1878 and 1973, with the landward parts of the marsh being colonized by scrub and woodland vegetation. However, the seaward borders of salt marshes in southern England, northern France, and New England, are generally cliffed and retreating, probably as the result of a rising sea level due to land subsidence (Guilcher, 1981). The same is true of the salt marshes of the Venice Lagoon, the seaward borders of which are receding in response to a rise of sea level relative to the land of about 27 cm between 1872 and 1985 (Pirazzoli, 1987).

The global sea-level rise predicted as a consequence of the greenhouse effect is likely to cause more rapid and extensive recession of the seaward margins of salt marshes and mangrove swamps (Fig. 9), except in areas where they can be maintained or enlarged by continuing rapid sedimentation or peat accumulation (Reed, 1990; Woodroffe, 1990; Ellison and Stoddart, 1991). This is supported by studies of the Holocene stratigraphy of mangrove areas (Woodroffe et al., 1985). Tidal creeks may be enlarged to accommodate greater inflow of sea water at high tides (Knighton et al., 1991), and marshland surfaces will be submerged, unless sedimentation rates are sufficiently rapid to maintain vertical accretion (Rethick, 1981). In the Chesapeake Bay, some marshy islands have diminished in extent and others have vanished altogether as a consequence of relative sea-level rise (Kearney and Stevenson 1991), but studies of New England salt marshes have suggested that aggradation has been sufficiently rapid to maintain them in recent decades (Redfield, 1972). In tropical areas, shoreward movement of muddy sediment from tidal flats may produce vertical accretion under mangroves as the sea level rises (Teh and Lim, 1993).

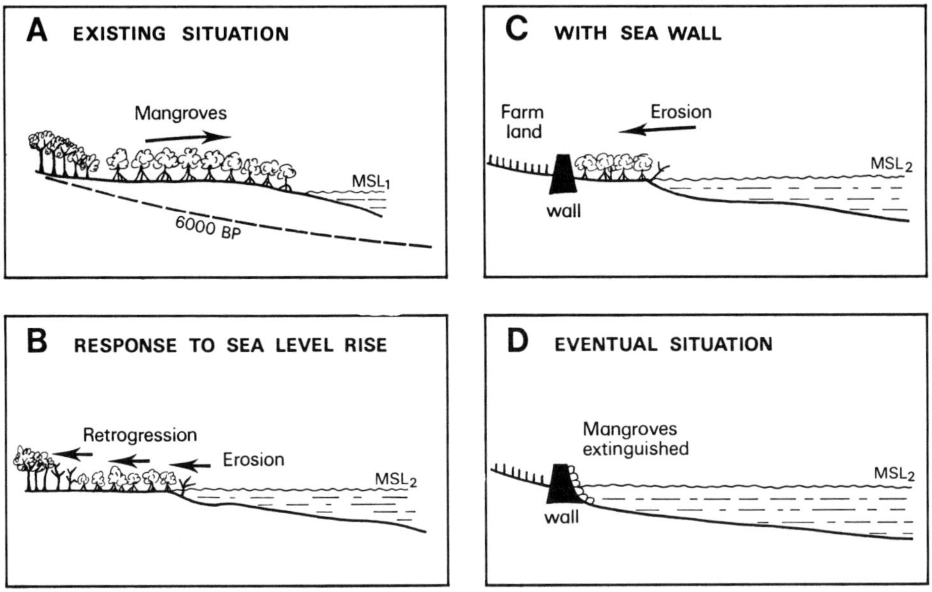

Fig. 9. Response of a mangrove-fringed salt marsh to a rise of mean sea level (MSL): above, where landward regression is possible; below, where such regression is prevented by a sea wall, so that the salt marsh (and eventually the mangroves) die out.

As submergence proceeds, the inner borders of mangrove swamps and salt marshes will tend to migrate landward at a rate determined by the transverse gradient. There will be little landward movement where the hinterland is steep or where embankments have been built to keep the sea out of reclaimed or developed land (Fig. 9) (Titus, 1986). It is difficult to find areas where such salt marsh regression has occurred because the landward borders almost invariably have been protected from marine incursion by sea wall construction, but examples of forest die-back at the landward edge of encroaching salt marshes have been documented in Maryland (Darmody and Foss, 1979) and behind the sounds of North Carolina (Stevenson et al., 1986). Phillips (1986) calculated that the areas of tidal wetlands on the Atlantic coast of the United States would shrink in situations where the seaward borders receded at more than 30 cm per year.

Salt marshes or mangrove swamps with well-defined zones of animal species or communities parallel to the shoreline are likely to show landward migration of these zones if sea-level rise exceeds the rate of sedimentary aggradation. In Corner Inlet, southeastern Australia, the mangroves that form a seaward fringe to salt marsh islands have been migrating back into the salt marsh in recent decades, a change which, accompanied by shoreline recession, appears to be a consequence of sea-level rise due to local tectonic subsidence in this area (Vanderzee, 1988). Vertical migrations of zoned marine organisms on rocky shores and pier supports are a predictable consequence of sea-level rise, and may provide evidence that supplements that from tide gauges (Bird, 1993).

Erosion of Developed Coastlines

There are many examples around the world of damage or destruction of developed property (buildings, roads, farmed land, forestry areas) as the result of regression of cliffs or erosion of beaches, deltas and swamps. Large amounts of money have been spent by governments and local authorities on the building and enlarging of sea walls and ramparts, or on the artificial replenishing of lost beaches as a protective feature (Walker, 1988). A global sea-level rise will greatly increase the problems of erosion of developed coastlines, and will certainly lead to demands for the improvement of existing anti-erosion projects and the installing of new protective structures (Stark, 1988). Many countries will find it necessary to revise their aims and attitudes in coastal development and management, and require planners and developers to avoid low-lying and erosion-prone coastlines (e.g. Gibb and Aburn, 1986; Psuty, 1986; Cocks et al., 1988).

The problems will be severe in areas where salt marsh and mangroves have been embanked and reclaimed, particularly on the coasts of southeast Asia, where brackish-water fishponds have been built extensively in former mangrove areas. On the northern coast of Java, for example, fishponds are quickly constructed on aggrading muddy areas around river mouths, but they are being fragmented by marine erosion on parts of the deltaic coastline no longer maintained by fluvial sedimentation, and thus are subject to land subsidence. A global sea-level rise of 1 m will accentuate this problem, inundating brackish-water fishponds, and causing flooding and salinity problems in the rice fields that generally lie to landward (Bird, 1993).

Conclusion

Erosion is already widespread around the world's coastline for a variety of reasons, including submergence and a near-shore sinking, which result from the sea-level rise on coasts where the land border has been subsiding. The global sea-level rise expected as a consequence of the greenhouse effect, likely to be of the order of 1 m over the next 100 to 150 years, will accelerate erosion on coastlines where it is already in progress, and initiate erosion in many areas that are at present stable or even expanding. Predictions of the extent of such erosion require detailed knowledge of coastal landforms and nearshore processes. A 1 m sea-level rise will enable the highest tides to reach levels of at least this much above their present limits, but on most coastlines the regression will extend well beyond this 1 m contour because of the initiation or intensification of erosion. As an initial approximation, the extent of erosional regression can be predicted by extrapolation of the existing nearshore cross-profile landward until it is restored in relation to the higher sea level. Refinements of this prediction will require assessment of wave energy regimes, as modified by changes in nearshore water depths, and expected sedimentation rates in the changing nearshore and coastal environments. Geomorphological and ecological research on the world's subsiding coastlines will thus contribute to analyses of expected changes in response to a global sea-level rise.

References

Barnett, T.P.,1984. The estimation of "global" sea level changes, *J. Geophys. Res.*, 89, 7980-7988.

Barth, M.G. and Titus, J.G.,1984. *Greenhouse Effect and Sea Level Rise*. Van Nostrand Reinhold, New York.

Bird, E.C.F., 1984. *Coasts*. Blackwell, Oxford.

Bird, E.C.F., 1985. *Coastline Changes*. Wiley Interscience, Chichester.

Bird, E.C.F., 1987. The world's disappearing beaches. in *Science and the Future Handbook*, Encyclopaedia Britannica, p. 115-127.

Bird, E.C.F., 1993. *Submerging Coasts: The Effects of a Rising Sea Level on Coastal Environments*. Wiley, Chichester.

Bird, E.C.F. and Ongkosongo, O.S.R., 1980. Environmental changes on the coasts of Indonesia. United Nations University, Tokyo, Japan. Bird, E.C.F. and Rosengren, N.J., 1984. The changing coastline of the Krakatau Islands, Indonesia. *Z. Geomorph.*, 28, 346-366.

Brunsden, D. and Jones, D.K.C., 1980. Relative time scales and formative events in coastal landslide systems. *Z. Geomorph.*, Supp. 34, 1-19.

Bruun, P., 1962. Sea level rise as a cause of shore erosion. J. Waterways and Harbors Div., *Proc. Amer. Soc. Civil Engnrs.*, 88, 117-130.

Carter, R.W.G., 1991. Near-future sea level impacts on coastal dune landscapes, *Landscape Ecology*, 6, 9-40.

Chen Jiyu, Liu Cangzi, and Yu Zhiying, 1985. China. In: E.C.F. Bird and M.L. Schwartz (eds.), *The World's Coastline*. Van Nostrand Reinhold, New York, 813-822

Cocks, K.D., Gilmour, A.J. and Wood, N.H., 1988. Regional impacts of rising sea levels in coastal Australia. In: G.I. Pearman (ed.), Greenhouse: Planning for Climatic Change, C.S.I.R.O. Division of Atmospheric Research, 105-120.

Darmody, R.G. and Foss, J.E., 1979. Soil-landscape relationships of the tidal marshlands of Maryland. *Soil. Sci. Soc. Amer. Journ.*, 43, 534-541.

Day, J.W. and Templet, P.H., 1990. Consequences of sea level rise: implications for the Mississippi delta. *Developments in Hydrobiolgy*, 57, 155-165.

Ellison, J.C. and Stoddart, D.R., 1991. Mangrove ecosystem collapse during predicted sea level rise: Holocene analogues and implications. *J. Coastal Res.* 7, 151-165.

Emery, K.O. and Aubrey, D.G., 1991. *Sea Levels, Land Levels, and Tide Guages*. Springer Verlag. New York.

Fisher, J.J., 1980. Shoreline erosion, Rhode Island and North Carolina coasts. In: M.L. Schwartz and J.J. Fisher (eds.), Proceedings of the Per Bruun Symposium, University of Rhode Island, 32-54

Gibb, J.G. and Aburn, J.H.. 1986. Shoreline fluctuations and an assessment of a coastal hazard zone along Pauanui Beach, Eastern Coromandel Peninsula, New Zealand. National Water and Soil Conservation Authority, Wellington, Technical Publication 27.

Gornitz, V. and Lebedeff, S., 1987. Global sea level changes during the past century. In: D. Nummedal, O.H. Pilkey and J.D. Howard (eds.), *Sea Level Change and Coastal Evolution*. Society of Economic Paleontologists and Mineralogists, Special Publication 41, 3-16.

Guilcher, A., 1981. Shoreline changes in salt marshes and mangrove swamps (mangals) within the past century. In: E.C.F. Bird and K. Koike (eds.), *Coastal Dynamics and Scientific Sites*, 31-53.

Hopley, D. and Kinsey, D.W., 1988. The effects of rapid short-term sea level rise on the Great Barrier Reef. In: G.I Pearman (ed.), Greenhouse Planning for Climate Change, C.S.I.R.O. Division of Atmospheric Research, 189-201.

Houghton, J.T., Jenkins, G.J., and Ephraums, J.J. (eds.), 1990. *Scientific Assessment of Climate Change*. Cambridge University Press.

Hutchinson, J.N., 1976. Coastal landslides in cliffs of Pleistocene deposits between Cromer and Overstrand, Norfolk. Laurits Bjerrum, Memorial Volume, University of Oslo, 155-182. Hulsbergen, C.H. and Schr?der, P., 1989. Republic of Maldives implications of sea level rise. Ministry of Economic Affairs, The Netherlands.

Kearney, M.S., and Stevenson, J.C., 1991. Island land loss and marsh vertical accretion rate evidence for historical sea level changes in Chesapeake Bay. *J. Coastal Res.*, 7, 403-415.

Knighton, A.D., Mills, K., and Woodroffe, C.D, 1991. Tidal-creek extension and saltwater intrusion in northern Australia. *Geology*, 19, 831-834.

Kolb, C.R. and Van Lopik, J.R., 1966. Depositional environments of the Mississippi deltaic plain. In: M.L. Shirley (ed.), *Deltas in Their Geologic Framework*, Houston Geol. Soc., Texas, 18-63.

Leatherman, S.P., 1990. Modeling shore response to sea level rise on sedimentary coasts. *Progr. Phys. Geogr.*, 14, 447-464.

Leontiev, O.K. and Veliev, K.A., 1990. Transgressive Coasts. [Caspian Sea]. International Symposium, Baku, USSR, 15-21 September 1990, 1-71. May, V.J. and Heeps, C., 1985. The nature and rates of change on chalk coastlines, *Z. Geomorph.*, Supp. 57, 81-94.

Nakano, T. and Takehisa, Y., 1983. Ground subsidence in the Niigata plain. International Symposium on Coastal Evolution in the Holocene, Niigata, 1983, 25-31.

Neumann, A.C. and Macintyre, I.G., 1985. Reef response to a sea level rise: keep-up, catch-up or give-up. *Proc. 5th Internat. Coral Reef Congress*, 3, 105-110.

Orlova, G. and Zenkovich, V.P., 1974. Erosion on the shores of the Nile delta, *Geoforum*, 18, 68-72.

Paskoff, R., 1981. *L'Erosion des C?tes*. Presses Universitaires de France, Paris.

Pernetta, J.C. and Sestini, G., 1989. The Maldives and the impact of expected climatic changes. UNEP Regional Seas Reports and Studies, 104 p.

Pethick, J.S., 1981. Long-term accretion rates on tidal salt marshes. *J. Sedim. Petrol.*, 51, 571-577.

Phillips, J.D., 1986. Coastal submergence and marsh fringe erosion, *J. Coastal Res.*, 2, 426-436.

Pirazzoli, P.A., 1986. Secular trends of relative sea level changes indicated by tide-gauge records. *J. Coastal Res.*, Special Issue 1, 1-26.

Pirazzoli, P.A., 1987. Recent sea level changes and related engineering problems in the lagoon of Venice. *Prog. Oceanogr.*, 18, 223-246. Pirazzoli, P.A., 1991. *World Atlas of Holocene Sea level Changes*. Elsevier Oceanography Series, 58.

Pringle, A., 1981. Beach development and coastal erosion in Holderness. In: J. Neale and J. Flenley (eds.) *The Ouaternary in Britain*, 194-205.

Psuty, N. P., 1986. Impacts of impending sea level rise scenarios: the New Jersey barrier island responses. *Bull. New Jersey Acad. Sci.*, 31, 29-36.

Pugh, D.T., 1987. The Global Sea Level Observing System. *Hydrographic J.*, 45, 5-8.

Ranwell, D.S., 1974. The salt marsh to tidal woodland transition. Hydrobiol. Bull., 8, 139-151.

Redfield, A.C., 1972. Development of a New England salt marsh. *Ecol. Monogr.*, 42, 201-237.
Reed, D.J., 1990. The impact of sea level rise on coastal salt marshes. *Progr. Phys. Geogr.*, 14, 465-481.
Richards, A.F., 1960. Rates of marine erosion of tephra and lava at Isla San Benedicto, Mexico. *Int. Geol. Congr.*, 21, 58-63. S.C.O.R. Working Group 89., 1991. The response of beaches to sea level changes: a review of predictive models. *J. Coastal Res.*, 7, 895-921.
Stark, K.P., 1988. Designing for coastal structures in a greenhouse age. In: G.I. Pearman (ed.), *Greenhouse: Planning for Climatic Change*, C.S.I.R.O. Division of Atmospheric Research, 161-176.
Stevenson, J.C., Ward, L.G. and Kearney, M.S., 1986. Vertical accretion in marshes with varying rates of sea level rise. In: M.S. Kearney (ed.), *Estuarine Variability*, Academic Press, New York.
Sunamura, T., 1992. *Geomorphology of Rocky Coasts*. Wiley, Chichester.
Teh Tiong Sa, 1993. Potential impacts of sea level rise on the permatang coasts of Peninsular Malaysia. *Malaysian J. Tropical Geogr.*, 24, 41-56.
Teh Tiong Sa and Lim Chow Hock, 1993. Impacts of sea level rise on the mangroves of Peninsular Malaysia. *Malaysian J.Tropical Geogr.*, 24, 13- 20.
Titus, J.G., 1986. The causes and effects of sea level rise. In: J.G. Titus (ed.), *Effects of Changes in Stratospheric Ozone and Global Climate*, Vol. 1 - Overview, 219-248.
Vanderzee, M.P., 1988. Changes in saltmarsh vegetation as an early indicator of sea level rise. In: G.I. Pearman (ed.), *Greenhouse: Planning for Climatic Change*, C.S.I.R.O. Division of Atmospheric Research, 147-160.
Verstappen, H.T., 1966. The use of aerial photographs in delta studies. In: *Scientific Problems in Humid Tropical Zone Deltas*, UNESCO, 373-379.
Wadsworth, A.H., 1966. Historical deltation of the Colorado River, Texas. In: M.L. Shirley (ed.), *Deltas in Their Geologic Framework*, Houston Geological Society, 99-105.
Walker, H.J., 1988. *Artificial Structures and Shorelines*. Kluwer Academic Publ.
Woodroffe, C.D., 1990. The impact of sea level rise on mangrove shorelines, *Progr. Phys. Geogr.*, 14, 483-520.
Woodroffe, C.D., Thom, B.G. and Chappell, J., 1985. Development of widespread mangrove swamps in mid-Holocene times in northern Australia. *Nature*, 317, 711-713.
Woodworth, P.L., 1991. The Permanent Service for mean sea level and the Global Sea Level Observing System. *J. Coastal Res.*, 7, 699-719. Zenkovich, V.P., 1967. *Processes of Coastal Development* (trans: O.G. Fry, ed. J.A Steers), Edinburgh. Zunica, M., 1990. Beach behaviour and defences along the Lido di Jesolo, Gulf of Venice, Italy. *J. Coastal Res.*, 6, 709-719.

CHAPTER 5

Land Subsidence in Bangkok during 1978–1988

Prinya Nutalaya, R.N. Yong, Thongchai Chumnankit, Somkid Buapeng

ABSTRACT: More than 4,550 km^2 of Bangkok was affected by land subsidence between 1960 and 1988; maximum local subsidence exceeded 160 cm. Most subsidence was due to soil compression at 10-200 m depth in response to the reduction of the aquifers' piezometric levels. Using aquifer characteristics, ground-water hydrochemistry and subsoil geotechnical properties, mathematical models have been developed to predict further subsidence and recharge response. To inhibit this subsidence, proper planning of land use, flood protection, waste water treatment, pollution control, and sound management of surface and ground-water resources are recommended.

Introduction

Situated on the Chao Phraya Basin, Bangkok is underlain by a thick layer of normally consolidated sediments of late Tertiary to Quaternary age. This layer exceeds 500 meters in thickness. Severe settling has occurred in recent years due to a combination of load-consolidation and subsurface water withdrawal. As a result, seasonal flooding of the populated areas during the rainy season has been common and has caused both property damage and loss of human life, in particular between 1983 and 1986.

The damage and inconvenience caused by flooding in the metropolitan area have prompted both short-term and long-term reparative and preventive measures. Several schemes have been proposed to alleviate flooding of the populated area, but none has been designed to inhibit the high rate of land subsidence. Two projects have been devoted to the evolution of issues surrounding land subsidence, e.g. the National Environment Board's (NEB) annual monitoring program of land subsidence in Bangkok and the study of the impact of Quaternary sediments on urban development and land use of the central plain of Thailand, jointly conducted by McGill University and the Asian Institute of Technology (AIT). From 1978 to 1982, in cooperation with the Royal Thai Survey Department and the Department of

J.D. Milliman and B.U. Haq (eds.), Sea-Level Rise and Coastal Subsidence, 105-130.

Mineral Resources and the AIT, the NEB conducted a comprehensive investigation of land subsidence caused by deep well pumping in the Bangkok area. Based on the results of the study, land subsidence was strongly linked to extensive deep well pumping (at a rate of about 1.3 million m^3 of water a day). Recharging the aquifers was recommended as one of the alternatives to alleviate the sinking of Bangkok. Since 1982, no major projects have been launched to directly study and/or prevent land subsidence problems, although there is a standing committee of the National Environment Board to monitor the subsidence and to review the overall situation. This committee submits its report to the cabinet every six months. In addition, a group has studied the feasibility of artificially recharging ground water.

The objective of this paper is to review the situation of land subsidence in Bangkok in the ten years since it was first studied in 1978, and to suggest some remedial measures.

Land Subsidence Prior to 1978

Bangkok ground surface elevations were first surveyed in the early 1930's. At that time, about eighty benchmarks were installed. In 1978, the Royal Thai Survey Department reviewed these benchmarks in order to evaluate the general land subsidence in Bangkok. Unfortunately, only eleven of the original benchmarks were found, but all of them showed a variable degree of subsidence, ranging from 20 to 85 cm (Fig. 1).

1978 to 1988

With the advent of the investigation entitled: *Land Subsidence Caused by Deep Well Pumping in the Bangkok Area* in 1978 and the subsequent monitoring program, the Royal Thai Survey Department ran a series of first-order leveling referenced to benchmarks on bedrock in Rachaburi. Within the subsiding area in Bangkok, 133 primary benchmarks and 514 secondary benchmarks were surveyed. The ground elevations along the surveyed line were calculated at 200 m intervals and plotted on a 1:50,000 scale map. The computed rate of subsidence is shown in Figures 2 and 3. At the same time, a network of subsidence observation stations was established, some of which had soil compression indicators with automatic recorders installed by the AIT and the NEB. These soil compression indicators yielded data on relative subsidence of the subsurface layers (Fig. 4).

The Department of Mineral Resources and the Division of Water Resources Engineering of the AIT also conducted a supplementary project on the ground-water resources in the Bangkok area during 1978 to 1982. A series of observation wells in various aquifers underneath Bangkok monitored the piezometric level changes (Figs. 5 to 8). In addition, the locations of extraction wells and the pumping rates were closely watched.

From these monitoring results the following observations can be made:

1) Total subsidence of Bangkok up to 1988 ranged from less than 20 cm to more than 160 cm. The magnitude of subsidence correlates well with the amount

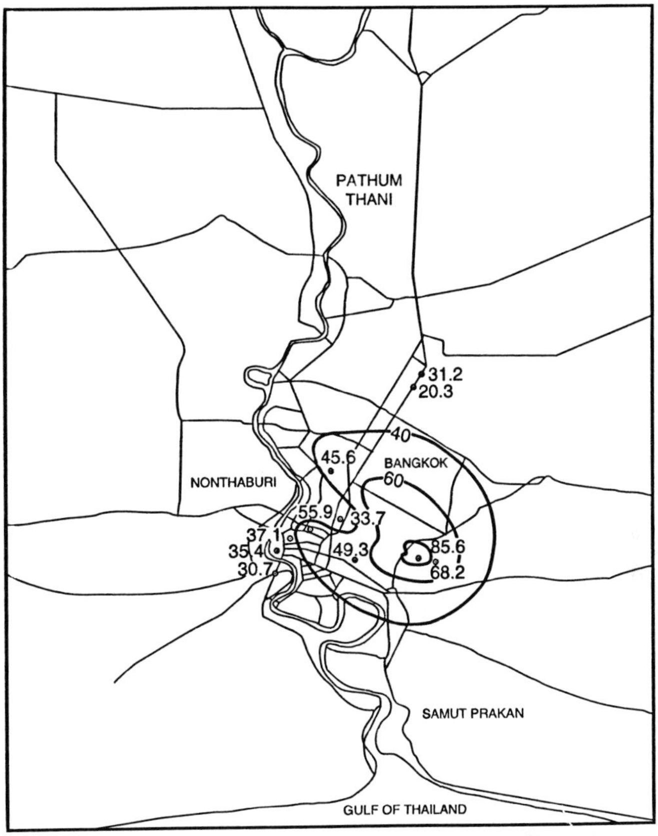

Fig. 1. Total land subsidence in Bangkok, 1933–1978.

of ground water withdrawn. As the rate of ground-water extraction follows city growth, so does land subsidence. The size of the subsiding area has increased markedly during the ten years, and the ground elevations in critical parts of Bangkok were already below the mean sea level (Fig. 9). In the inner part of the city, where there is sufficient water supply and the ground-water pumping is prohibited, the subsidence decreased considerably. At some places, the ground level even showed a slight rebound.

2) The total area affected by land subsidence was approximately 4,550 km^2. Land with more than 10 cm/year subsidence shifted toward the east and north, but new subsidence also developed to the northwest and west of Bangkok.

3) The soil compression indicators show that only 5–10% of total subsidence occurred in the first ten meters of depth, related to seasonal aridation and flooding, natural consolidation of the clays, load of buildings and land fills, and traffic vibration. The major subsidence took place at the depth ranges of 10–200 m, and was directly related to the piezometric heads decline in the Bangkok, Phra Pradaeng, Nakhon Luang and Nonthaburi aquifers.

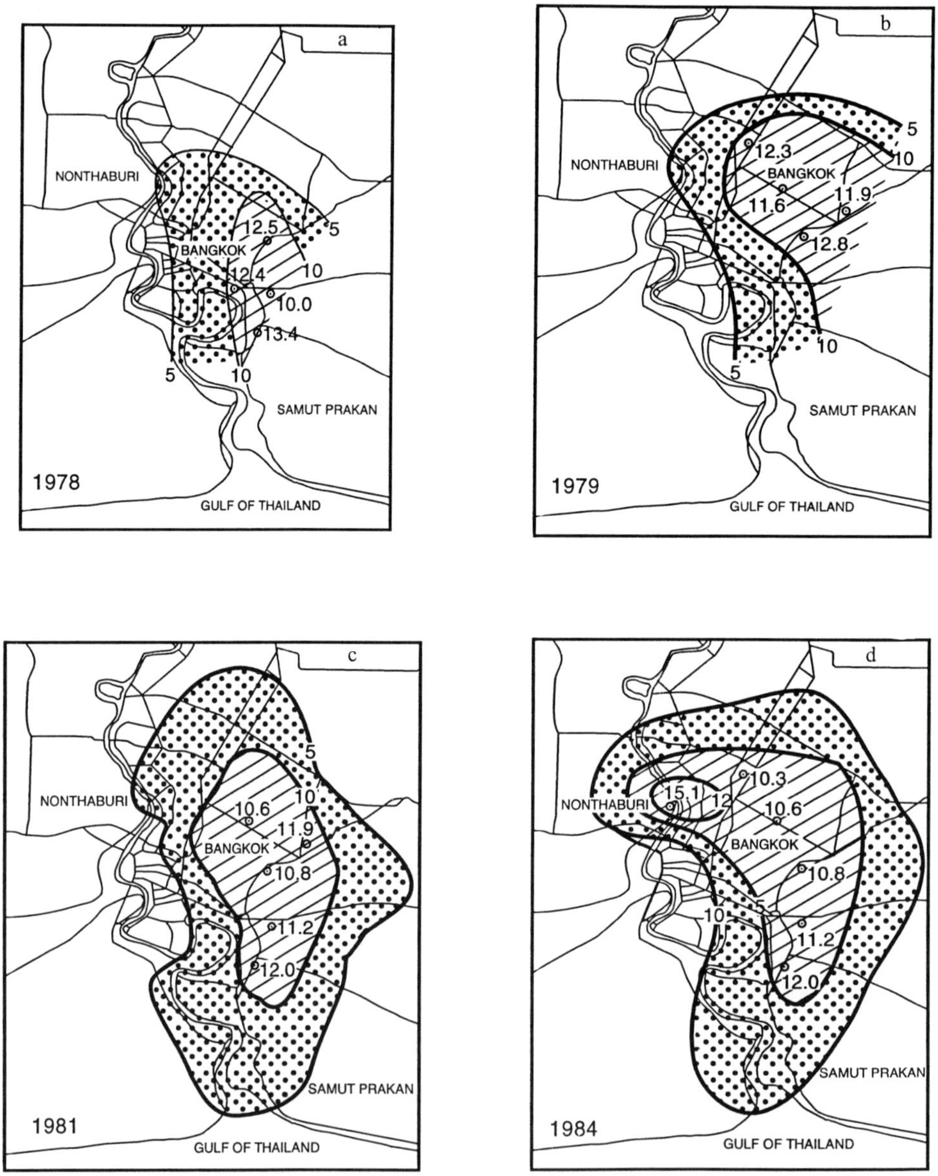

Fig. 2 a-d. Land subsidence (in cm) in 1978, 1979, 1981 and 1984.

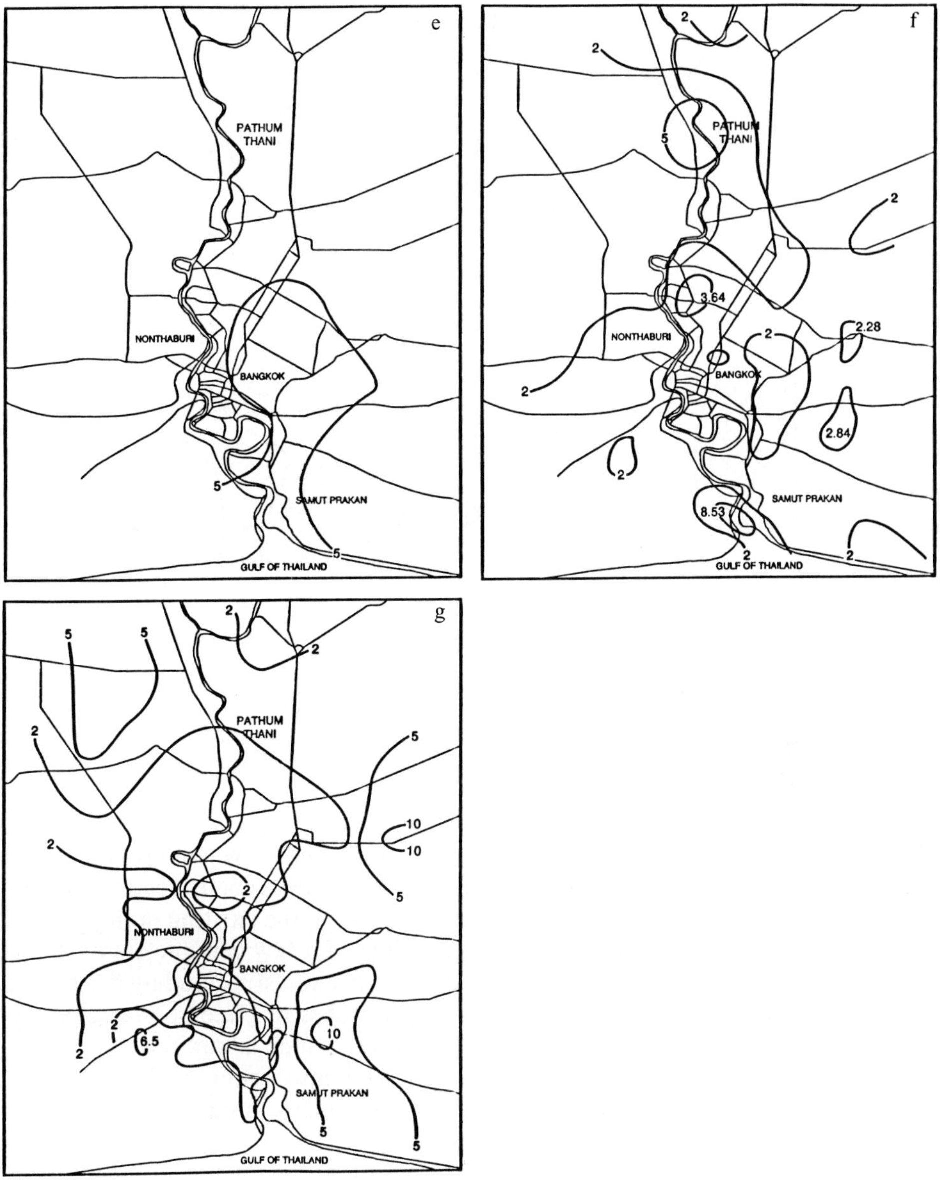

Fig. 2 e-g. Land subsidence (in cm) in 1985–1987.

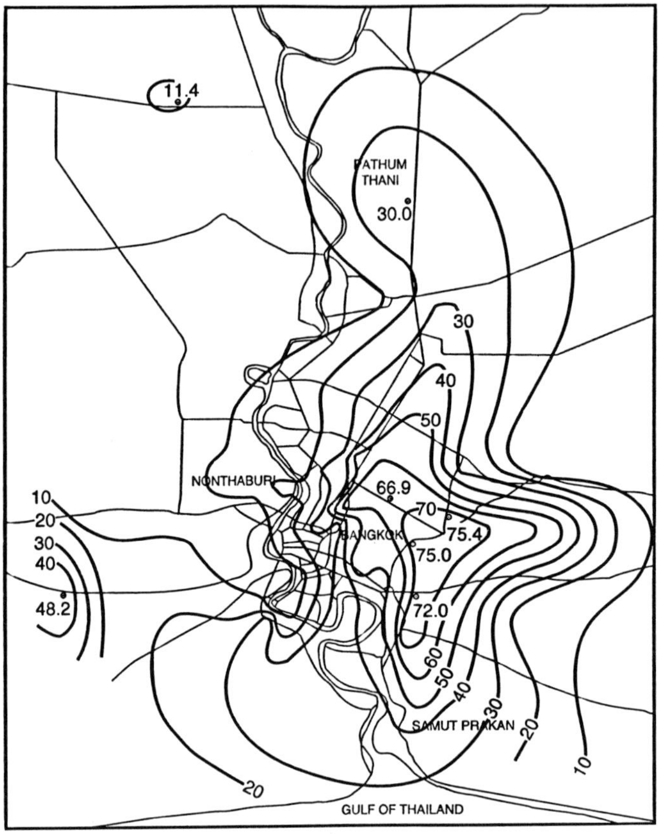

Fig. 3. Total subsidence (in cm) 1978-87.

Effects of Land Subsidence

Bangkok is situated on an extremely flat and low-lying area where ground elevations range from only 0–1.5 meters above mean sea level. Floods occur regularly during the rainy season and seasonally high tides. With the existing drainage system of Bangkok, a 30 mm rain can cause flooding of a few centimeters on some streets for up to six hours. If the rain exceeds 60 mm, extensive flooding can occur, requiring about 6 to 24 hours to drain. During October to December, seasonally high tides rise up to 1.35 m above the mean sea level, which in turn raises the Chao Phraya River water level, and the water flows back into sewers and canals. The low-lying and coastal areas, where ground elevation is lower than 1.35 m, will be affected if there is no flood protection system.

Because of differing rates of subsidence at different depths and/or locations, the settling of building foundations, underground pipe lines, roads, walkways and bridges occur at different rates if placed at different depths or at different locations. This difference in settlement has caused cracking and bending of pipes, floor slabs, concrete walk-ways and steps, and detachment of septic tanks and sidewalks or

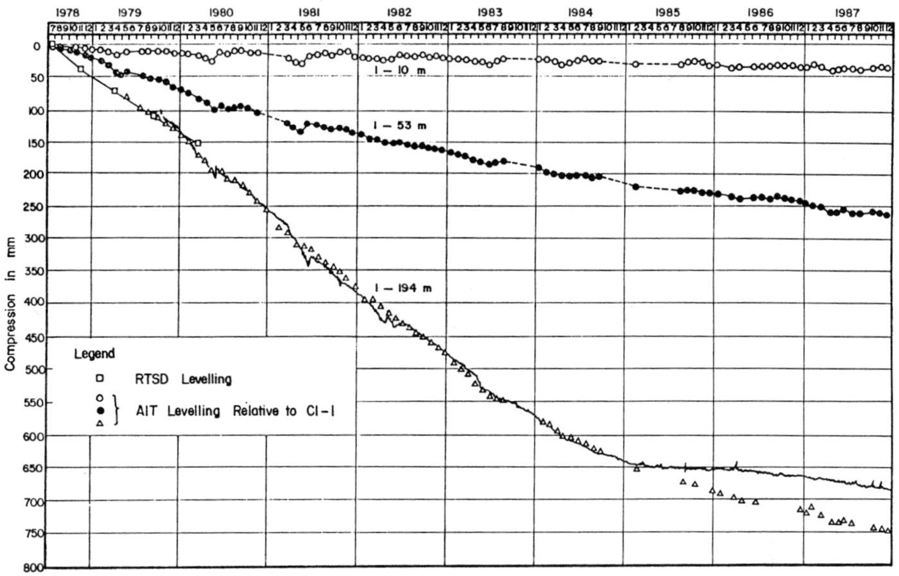

Fig. 4. Subsidence of Station 21 (Bang Na) as measured by an automatic recorder and levelling.

steps from buildings. Horizontally buried pipe lines can be lifted above ground by protrusion of ground-water wells. Buildings with shallow foundations can sink with the ground, and when the roads are resurfaced or elevated they may be partially buried.

Every benchmark in the subsidence area is not reliable because they, too, are susceptible to subsidence. In Bangkok, the Chao Phraya River water levels are reported relative to the benchmark at the Memorial Bridge pier. None of these data can be assumed to be true unless the actual subsidence at that point is known. The same problem is true of the road surface leveling and flood protection dikes. As a result, *no* benchmark in Bangkok can be used as a reference point.

Overdraw of ground water without replenishment causes a decline in piezometric levels and pore pressures. In the area where an aquifer is connected to the sea, salt water gradually will replace the fresh water. The deterioration of ground-water quality is also caused by the migration of the mineralized water from the higher pore pressure areas in the clay layers into the sand layers. Ground water in the Bangkok aquifer is now heavily polluted to the point that it is no longer potable. Several ground-water wells that tap the lower aquifers also are facing this problem.

If the land surface is allowed to subside farther, the flooding problem will become more acute. Flooding will cover much wider areas and last over a longer period. Draining of flood water by gravity will be reduced and costs of pumping will increase. The areas already flooded will be under deeper water. As a result, septic tanks in the flooded area will become water logged and sewage and associated diseases become potentially lethal health hazards. Ground-water wells and even the raw-water supply canal will be contaminated. The transportation system would

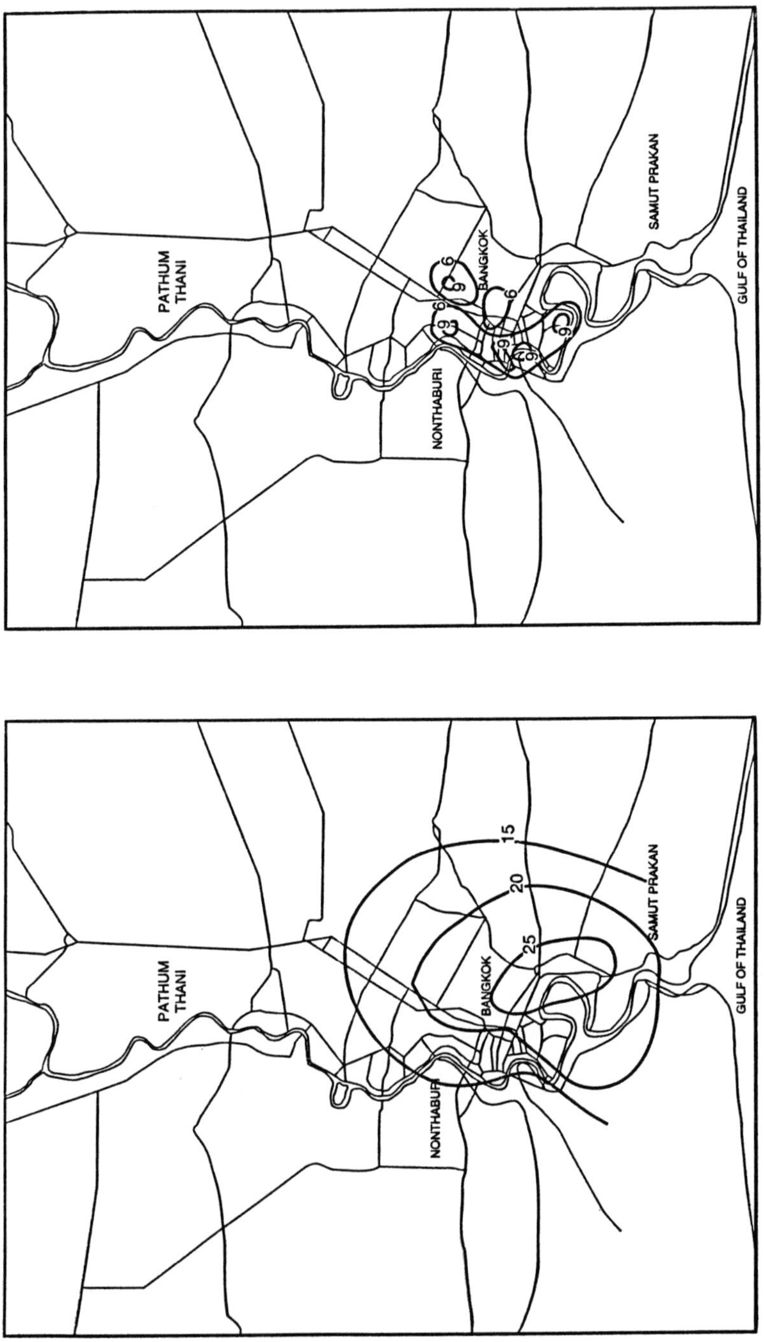

Fig. 5. *Left:* (a) Piezometric level of Bangkok aquifer, 1979. *Right:* (b) Piezometric levels of Phra Pradaeng and Nakhon Luang aquifers, 1958-59.

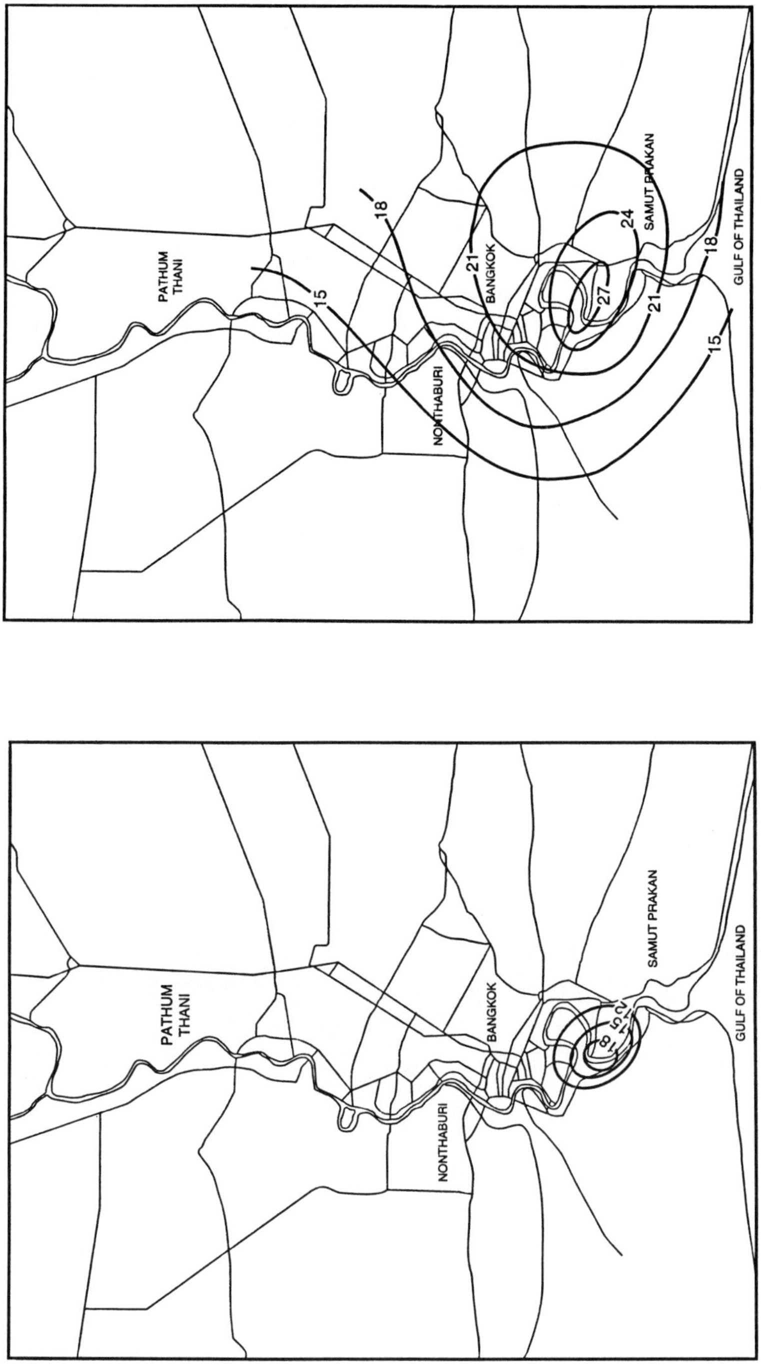

Fig. 6. *Left:* (a) Piezometric level of Phra Pradaeng aquifer, 1969. *Right:* (b) Piezometric level of Phra Pradaeng aquifer, 1974.

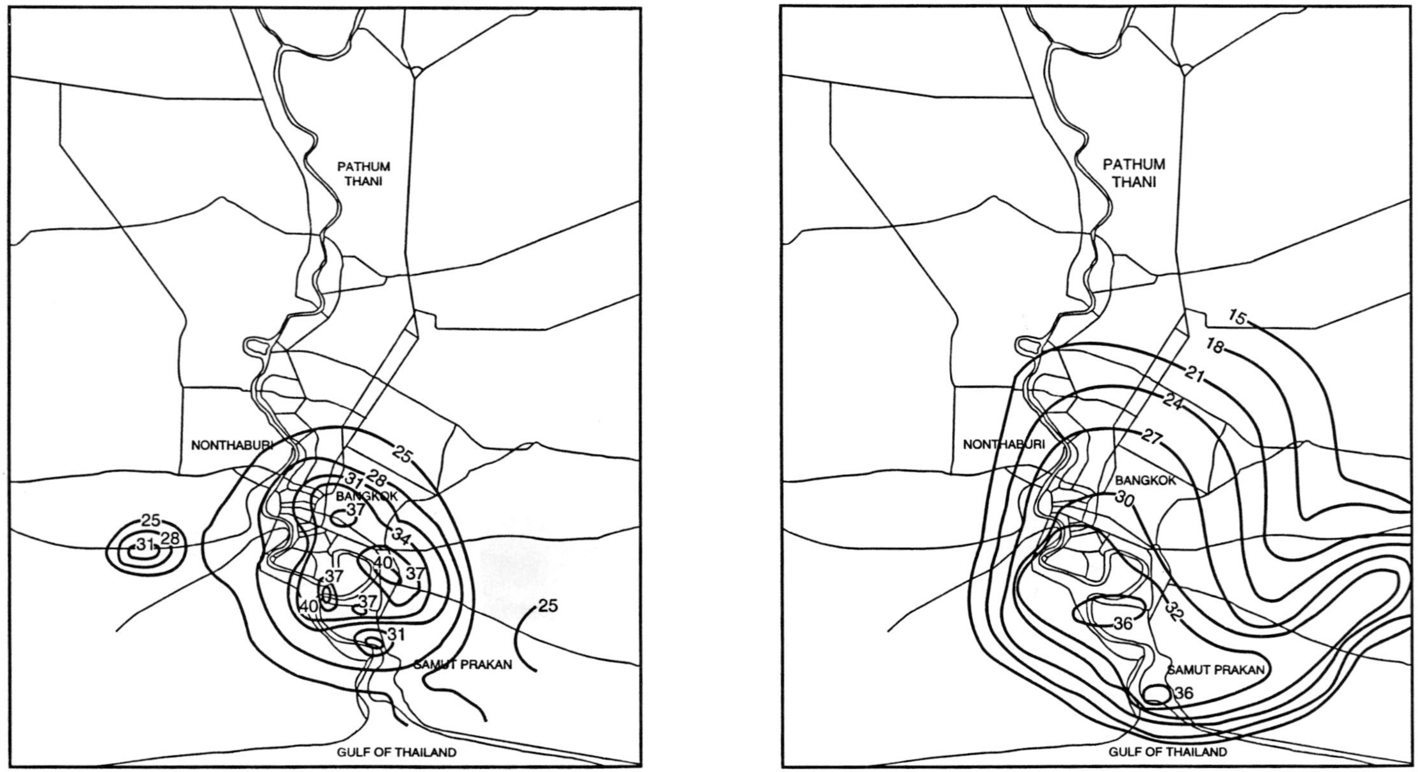

Fig. 6. *Left*: (c) Piezometric level of Phra Pradaeng aquifer, 1979. *Right*: (d) Piezometric level of Phra Pradaeng aquifer, 1982.

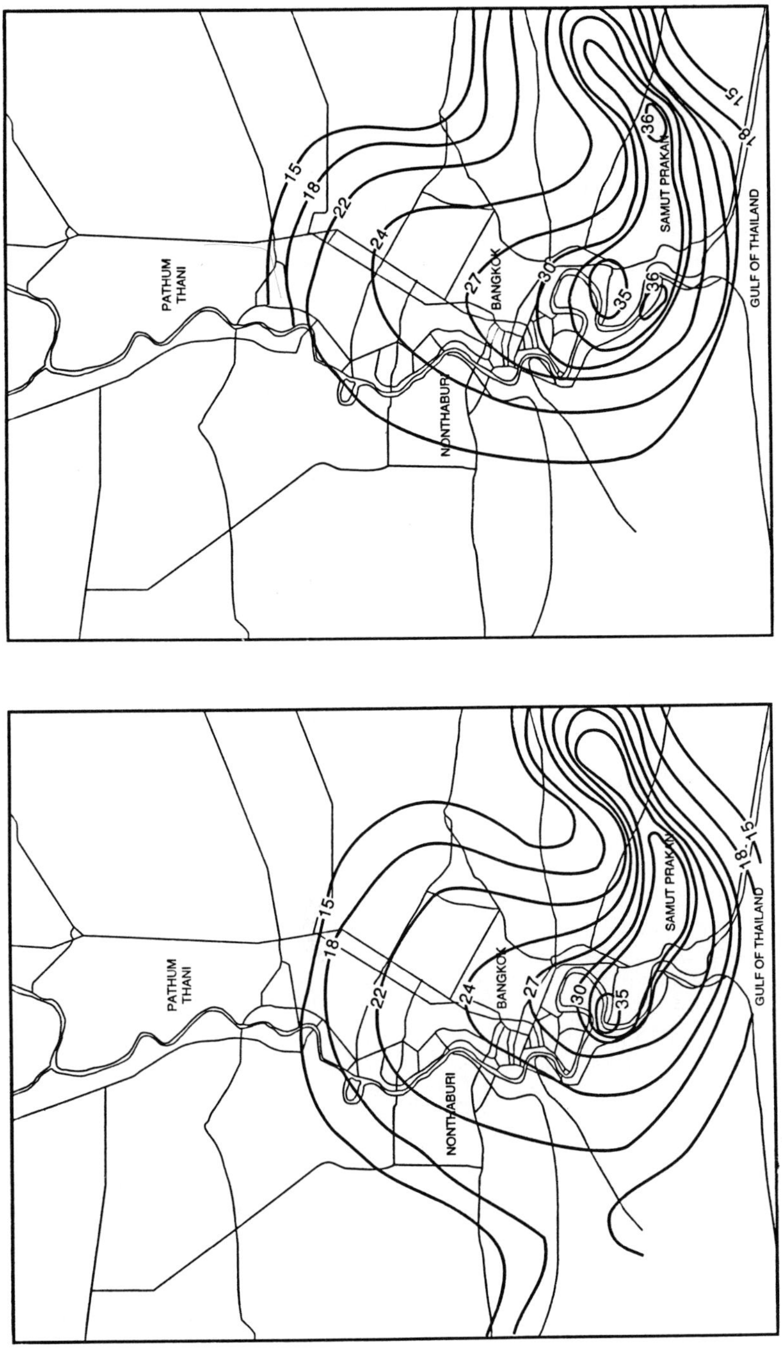

Fig. 6. *Left:* (e) Piezometric level of Phra Pradaeng aquifer, 1985. *Right:* (f) Piezometric level of Phra Pradaeng aquifer, 1986.

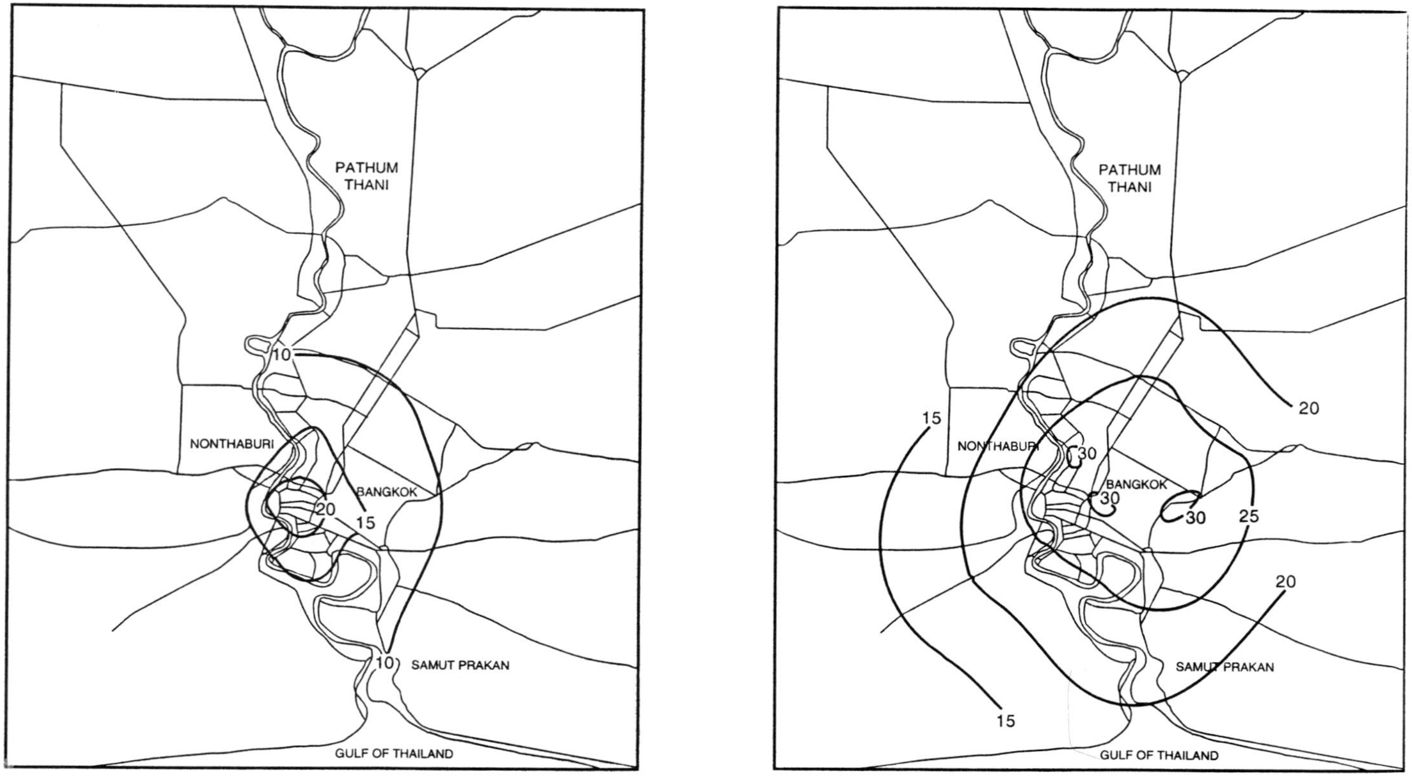

Fig. 7. *Left*: (a) Piezometric level of Nakhon Luang aquifer, 1969. ***Right***: (b) Piezometric level of Nakhon Luang aquifer, 1974.

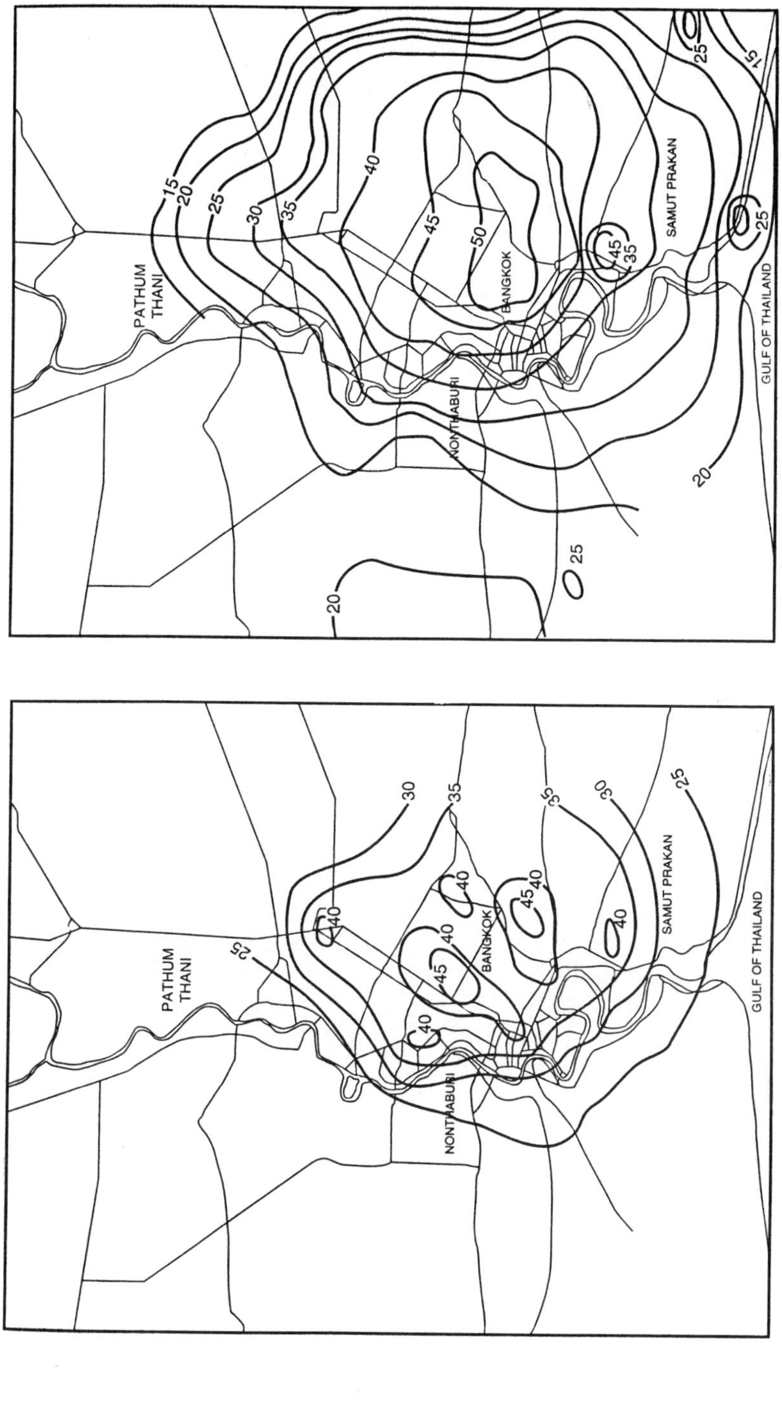

Fig. 7. *Left:* (c) Piezometric level of Nakhon Luang aquifer, 1979. *Right:* (d) Piezometric level of Nakhon Luang aquifer, 1982.

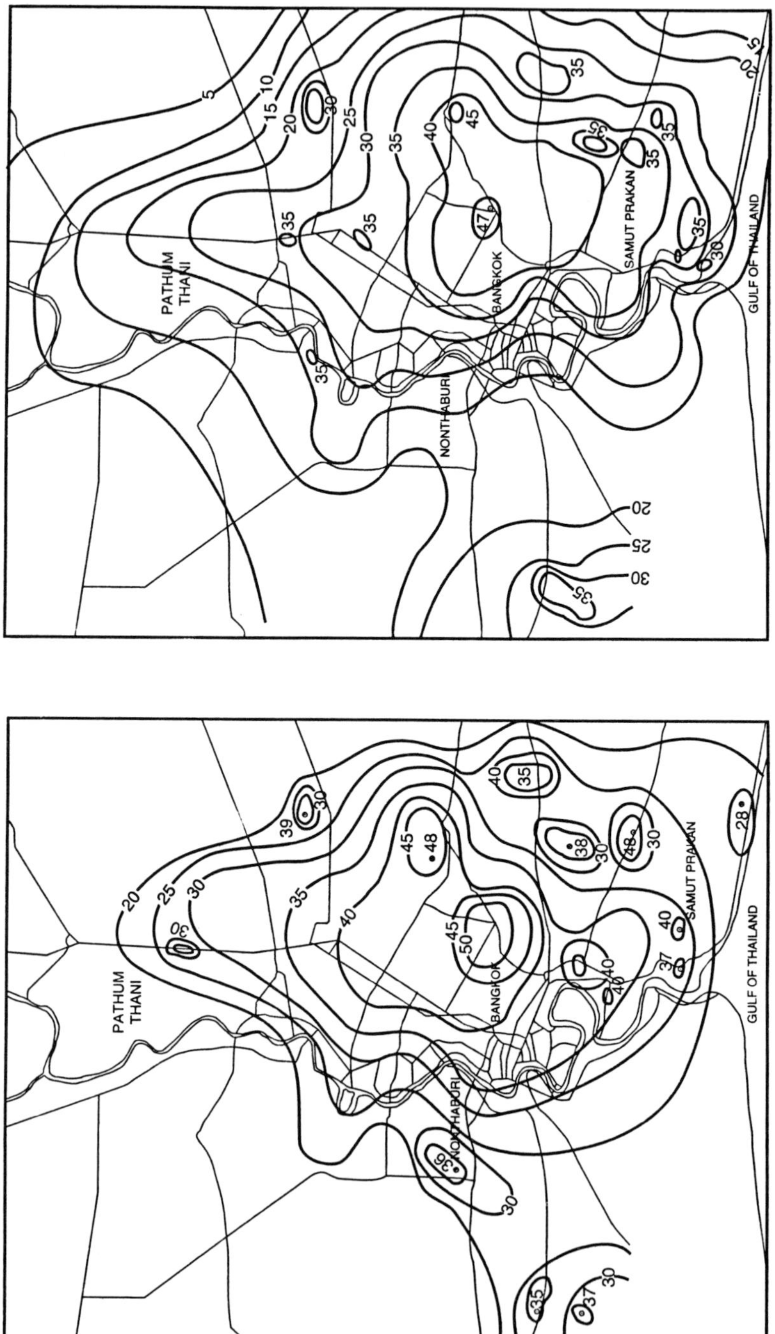

Fig. 7. *Left*: (e) Piezometric level of Nakhon Luang aquifer, 1985. *Right*: (f) Piezometric level of Nakhon Luang aquifer, 1986.

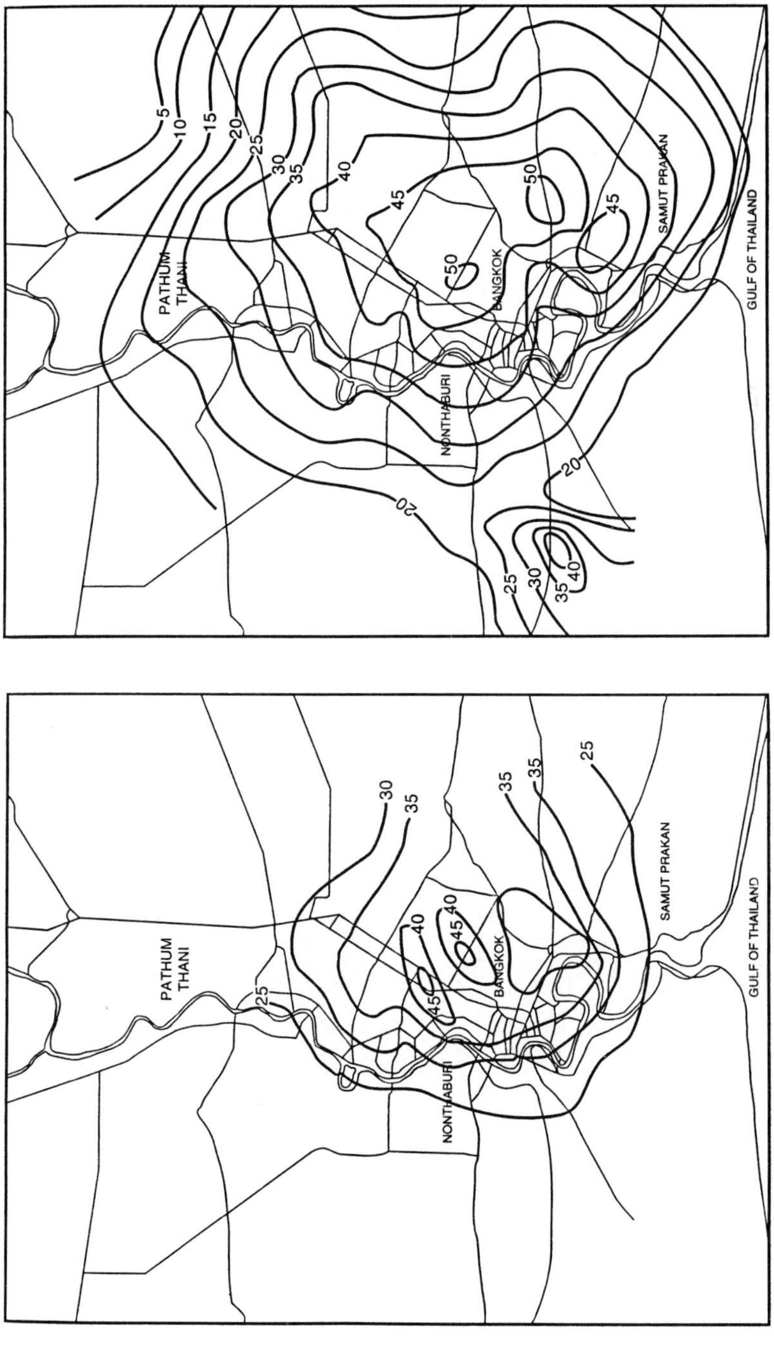

Fig. 8. *Left:* (a) Piezometric level of Nonthaburi aquifer, 1979. *Right:* (b) Piezometric level of Nonthaburi aquifer, 1982.

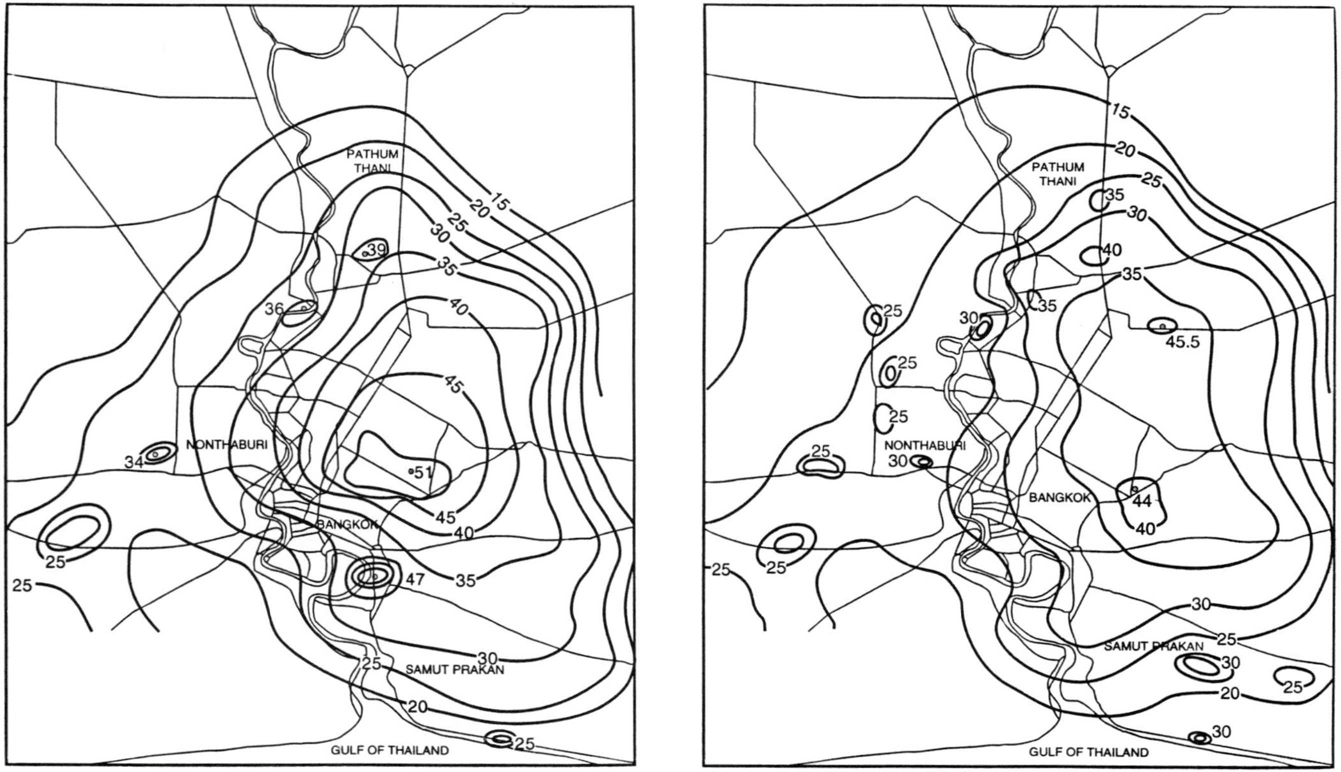

Fig. 8. *Left*: (c) Piezometric level of Nonthaburi aquifer, 1985. *Right*: (d) Piezometric level of Nonthaburi aquifer, 1986.

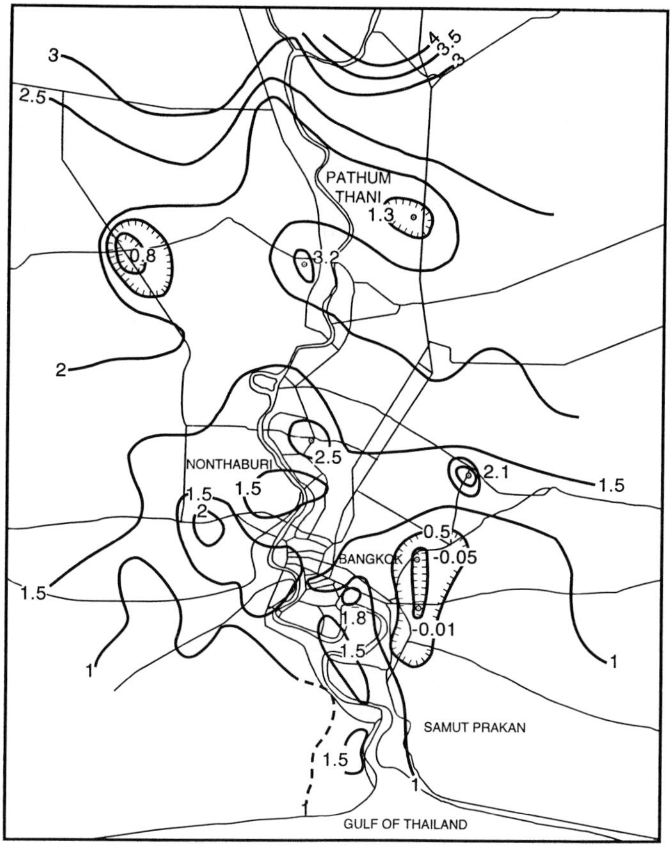

Fig. 9. Ground elevation of Bangkok, 1987.

be disrupted as would the supplies such as food and medicine. The effects of these wastes on the environment would be even more tenacious then on humans.

Aquifer Delineation

The subsurface formations of the Chao Phraya Basin consist of a complex sequence of unconsolidated, deltaic, shallow marine, and alluvial sediments of late Tertiary and Quaternary ages lying below the soft Bangkok clay of Holocene (Recent) age. At certain places the thickness of this unconsolidated sediment is more than 2000 m, and in some places the depth to bed rock may be as much as 3,300 m (AIT, 1981a).

Several thousand ground-water wells have been drilled through the sediments of the Lower Central Plain over the past three decades. Unfortunately, the lithologic logs of most of these ground-water wells are not available, and no systematic collection of this information has been attempted. The north-south and east-west stratigraphic sections of the lower central plain of Thailand were prepared by

Piancharoen & Buapeng (1973) using the well-log data and are presented in Figure 10.

The geohydrology of the Chao Phraya Basin with respect to subsidence was studied by Sodsri (1978) using the variation of aquifer characteristics, general sequence of strata, and thickness and types of sediments. She found that the boundary of the Bangkok sub-basin covers an area spanning both sides of the Chao Phraya River. The western limit is fault-bound along the Tha Chin River, the eastern limit is east of Pathum Thani and west of Chachoengsao Provinces. The northern boundary is south of Ayudhya (Fig. 11).

Source and Age of Ground Water

Stable isotope analysis (^{18}O and ^{2}H) indicates that the shallow ground water is recharge water of modern meteoric origin infiltrated through the alluvial deposits. The ground-water flow patterns established on the basis of the ^{18}O enrichment suggest that the recharge water in the Mae Klong River basin apparently moves towards the Chao Phraya River basin and the recharge water in the northern region moves south along a flow path similar to the Chao Phraya River (Yong and Nutalaya, 1988).

Isotopic ages of ground water determined by ^{14}C analysis indicate that the majority of ground water in the lower central basin is considerably younger than the corresponding sediments of the basin and older than the last regression in the area. These data suggest that the formation water of the sediments was flushed out by recharge water during deep ground-water circulation and has evolved to the present water type. Two major ground-water circulations probably occurred in the period of 10 to 20 ka and 30 to 40 ka. The ground water in the central region along the Chao Phraya River appears to have circulated during the period of 10–20 ka. The major hydrochemical processes that influence the composition of the present ground-water composition include carbonate dissolution, ion exchange, and the mixing of fresh and saline waters.

Geotechnical Properties of Soils

Laboratory tests to determine basic soil properties were performed on undisturbed soil samples obtained from bore holes. The basic properties included natural water content, Atterberg limits, total unit weight, compressive strength from pocket penetrometer test, salt content and pH, and grain-size distribution. A summary of the basic soil properties at the two locations is given in Figure 12.

The permeability test was conducted on both clay and sand samples. The results showed that the permeability of sand strata affects the rate of ground-water flow toward a pumping well while the permeability of clay layers controls the amount of vertical seepage between layers.

The typical permeability value of sand samples is 10^{-5} to 10^{-4} cm/sec while the permeability of clay samples varies from 5×10^{-8} to 10^{-6} cm/sec. Generally, the permeability is low for the samples that have a relatively low sand content. For

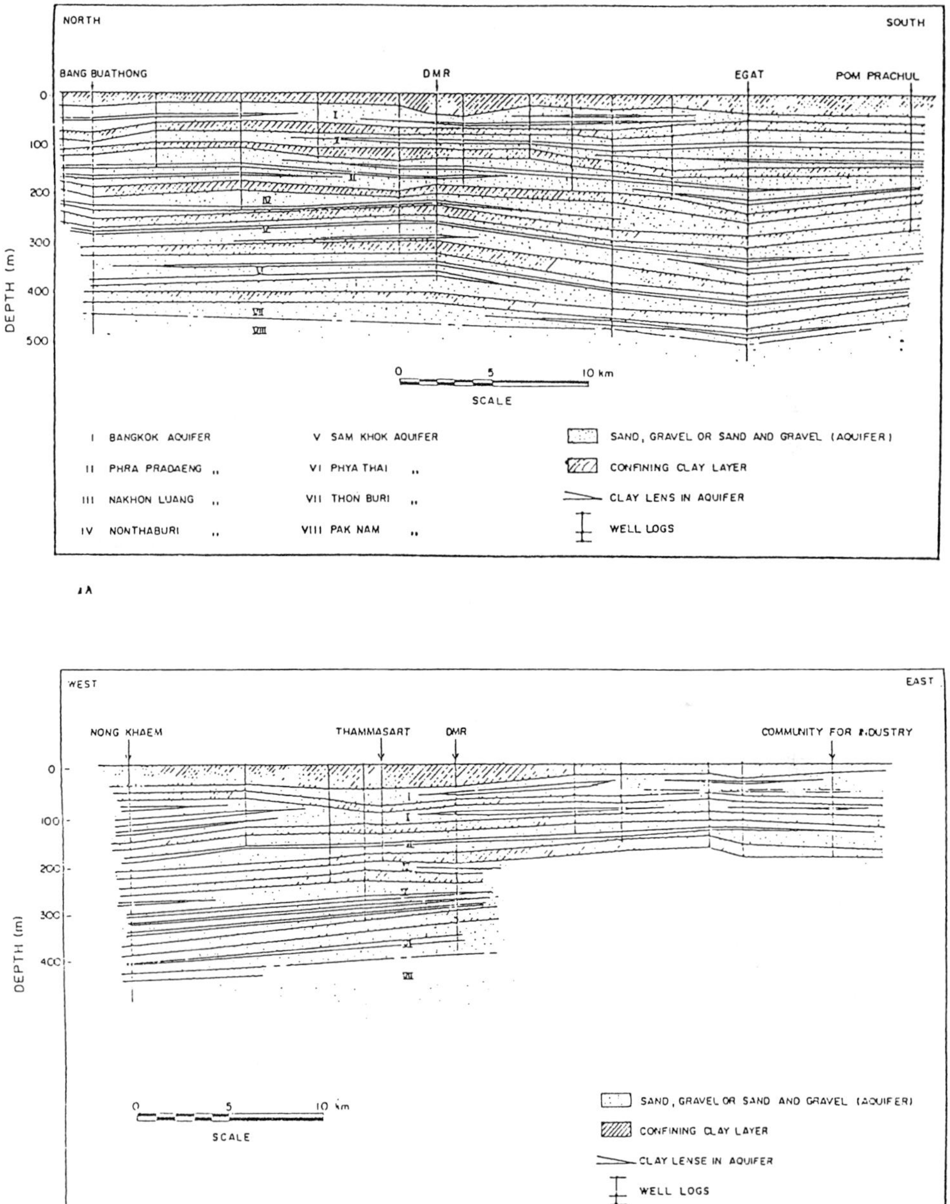

Fig. 10. *Top*: (a) North-south stratigraphic section of central plain of Thailand. *Bottom*: (b) East-west stratigraphic section of central plain of Thailand (after Piancharoen and Buagpeng, 1973).

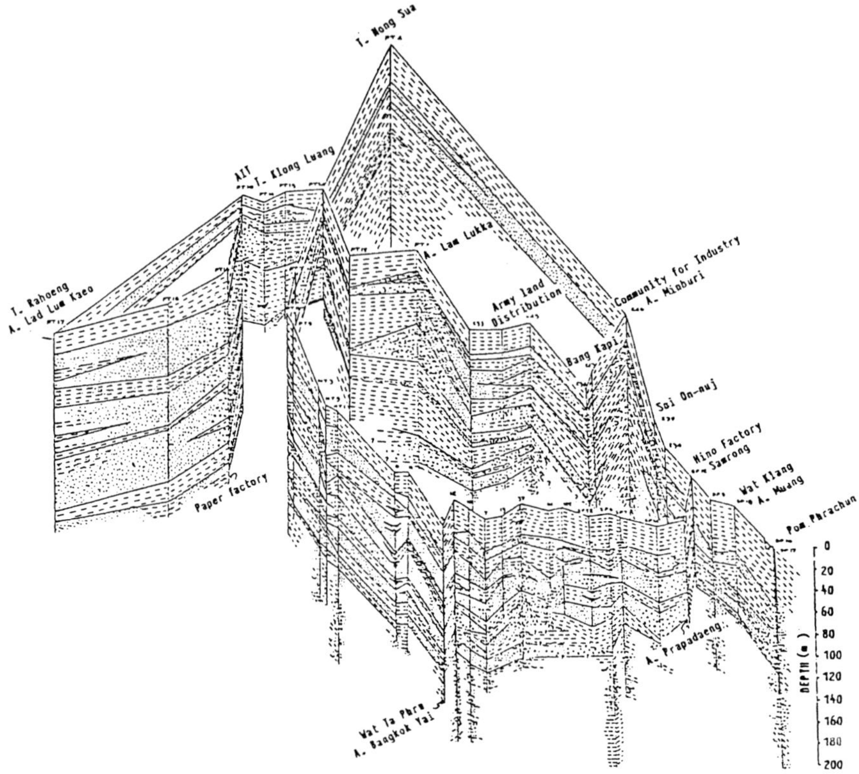

Fig. 11. Fence diagram of Bangkok subsoils (After Sodsri, 1978).

samples with a sand content less that 90%, the permeability is low, approaching the typical permeability of clay (i.e. about 10^{-7} cm/sec).

Even though the deep strata consist of hard clay and dense sand and are much less compressible than the upper clay layer, their consolidation still contributes to substantial proportion of the total surface subsidence since their total thickness is so great (about 300 meters) and, at a depth of 200 meters, a large ground-water pressure drop (more than 40 t/mo) occurs.

Summary of compression characteristics of soil sample are given in Figures 13–15. Generally, it was found that the compressibility of sand samples is very low. However, for the sand samples that has large clay content (more than 10%) they show high compressibility.

Mathematical Modeling to Predict Subsidence and Recharge Response

Two mathematical models were developed: a finite difference model to simulate the piezometric head changes in the top three aquifers with the imposition of various ground-water activities, and a finite difference linear model to find the hydrodynamically optimal locations and corresponding rates of artificial recharging for an

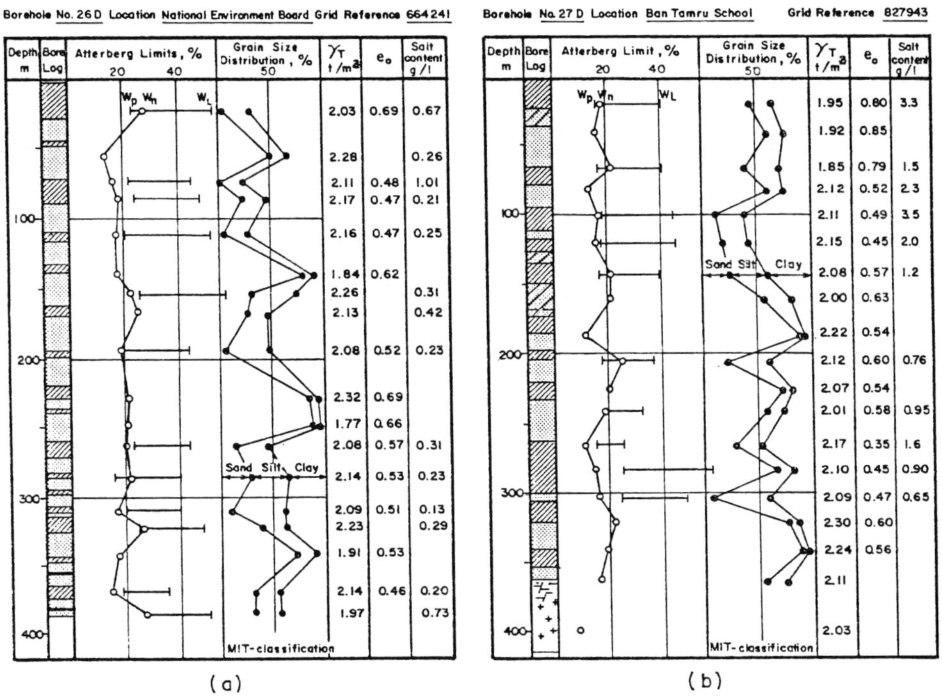

Fig. 12. Geotechnical properties of soils.

aquifer. In the first model (multi-aquifer model), leakage between aquifers across the intervening aquitard was taken as directly proportional to the piezometric head gradient across the aquitard. The model was calibrated for the period of 1979–1982. Monthly piezometric head observations of the three aquifers at locations distributed over the area have been recorded during this period (AIT, 1981a,b). These observations were used to estimate the rate of the piezometric head drop in aquifers, and those rates were taken as the first criterion for calibration of the multi-aquifer model.

In the estimation of optimal recharge rates, several recharge schemes were used. In each scheme, recharge water was distributed according to the optimal pattern, but the total rate of recharge was changed. Twenty, forty, fifty, sixty and eighty percent of total optimal recharge rates were considered. Behavior of the system without any recharging also was investigated.

Figure 16 shows the piezometric heads at the end of the simulated recharging period, along an east-west grid in the Bangkok aquifer with different recharging schemes. Similar variations of piezometric head were observed in the Phra Pradaeng and Nakhon Luang aquifers. The multi-aquifer solution shows that the piezometric head in each aquifer is above the target level, corresponding to the optimal recharge rate. The reason is that the leakage from the water table and between

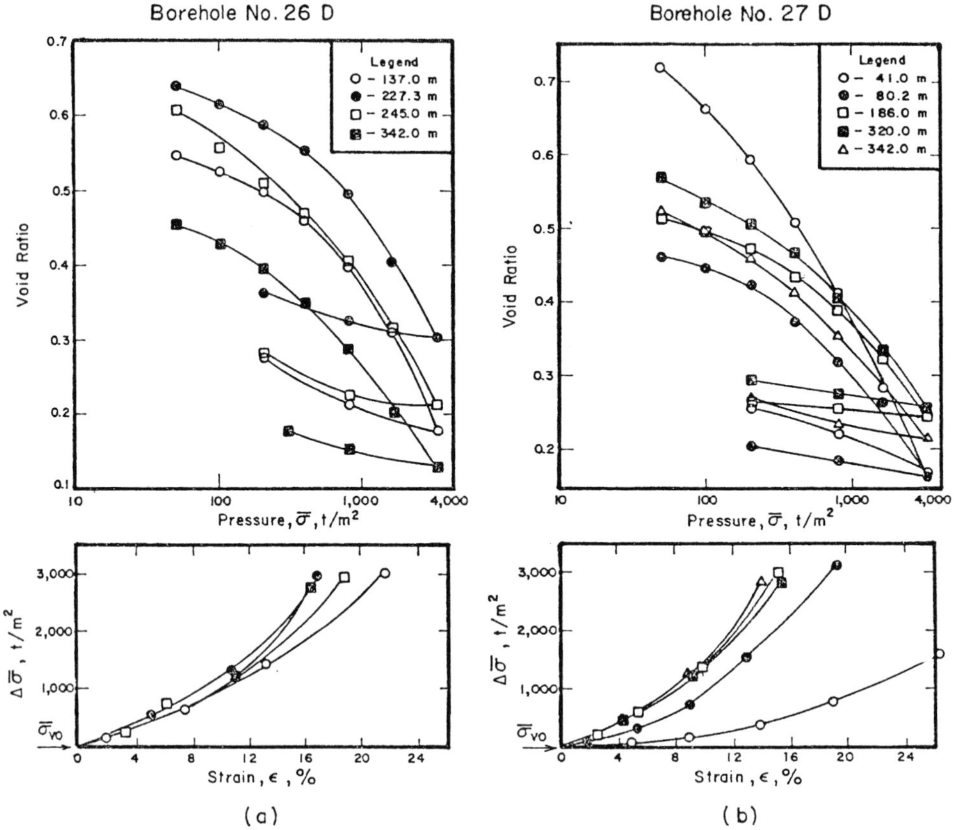

Fig. 13. Compression characteristics of sands.

aquifers was ignored when optimal recharge rates were estimated, but included in the multi-aquifer model, thus giving some additional piezometric head gain. Application of eighty percent of optimal recharge rate was sufficient to achieve the objective — i.e. the piezometric heads in the aquifers rise above or reach the target level. Variation of the piezometric head with time at a typical node, near the center of the cone of depression is given in Figure 18. The results show that, if withdrawal is constant, the rate of piezometric head drop diminishes and the head values reach a stable level within about the recharge period without any artificial recharging. This is an optimistic result caused by the assumption of constant head boundaries. The piezometric head in the aquifers under actual conditions might be lower than the predicted values. However, if the natural replenishment of aquifers is sufficient to maintain the piezometric level assumed at boundaries, the above result shows that the aquifer system will reach a steady state within about two years without further significant drop in piezometric head, if new ground-water withdrawals are banned. Recharging of the aquifers at a rate of 20% of the optimal

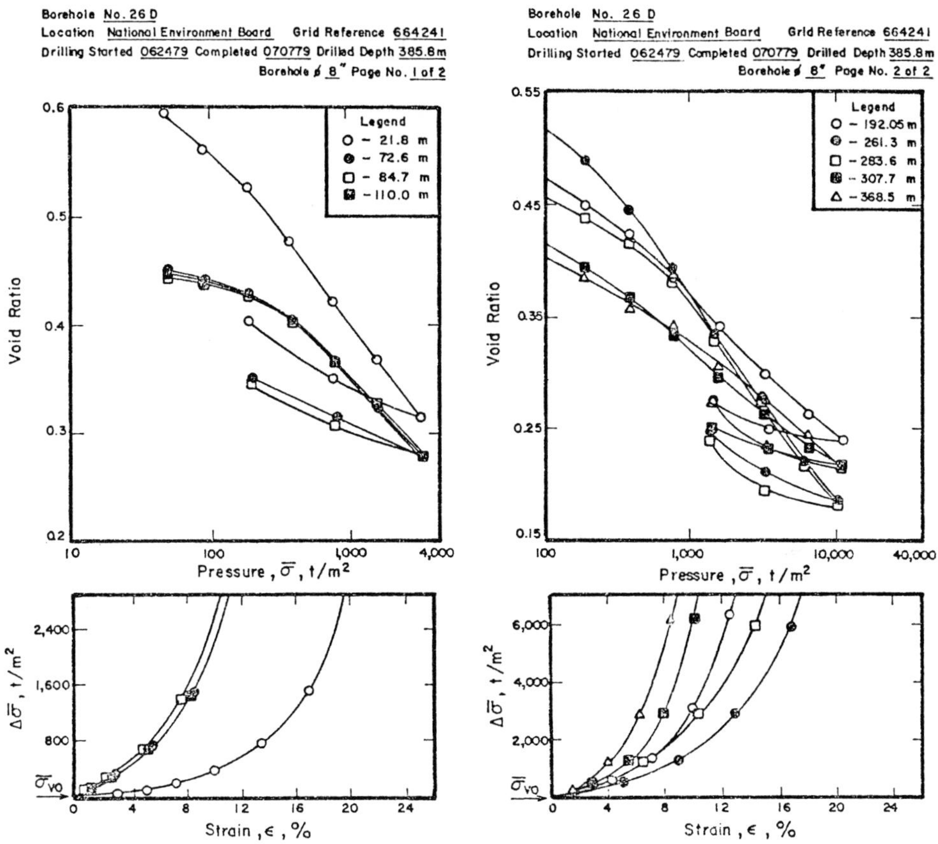

Fig. 14. Compression characteristics of clays (from National Environment Board).

value helps maintain the piezometric head in the aquifer at or a little above the levels in the beginning of recharging (Fig. 17).

Recommendations

The complex land subsidence problem of Bangkok greatly affects most of the communities that make up the Bangkok metropolis, causing millions of bahts of loss annually. Since the ground-water pumping will increase as the city grows, land subsidence problem will accelerate unless a solution is found and implemented. The shear size of the problem in terms of the area, the number of people it affects, the money lost and the money needed to remedy the situation require an integrated method of arriving at a set of solutions.

Many problems related to ground-water extraction can be avoided by reliable management planning to satisfy future water demands and stablize the water table, which in turn can help solve specific problems such as land subsidence,

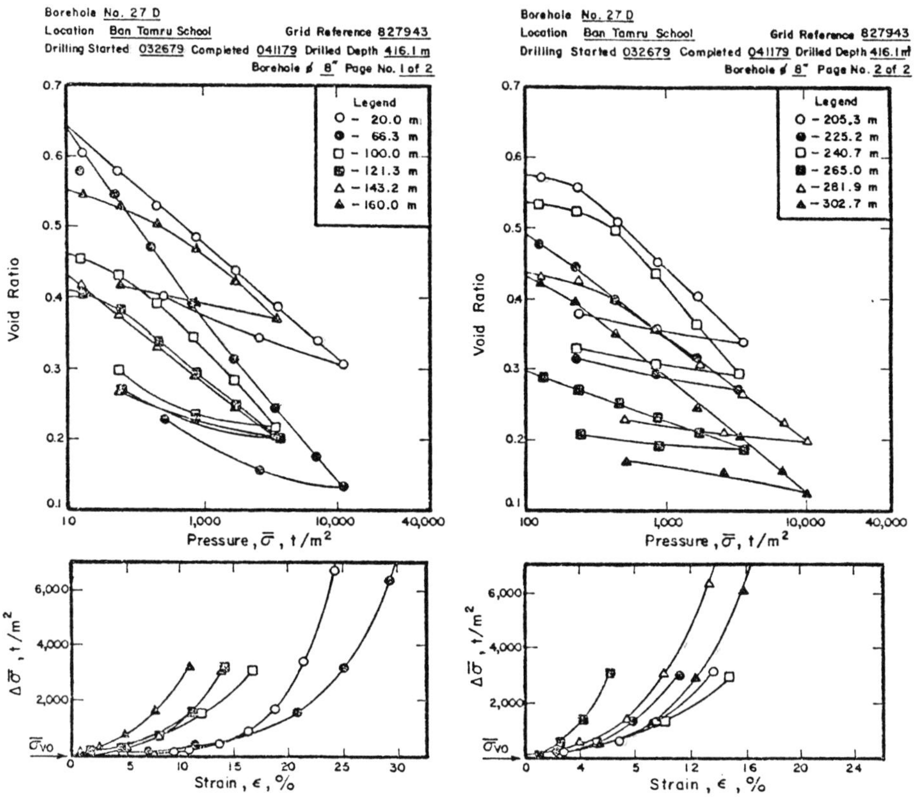

Fig. 15. Compression characteristics of clays (from Ban Tamru School).

deterioration of water quality and lowering water level in the ground-water basin. The problems in ground-water development are caused generally by improper or unplanned development of ground-water resources and to a lesser extent by such natural conditions as poor-quality water and low-yielding aquifers. Continuous extraction of water from aquifers in excess of recharge results in a continuous lowering of water levels in the Bangkok area, which in turn leads to land subsidence and contamination of the ground-water resources. Remedial measures for these problems, which are difficult to solve independently, should be avoided as short term solutions on a one to one basis, and instead should be incorporated into a sound ground-water management plan.

Proper planning and management of ground-water resources are urgently needed. Ground-water management programs in Bangkok should be planned on the basis of a regional system and long-term solutions, and should be conducted in parallel with the managements of surface-water hydrology, flood protection, land-use regulations and waste-water treatment and pollution control.

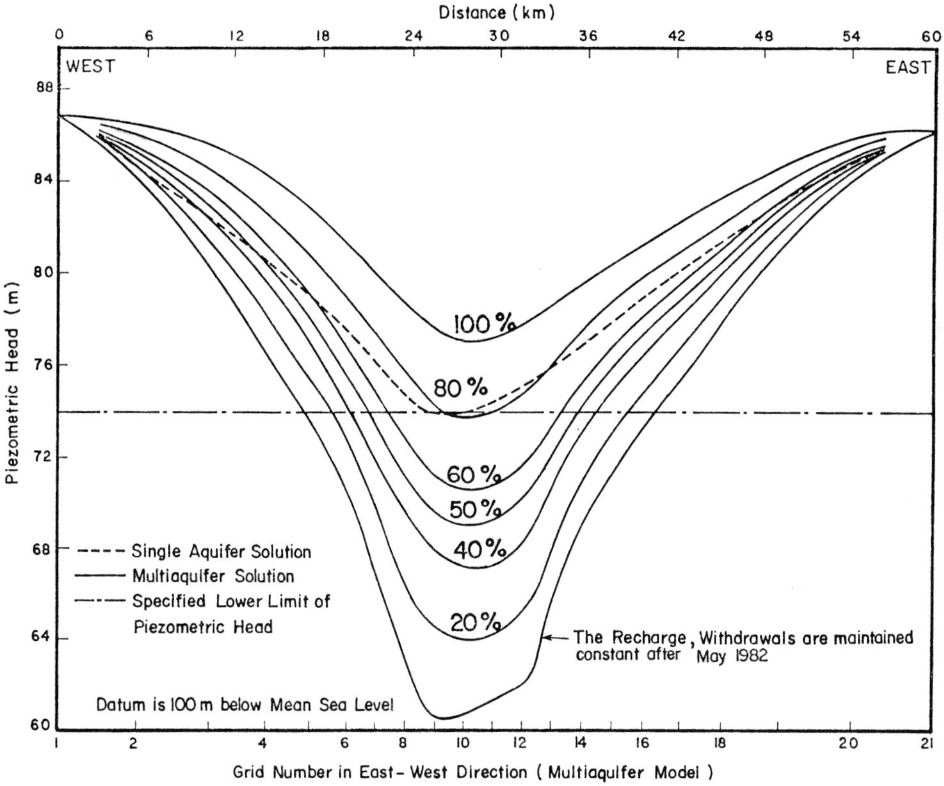

Fig. 16. Piezometric head along an east-west section of the Bangkok aquifer at the end of recharge, with application of different percentages of optimum rate of recharge.

Acknowledgements

The authors wish to express their sincere gratitude and appreciations to Miss Kasama Narkthong who typed the manuscript and to Mr. Wanchai Sophonsakulrat who drafted figures.

References

Asian Institute of Technology, 1978. Investigation of Land Subsidence Caused by Deep Well Pumping in the Bangkok Area, Phase I Final Report, A.I.T. Research Report No. 87, Bangkok, 222 pp.

Asian Institute of Technology, 1980. Investigation of Land Subsidence Caused by Deep Well Pumping in the Bangkok Area, Phase II Final Report, A.I.T. Research Report No. 89, Bangkok, 196 pp.

Asian Institute of Technology, 1981a. Investigation of Land Subsidence Caused by Deep Well Pumping in the Bangkok Area, Comprehensive Report 1978-1981, A.I.T. Research Report No. 91, Bangkok, 365 pp.

Asian Institute of Technology, 1981b. Ground water Resources in Bangkok Area, Comprehensive Report 1987-1982, A.I.T. Research Report No. 137, Bangkok, 333 pp.

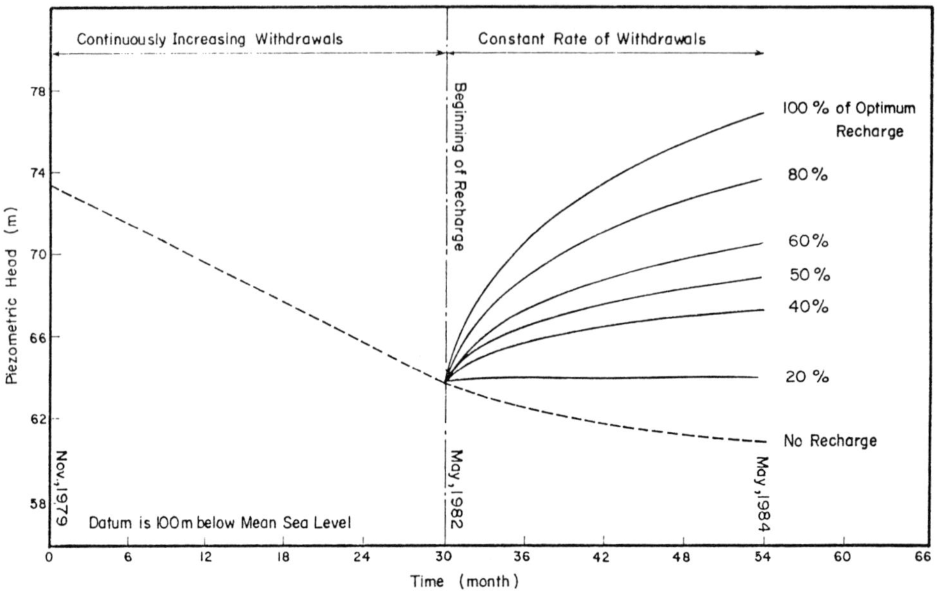

Fig. 17. Typical piezometric head variation with time at a point in the Bangkok aquifer, with application of different percentages of optimum rate of recharge.

Asian Institute of Technology, 1982. Investigation of Land Subsidence Caused by Deep Well Pumping in the Bangkok Area, Phase IV Final Report, A.I.T. Research Report No. 161, Bangkok, 98 pp.

Chuamthaisong, C. and Yuthamanop, A., 1980. Relative Age Dating of Ground water and the Velocity of Ground water Flow in the Vicinity of Bangkok Area, Preliminary Report, Department of Mineral Resources, Bangkok.

Hunter, J.A., Pullan, S.E. and Whiteley, R.J., 1986. High Resolution Seismic Reflection Surveying in the Chao Phraya Basin, Bangkok, Thailand – Initial Results. Unpublished Report.

Phiancharoen, C. and Buapeng, S., 1973. Hydrogeology of the Lower Chao Phraya Delta, Ground water Bulletin No. 4, Department of Mineral Resources, Bangkok.

Piyasena, W.M., 1982. Rehabilitation of a Depleted Aquifer System through Artificial Recharge – An Application to Bangkok, A.I.T. D. Eng. Dissertation No. GT-82-1-, Bangkok, 327 pp.

Ramnarong, V. and Buapeng, S., 1978. Ground water Level Map of Bangkok, Mitigation of Ground water Crisis and Land Subsidence in Bangkok Metropolitan Area Project, Department of Mineral Resources, Bangkok.

Royal Thai Survey Department, 1988. Report of Investigation of Land Subsidence in the Bangkok Area from February to July 1988, Royal Thai Survey Department, Bangkok.

Sodsri, S. (1978) . Geohydrology of the Chao Phraya Basin with Respect to Subsidence. A.I.T. M.Sc. Thesis No. 1296, Bangkok, 154 pp.

Yong, R.N. and Nutalaya, P., 1988. Impact of Quaternary Sediments on Urban Development and Land Use of the Central Plain of Thailand, Final Report Submitted to the International Development Research Centre (IDRC), Ottawa 242 pp.

CHAPTER 6

Coastal Erosion in the Gulf of Thailand

S. Vongvisessomjai, R. Polsi, C. Manotham, D. Srisaengthong and S. Charulukkana

ABSTRACT: Shorelines in the Gulf of Thailand that have undergone extreme change (such as, the Chao Phraya River mouth and the east coast of the peninsula) have been studied through the analysis of available navigation and topographic maps, aerial photographs and satellite images to determine areas of erosion and accretion. Severe shoreline erosion can be found on the west of the Chao Phraya River mouth (−500 m), Phetchburi (−200 m) and Hua Hin (−100 m), while accretion can be found on the east of Chao Phraya River mouth and south of Songkhla Lagoon inlet. Active sand-spit migration at Bang Nara River mouth poses severe problem for navigation of fishing boats.

Introduction

Locations of shorelines selected for this study (Fig. 1) are at the head of the Upper Gulf of Thailand and the east coast of the peninsula. These shorelines have been formed by sediments carried by rivers draining into the Gulf of Thailand: the head of the Upper Gulf by the Chao Phraya and Meklong Rivers; Bandon Bay by the Tapi River; Nakorn Si Thammarat Bay by the Pakpanang River; and Pattani Bay by the Pattani River. The southernmost shoreline of Thailand receives huge sediment supply from Malaysian rivers, namely the Kolok and Kelantan. Other sources of sediment along these shorelines are from the small streams draining into the Gulf of Thailand.

Most of the shorelines are sandy between rock outcrops, or muddy near the mouths of the big rivers mentioned above (Fig. 2). Breaking waves are the major factor in transporting sediment along the coastline. Bays and river mouths serve as depositional sites for these sediments. Erosion prevails downcoast of the bays and river mouths when the incoming sediments are less than the outgoing sediments. Jetties constructed at river mouths generally facilitate accretion upcoast but increase erosion downcoast.

Shoreline accretion and erosion in turn depend on wind speed and direction and available fetch lengths. Winds over Thailand generated by northeast and southwest

J.D. Milliman and B.U. Haq (eds.), Sea-Level Rise and Coastal Subsidence, 131-150.

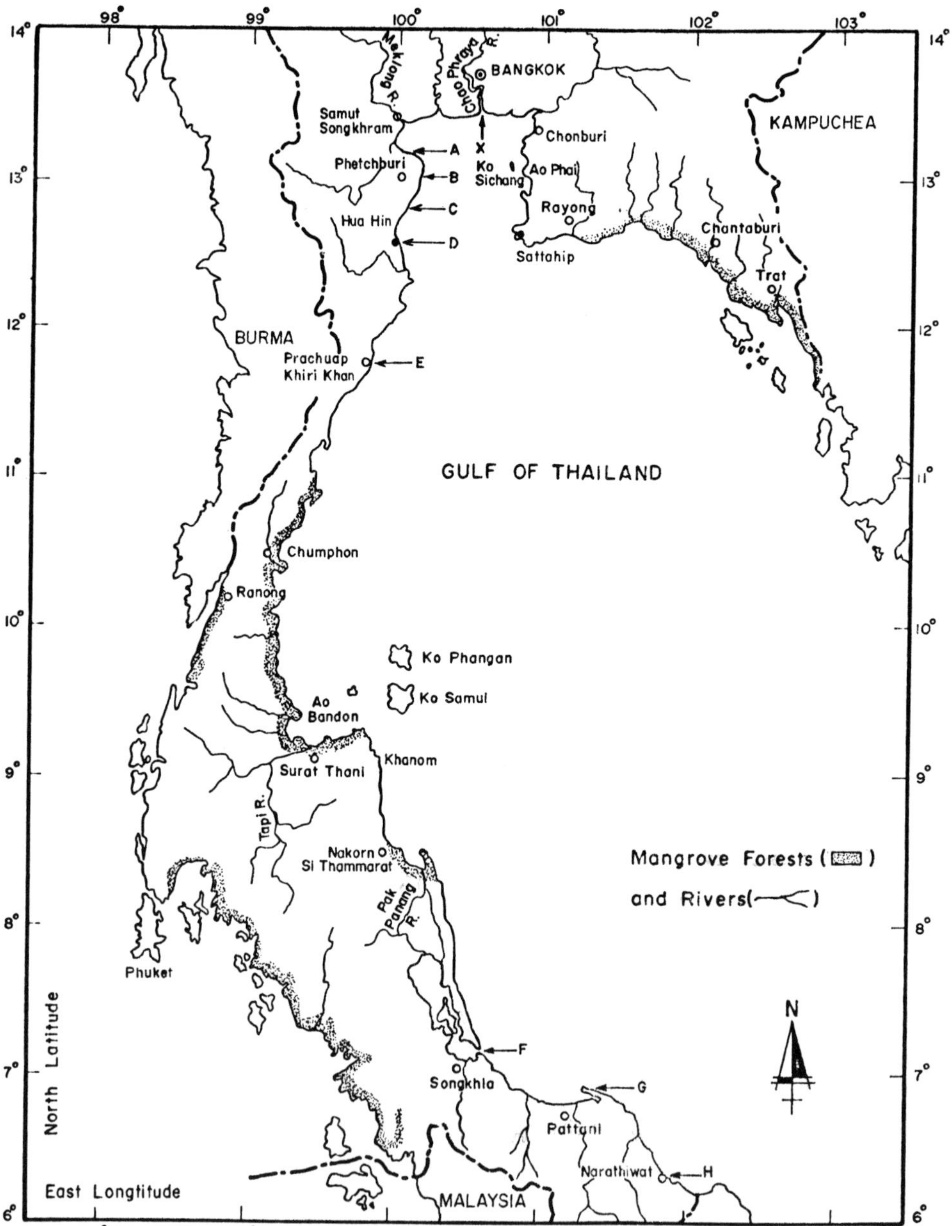

Fig. 1. Location map showing mangrove forests and rivers.

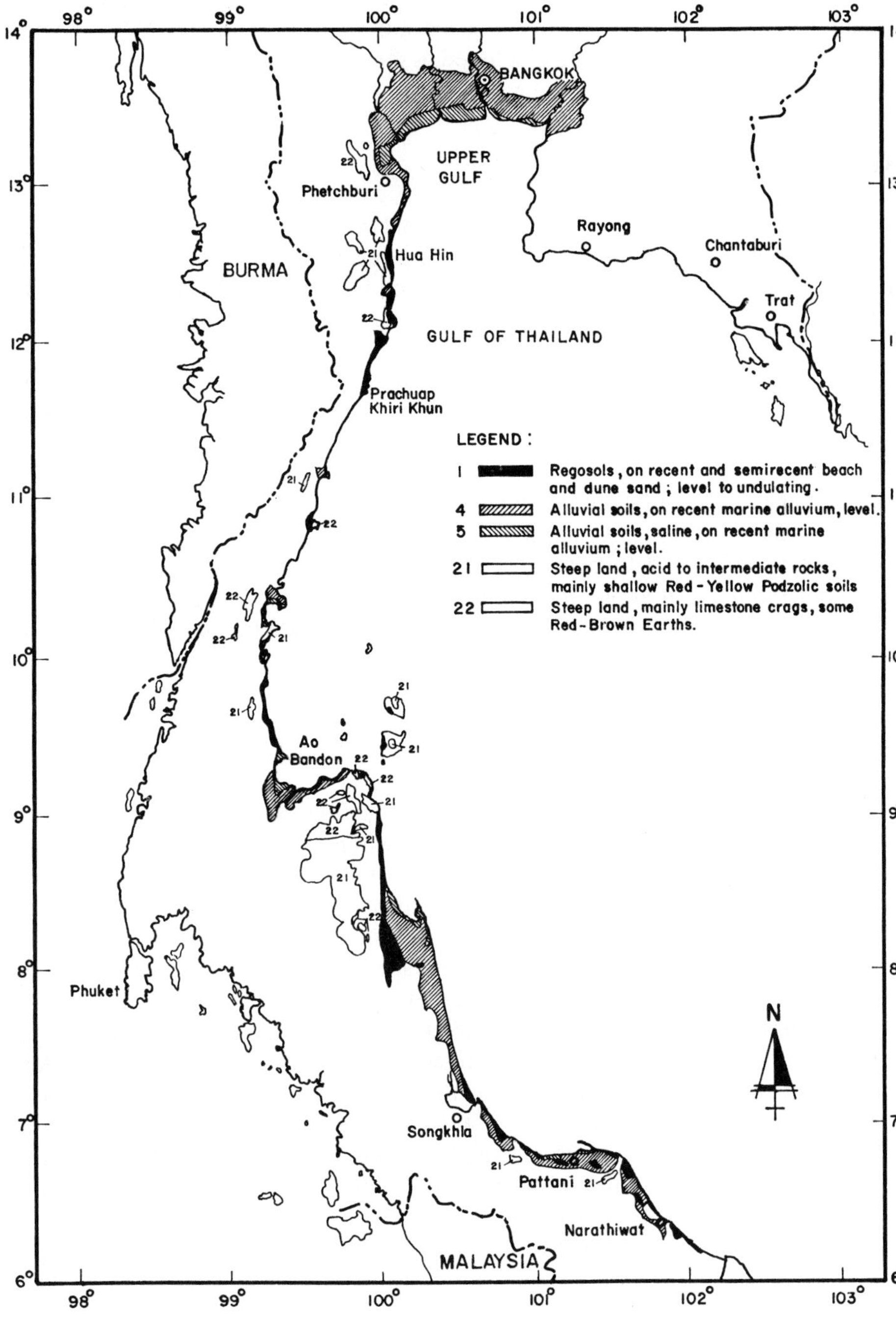

Fig. 2. Soil map of Thai coastline (from Moorman and Rojanasoonthon, 1972).

monsoons and cyclones from the South China Sea are used to calculate the waves in deep water using the method of Silvester and Vongvisessonijai (1970). The resulting coastal morphologies are discussed in the section following oceanographic conditions.

Oceanographic Conditions

Thailand's tropical climate is characterised by two major seasons. The southwest monsoon from mid-May through September, during which hot moist equatorial air masses traverse the country, is accompanied by moderate to heavy rains. The southwest winds generate moderate waves along the eastern seaboard of the Gulf of Thailand and along the west coast of the southern peninsula in the Andaman Sea. The northeast monsoon, from November to mid-March, brings a reversal of air movement. The northeast winds generate waves along the east coast of the southern peninsula of Thailand. Waves along the eastern seaboard of the Gulf of Thailand are very small during this period. Retreat of the southwest monsoon in September and October is frequently accompanied by peak wind and wave intensity caused by the passage of cyclones generated in the South China Sea.

Cyclone tracks tend to pass north of Thailand in the first half of the year, across the land mass in September, and over the Gulf on a westerly course during October–December, moving farther south as the year progresses. These tropical storms cause the extreme wind, sea and current conditions.

WINDS

The prevailing winds can be summarized as follows:

1. Prachuap Khiri Khan and Northward (including coastal areas in the east): During the northeast monsoon, the prevailing winds are from the north and east during the early season, later becoming more variable and southerly. Wind speeds are between 4–8 kts. Winds in the spring transition are variable, with about the same wind speed. In the southwest monsoon season, the prevailing winds are more persistent from the south and west. The wind speeds off Sattahip, Ko Sichana and Chonburi are stronger in this than in any other season.

2. South of Prachuarp Khiri Khan: Winds are generally from the east during the northeast monsoon season, with stronger winds during the season's end. During summer, wind directions are similar to those in winter, but speeds decrease. During the southwest monsoon season, winds increase along the northern portion of this region, but decrease south of Sonakhla. Wind speeds are mostly 5 kts and, except in Narathiwat, from the south and west. In the fall transition, winds begin to be variable and light, but speeds begin to increase as the northeast monsoon progresses. Surat Thani experiences weak winds (2–3 kts) throughout the year while Songkhla experiences stronger winds, up to 11–12 kts during the late northeast monsoon season. Figs. 3–6 show monthly wind-roses of Sattahip, Hua Hin, Surat Thani and Songkhla, respectively.

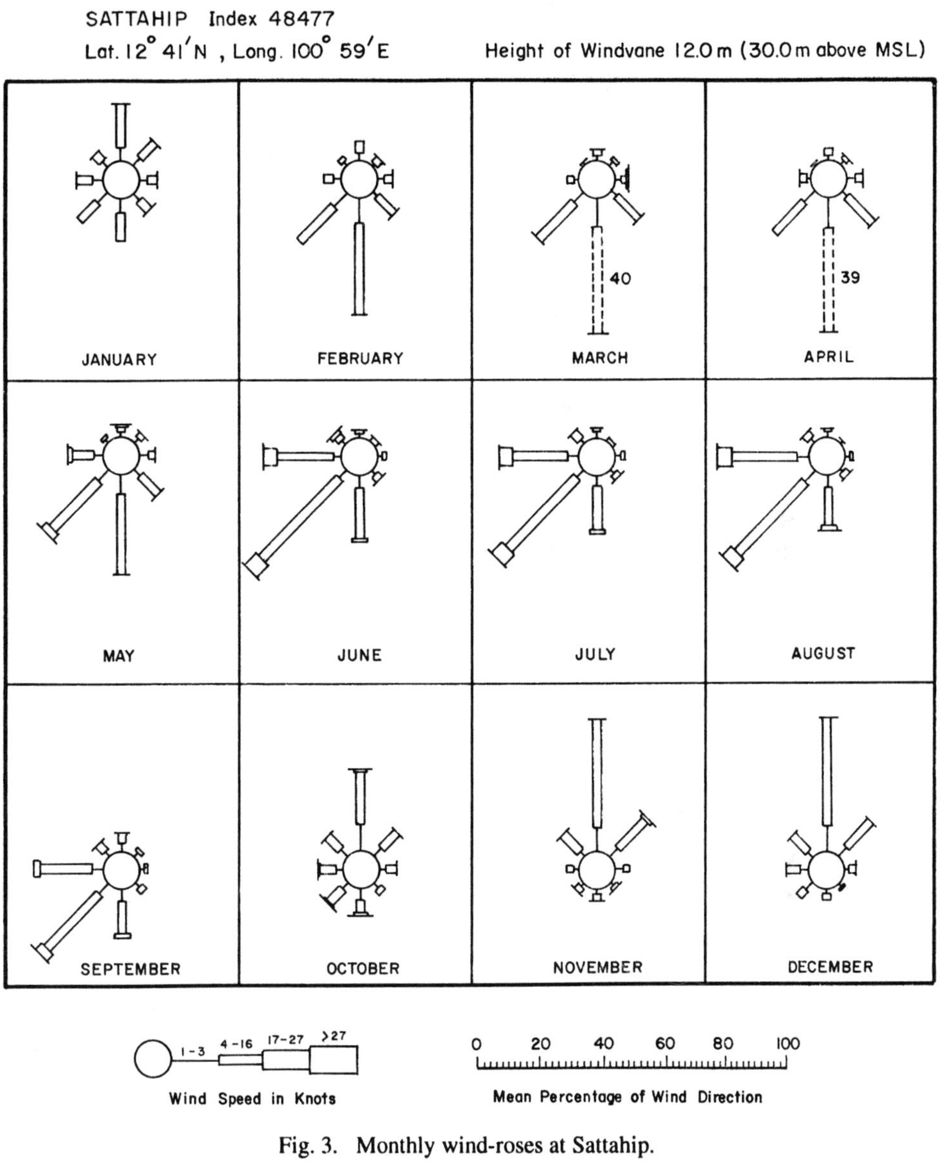

Fig. 3. Monthly wind-roses at Sattahip.

WAVES

The longest fetch length in the Gulf of Thailand is almost 1000 km from the southeast. The width of the Upper Gulf of Thailand between Sattahip and Hua Hin is about 100 km (Fig. 1). Wave heights are generally moderate in all areas, and even the most exposed Gulf areas generally have waves less than 0.25 m in height. Most wave data are computed from wind data using the method proposed by Silvester and Vongvisessonijai (1970). Fig. 7 shows the probabilities of occurrence

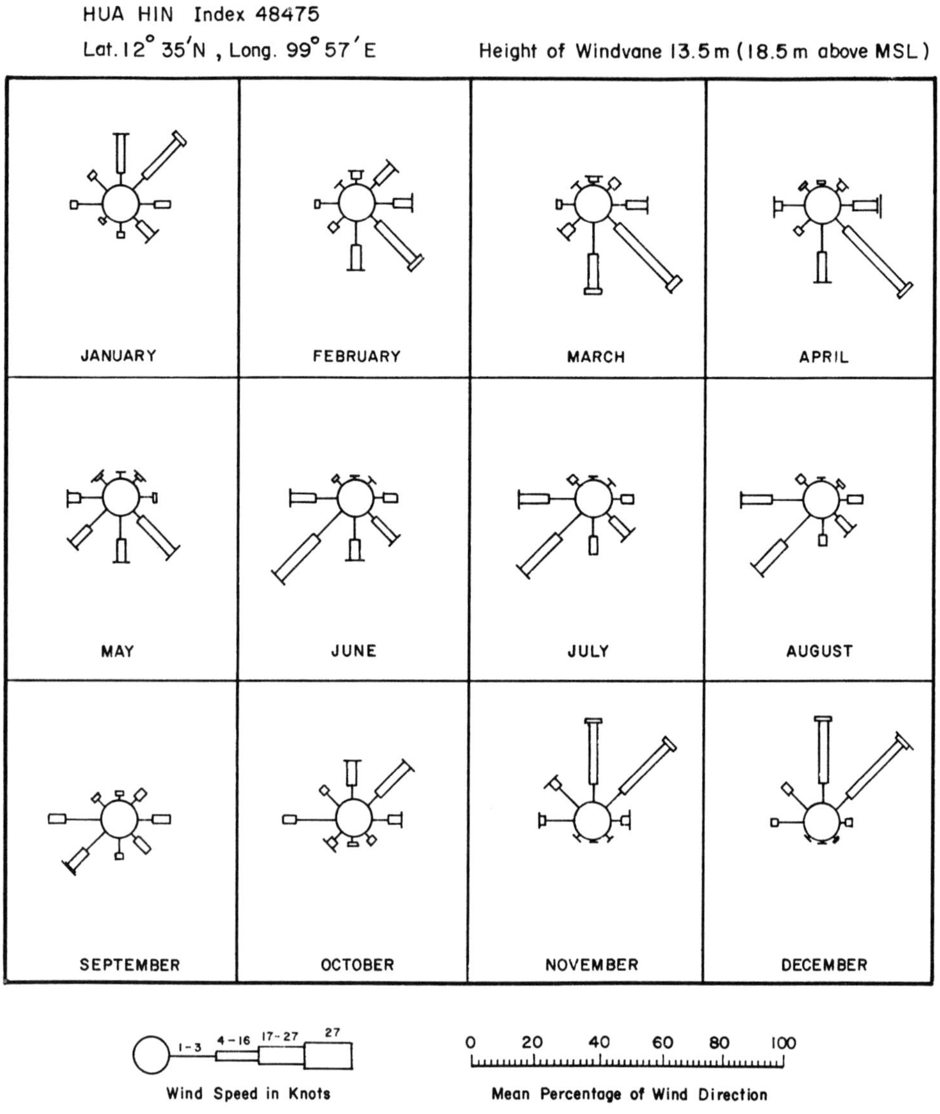

Fig. 4. Monthly wind-roses at Hua Hin.

of significant wave height and wave period along the eastern seaboard at Ao Phai or Ko Sichang and Sattahip, while Fig. 8 shows conditions for the east coast of the southern Thailand near Surat Thani (Khanom and Ko Samui). The waves in the Upper Gulf of Thailand (Fig. 7) are the smallest due to the short fetch and weak southwest winds. Waves near Surat Thani (Khanom and Ko Samui in Fig. 8) are larger due to stronger northeast winds and greater fetch. The waves in the South

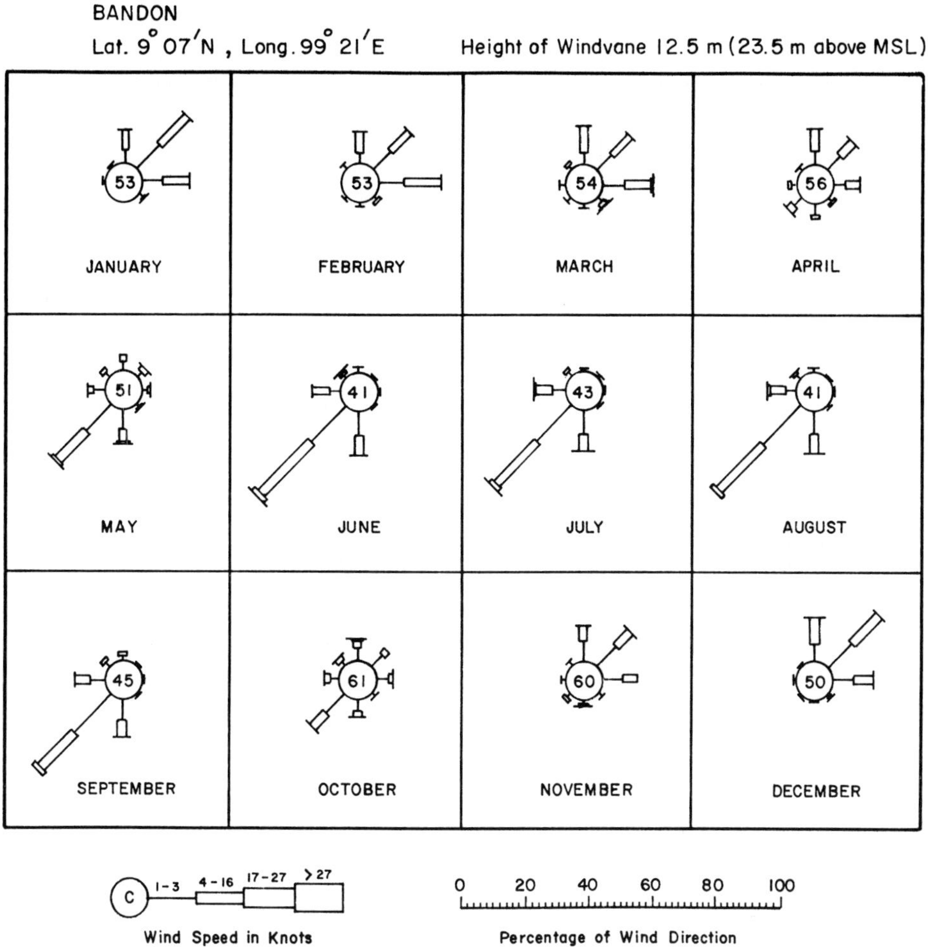

Fig. 5. Monthly wind-roses at Bandon, Surat Thani.

China Sea coming to Songkhla, Pattani and Narathiwat are more severe due to stronger northeast winds and much longer fetch.

LITTORAL DRIFT

When breaker characteristics are known, the longshore transport rate Q_s can be computed from the longshore component of energy flux in the surf zone P_{ls} (Coastal Engineering Research Center, 1984):
Immerged weight transport rate $I_{ls} = 0.77P_{ls}$, where $I_{ls} = (\rho_s - \rho)ga'Q_s$ and $P_{ls}(\rho g/16)H_b^2C_g \sin 2\alpha_b$ in which ρ_s = density of sand; ρ = density of seawater; g = gravitational acceleration; a' = correction factor for pore space = 0.6; H_b = breaker height; C_g = wave group celerity; and α_b = breaking angle.

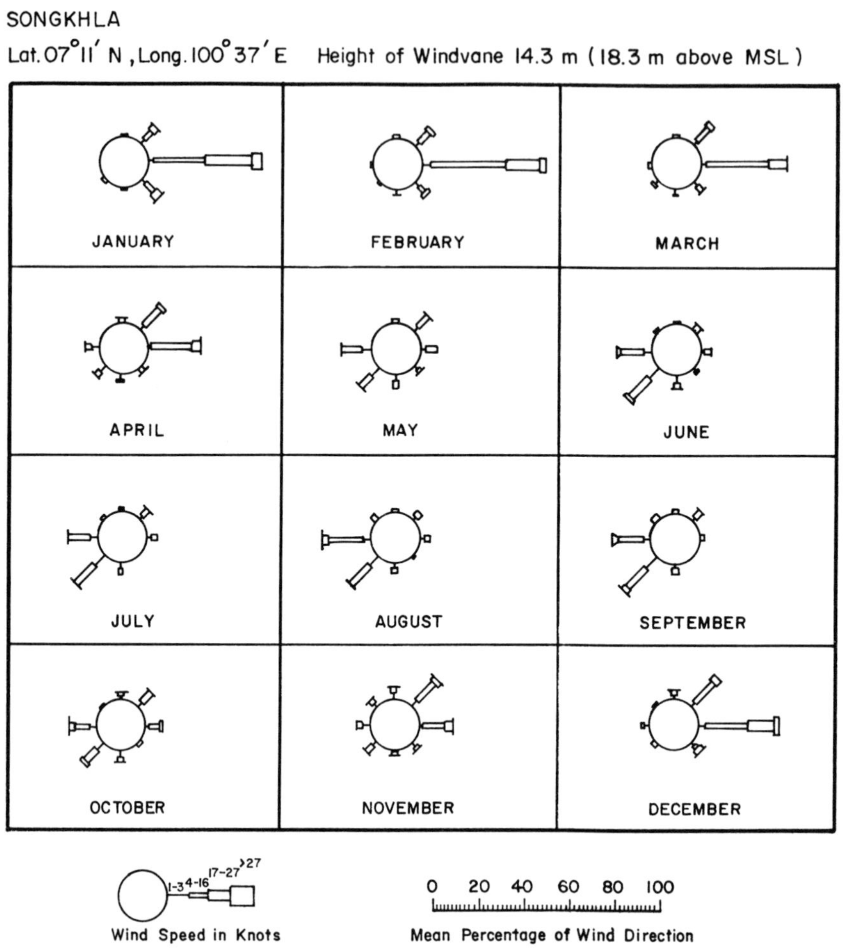

Fig. 6. Monthly wind-roses at Songkhla.

The breaker characteristics (H_b and α_b) depend on deep-water wave characteristics (H_0 and α_b) and bottom topography or beach slope. As waves travel from deeper water, they change height and direction due to refraction, shoaling, and the bottom friction. Refraction is the bending of wave crests due to the slowing down of that part of the wave crest in shallower water. As a result, refraction tends to decrease the angle between the wave crest and the bottom contour. Thus, for most coasts, refraction reduces the breaker angle and spreads the wave energy over a longer crest length. Shoaling is the change in wave height due to conservation of energy flux. As a wave moves into shallower water, the wave height first decreases slightly and then increases continuously to the breaker position, assuming friction and refraction effects are negligible. Bottom friction is important in reducing wave height where waves must travel long distances in shallow water.

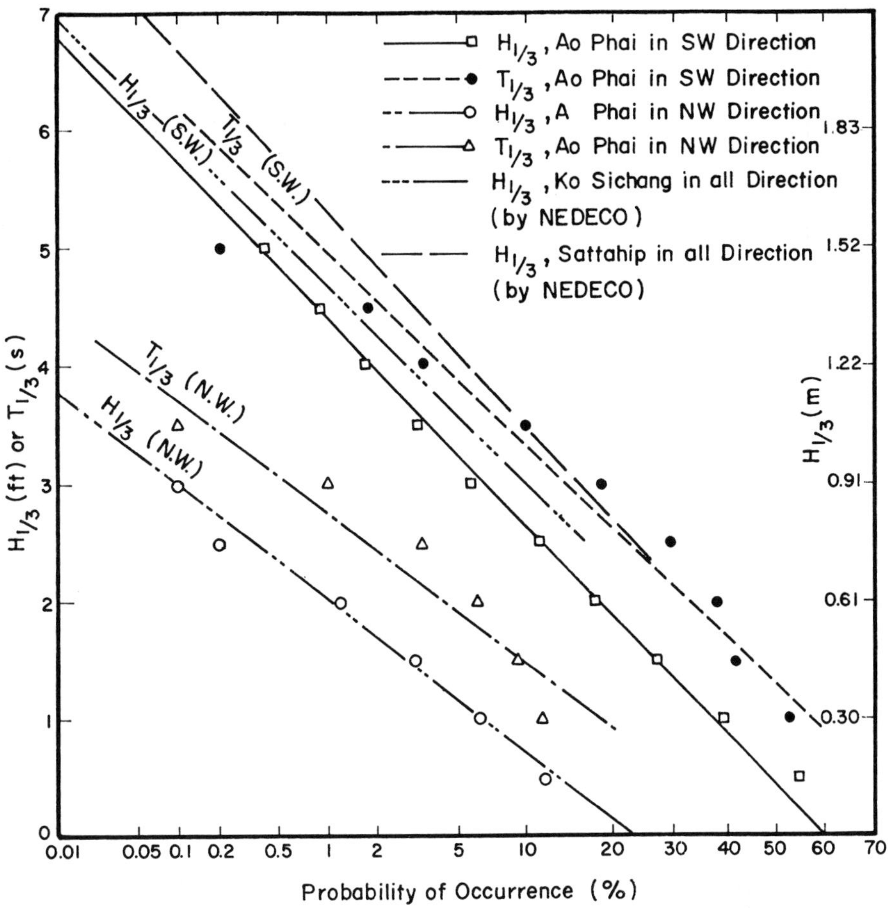

Fig. 7. Probability of occurrence of significant wave heights and wave periods along the eastern seaboard.

When breaker characteristics are not known, the longshore transport rate Q_s can be computed from the deep-water wave characteristics. The following formula can be used for open coastlines with rather uniform topography. (Coastal Engineering Research Center, 1984):

$$O_s\ (\mathrm{m}^3/\mathrm{month}) = 169,300\, f H_0^{5/2} F(\alpha_0),$$

where $F(\alpha_0) = [\cos\alpha_0)^{1/4} \sin 2\alpha_0]$ is the direction term, f is the decimal frequency of the wave, H_0 is the deep-water wave height in metres, and α_0 is the deep-water angle between wave crest and shoreline.

Coastal Morphology

Locations of the shoreline used in this study are the Chao Phraya River mouth (X in Fig. 1), the east coast of the peninsula from Samut Songkhram province to

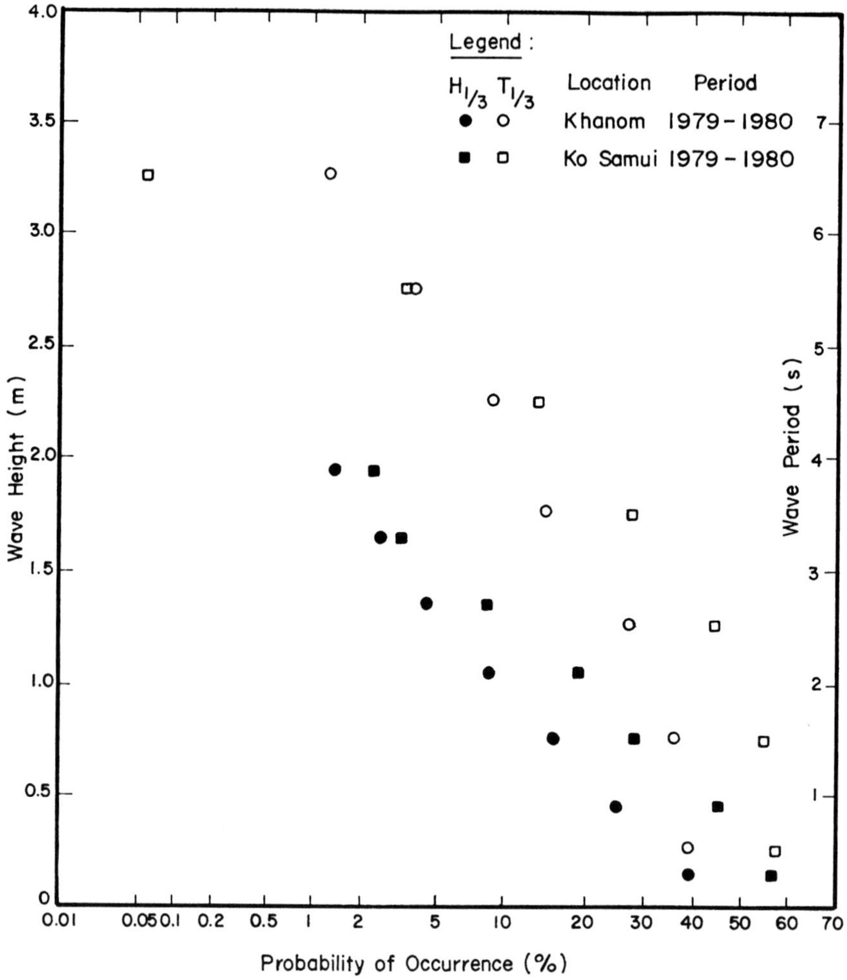

Fig. 8. Probability of occurrence of significant wave heights and wave periods at Khanom and Ko Samui (Surat Thani).

Phetchburi Province (A), Ban Bang Kaew (B), Hua Hin (D), Prachuap Khiri Khan (E), Songkhla Lagoon inlet (F), Laem Tachi of Pattani (G), and Bang Nara River mouth of Narathiwat (H). Two large sand spits can be seen in the south at Nakorn Si Thammarat and Pattani. All four shorelines in area X are taken from satellite imageries; three shorelines were analysed each for areas A to E, the first two from aerial photographs and the last one from satellite imagery. Shorelines from areas F to H were taken from navigation maps.

CHAO PHRAYA RIVER MOUTH

The lower Chao Phraya plain is classified as a geologic depression with mountainous borders to the west and northeast. This depression zone stretches from the Gulf of Thailand to as far north as Sukothai Province. The depression is filled with brackish and marine alluvium between 500 and 2,000 m in thickness. There are at least six underlying aquifers, which are interspersed with layers of clay. The thickness of each clay layer is about 10 to 30 m. The uppermost stratum is a clay deposit, known as the Bangkok clay, about 25 m deep. This uppermost layer can be subdivided into the weathered clay layer from 0 to 4 m, the soft clay layer from 4 to 15 m deep, and a more consolidated, stiff clay layer from about 15 to 25 m deep.

Bangkok Port, the most important port of Thailand, is located on the east bank of the Chao Phraya River about 27 kilometres from the river mouth (Fig. 1). The navigation channel is maintained by dredging to a depth of 10 m below mean sea level. Most dredged material is draped along the eastern coast resulting in artificial accretion along the shoreline. Fig. 9 shows the western and eastern shorelines of the Chao Phraya River mouth in 1969, 1973, 1979 and 1987. The western shoreline lost 7.2×10^6 m^2 between 1969 and 1987, with a maximum erosion of 500 m. The eastern shoreline, however, was replenished by the dredged material, resulting in a net accretion area of 3.4×10^6 m^2 from 1969 to 1987.

SAMUT SONGKHRAM-PHETCHBURI

This section of shoreline (Fig. 11) is supplied by sand and silt from the Meklong River in the north and the Phetchburi River in the middle section. Therefore, most of the shoreline is accreting. From 1954 to 1974, the accretion area was 1.5×10^6 m^2 and the eroded area was 0.2×10^6 m^2, a net accretion of 1.3×10^6 m^2.

BAN BANG KAEW

The shoreline of Ban Bang Kaew (Fig. 11) has suffered from severe erosion in the past due to northeast monsoon waves. From 1954 to 1974, much of the shoreline retreated about 200 m, with the corresponding eroded area of 1.35×10^6 m^2 and accretion area of 1.25×10^6 m^2. The Department of Land Development requested the Asian Institute of Technology and the Royal Irrigation Department of Thailand to design and implement a protection scheme for the shoreline. A series of 14 offshore breakwaters 30 m long and 120 m apart were constructed in 1985 to protect 2,000 m of shoreline from Komnarani Temple to Bang Kaew Canal. After 3 years, the 1988 shoreline had accreted up to 150 m from the 1974 shoreline, and the previous erosion had been reversed.

HUA HIN

Similar to the shoreline of Ban Bang Kaew, Hua Hin (Fig. 12) has suffered from severe erosion. From 1954 to 1988, the shoreline retreated an average of 100 m, with a total eroded area of 2.15×10^6 m^2.

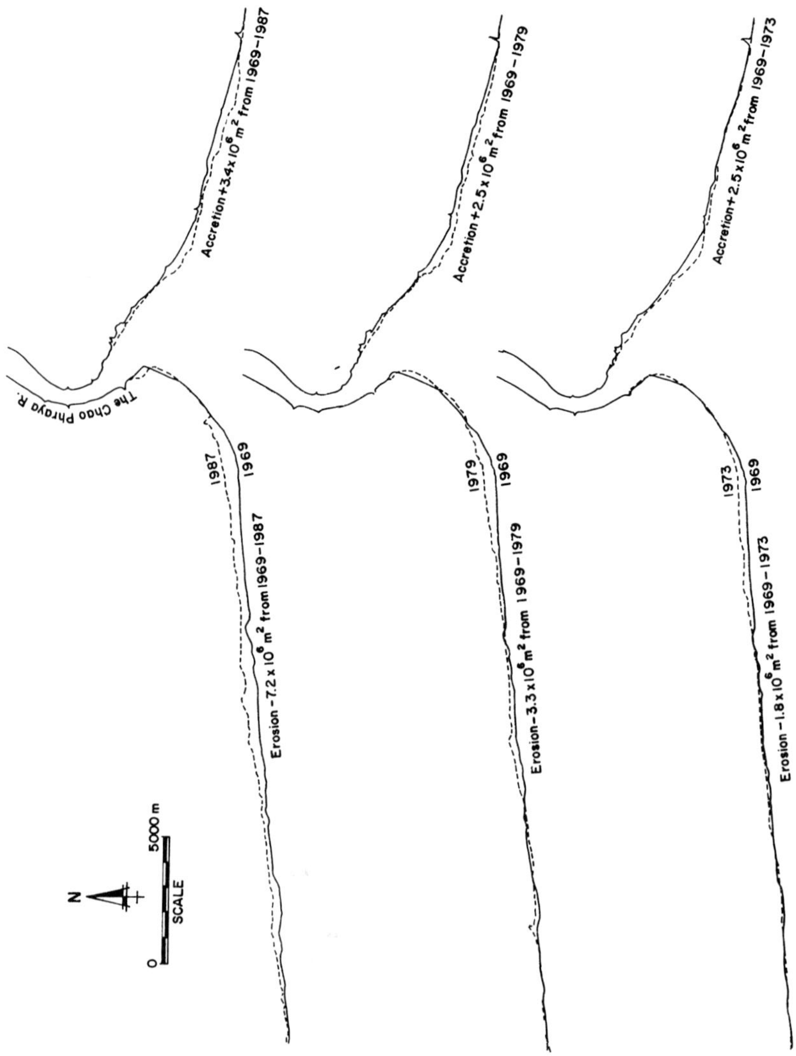

Fig. 9. Changes in the configuration of the Chao Phraya River mouth, 1969–1987.

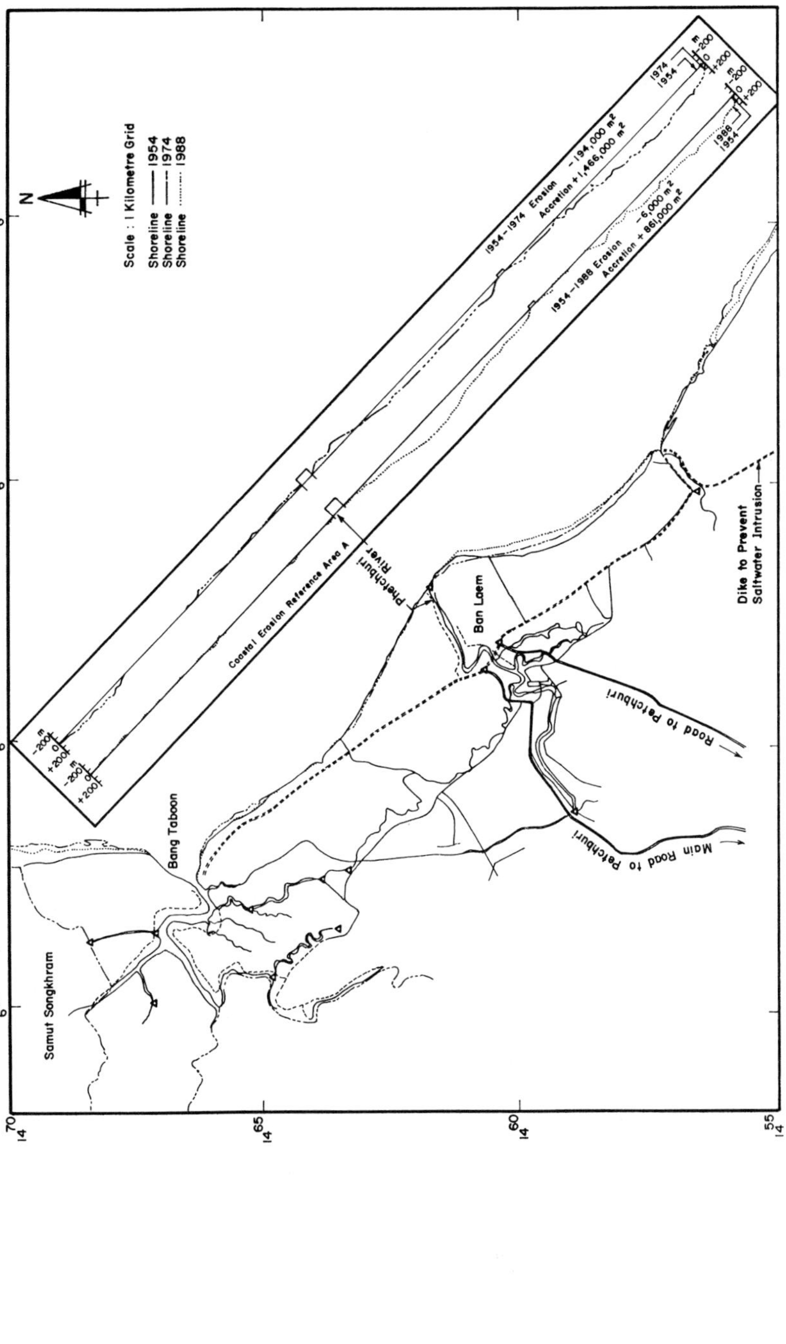

Fig. 10. Erosion and accretion (1954–1988) of the coastline in Area A - Samut Songkhram to Phetchburi.

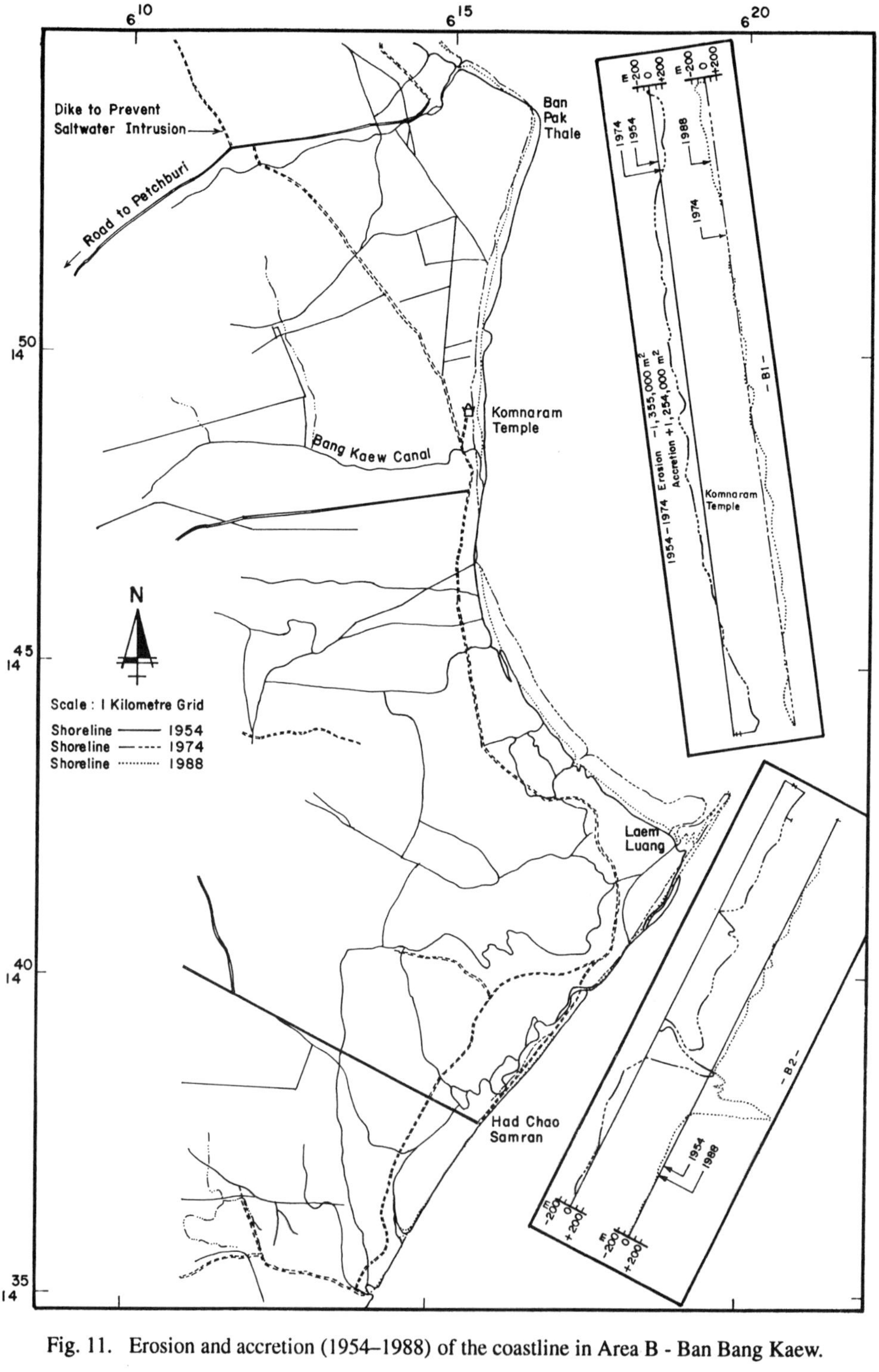

Fig. 11. Erosion and accretion (1954–1988) of the coastline in Area B - Ban Bang Kaew.

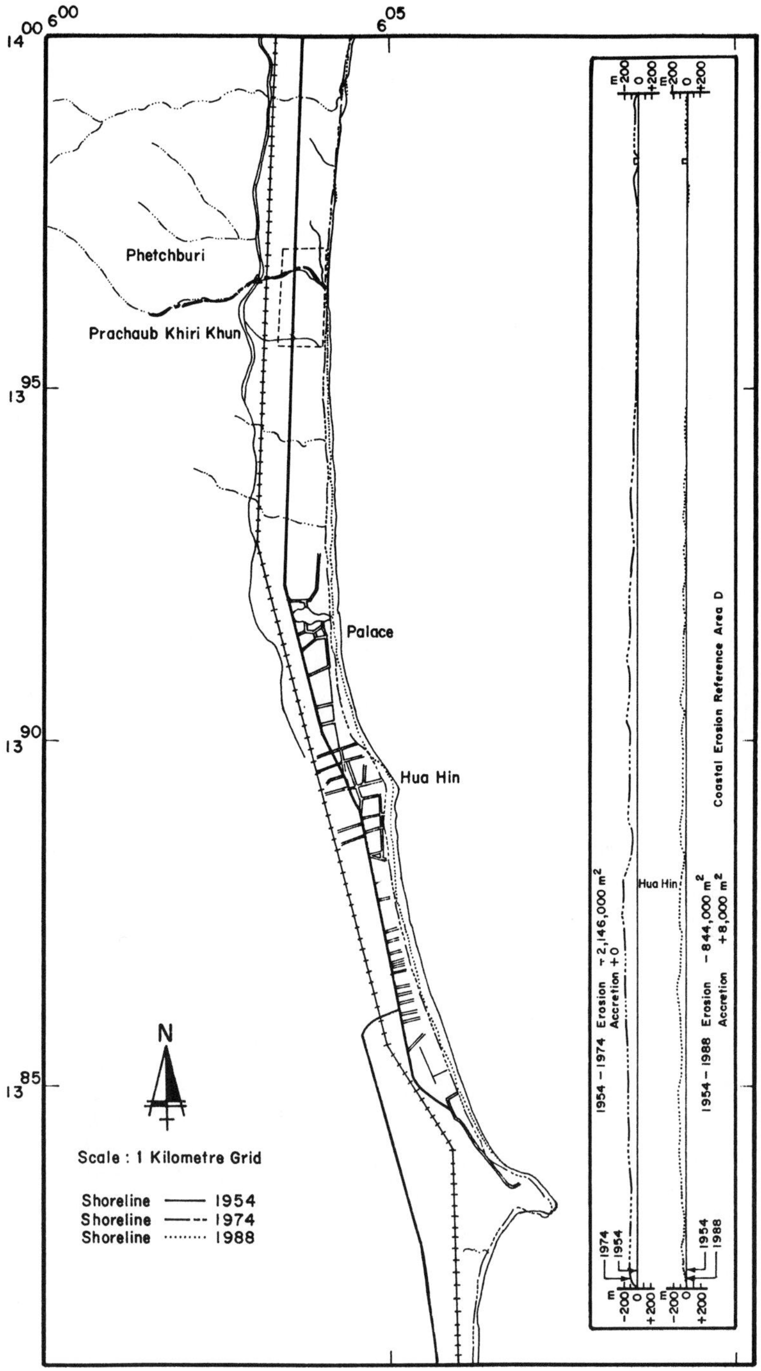

Fig. 12. Erosion and accretion (1954–1988) of the coastline in Area D - Hua Hin.

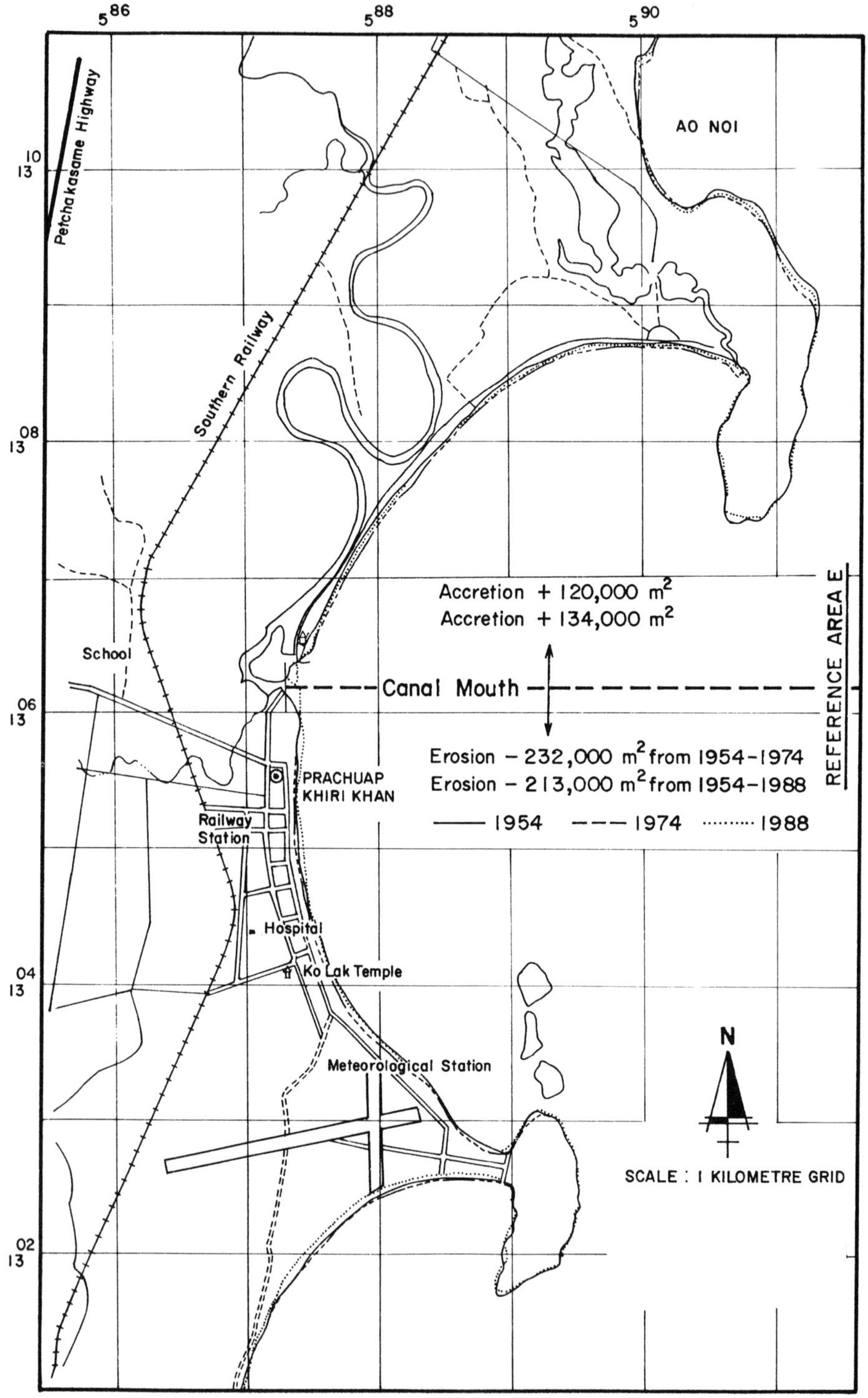

Fig. 13. Erosion and accretion (1954–1988) of the coastline in Area E - Prachuap Khiri Khan.

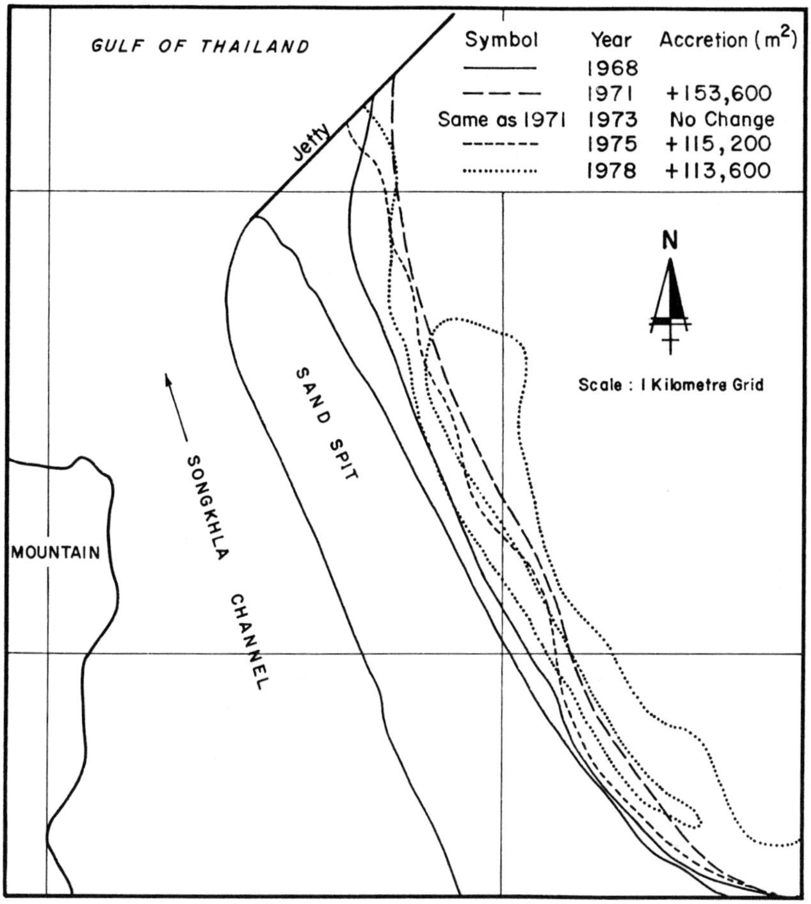

Fig. 14. Changes in the coastline at Songkhla Lagoon inlet, 1968–1978.

PRACHUAP KHIRI KHAN

Prachuap Khiri Khan Bay (Fig. 13) is located between two rock outcrops. Because it is almost symmetric in the north and south, the bay is influenced by both north and south littoral drifts. A canal in the middle of the coast serves as a dividing point of accretion in the north and erosion in the south. From 1954 to 1974, the northern shoreline of the bay accreted in area by 0.12×10^6 m^2, while the southern part eroded by 0.23×10^6 m^2.

SONGKHLA LAGOON INLET

Songkhla Lagoon has been formed by a sand spit (Fig. 14). About ten small rivers drain into the lagoon, and the lagoon serves as an important freshwater resource in the south as well as a breeding ground for shrimp and fish. The high freshwater discharge from the lagoon helps maintain the inlet. A 700 m jetty was constructed in 1968 at the south of the inlet to trap littoral drift from the south.

As seen in Fig. 14, the shoreline south of the jetty has advanced significantly. From 1968 to 1971 accretion amounted 0.15×10^6 m^2. There was little change between 1971 and 1973; 0.11×10^6 m^2 accretion from 1973 to 1975; and 41×10^6 m^2 accretion from 1975 to 1978. In 1978 the shoreline prograded more than 400 m, and because the water depth at the tip of the jetty on the beach side is rather shallow (about 1 m below mean sea level), the littoral drift from the south started to bypass the short jetty, resulting in a slow accretion of the shoreline.

Laem Tachi of Pattani

The Tachi sand spit, formed by northerly littoral drift, provides shelter for Pattani Bay from northeast and east waves. The bay is rather shallow due to deposition of silt carried by the Pattani River (Fig. 1). Between 1981 and 1985 the termination of the sand spit accreted southwestward ca. 0.4×10^6 m^2 (Fig. 15).

Bang Nara River Mouth

The Bang Nara River mouth (Fig. 16) is bounded on both sides by active sand spits. These spits change every year due to active littoral drift. Spit migration is generally in a northwesterly direction; when it becomes too long, it is cut by waves. Because of the spit migration and the high rate of sedimentation at the river mouth, fishing and cargo boats often experience navigation problems.

Summary

The upper Gulf of Thailand shoreline in the vicinity of the Chao Phraya River mouth has suffered from wave attack from the south, with a maximum shoreline regression of 500 m west of the river mouth. The shorelines of Phetchburi and Prachuap Khiri Khan are eroded by storm waves during the northeast monsoon and are accreted by swell and waves during the southwest monsoon and transition period.

Further south at Songkhla, Pattani and Narathiwat, the waves and littoral drifts are more severe. Littoral drift along these shorelines is predominantly from the south, creating several long sand spits at Laem Tachi (Pattani), the southern part of Songkhla Lagoon inlet, and Laem Talumpuk (Nakorn Si Thammarat).

Acknowledgements

This paper presents the preliminary results of a more detailed study on the coastal morphology of Thailand. The financial support from the National Research Council of Thailand made this project possible.

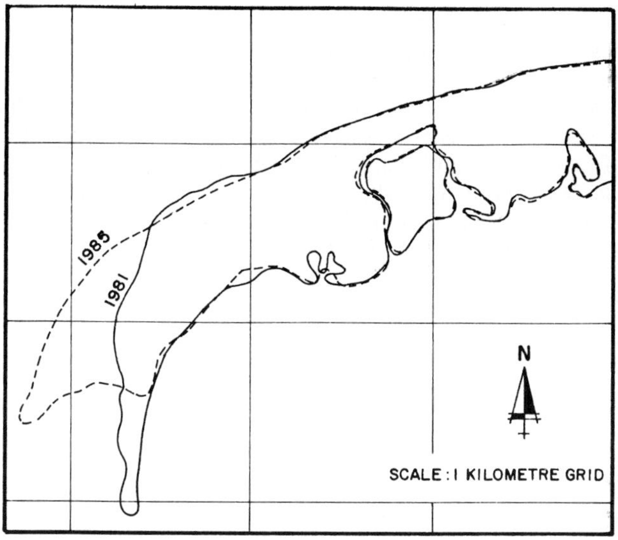

Fig. 15. Sand spit changes at Laem Tachi (Pattani).

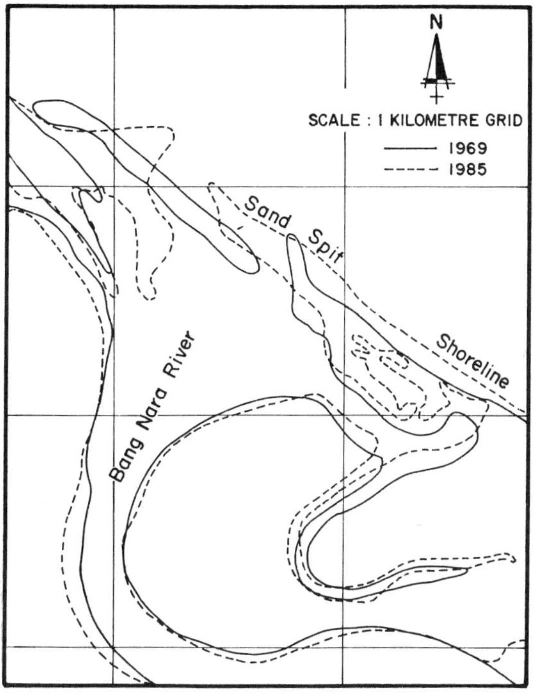

Fig. 16. Coastline changes at the Bang Nara River mouth (Narathiwat).

References

Coastal Engineering Research Center, 1984. Shore Protection Manual, Department of US Army, Waterways Experiment Station, Corps of Engineers, Vicksburg, Mississippi, vol. 1.

Moorman, F. R. and Rojanasoonthon, S., 1972. The Soils of the Kingdom of Thailand - Explanatory Text of the General Soil Map. Report SSR-72A, Soil Survey Division, Department of Land Development and Food and Agriculture Organization (FAO) of the United Nations, Ministry of Agriculture and Cooperatives, Kingdom of Thailand.

Silvester, R. and Vongvisessomjai, S., 1970. Energy distribution curves of developing and fully arisen seas. *J. Hydraulic Res.*, 8, 493-521.

CHAPTER 7

Nature of Sediment Load in the Ganges-Brahmaputra River Systems in India

V. Subramanian and A. L. Ramanathan

ABSTRACT: No generalizations can be made from the existing data about the sediment load transported by the Ganges-Brahmaputra River systems due to wide diurnal, seasonal and annual variations in the sediment-carrying capacity of these rivers. Estimates of the sediment load are highly variable, ranging from 402 to 710 × 10^6 tonnes/year for the Brahmaputra River and from 403 to 660 × 10^6 tonnes/year for the Ganges River. Both these rivers carry predominantly coarse silt to sand-size particles. Human impact on the natural processes of sediment erosion and deposition are evident in both river systems although at different levels. Detrital sands and illitic clays dominate the mineral composition of the river loads. In comparison to the global averages, the denudation rates in the Himalayan river system are four-fold higher.

Introduction

Rivers are the major agents of land sculpture in various environments and are capable of producing enormous variation in land forms. Measurement of river load occupies an important position in the study of fluvial-denudation systems, at the scale of the small catchment as well as in larger river basins. Of the 15 × 10^9 tonnes/year of global sediment flux to the oceans, Asian rivers have been suggested to contribute half (Milliman and Meade, 1983), and of this, the Himalayan rivers may contribute as much as 90%.

The Himalayan region of Indian Subcontinent is drained by three major river systems, the Indus (not discussed here), Ganges and the Brahmaputra. Together, they drain some of the most intensely utilised land areas in the world. These two rivers are characterized by heavy sediment load, high-altitude and tectonic-prone drainage areas, and a high but irregular rate of downstream deposition. Study of these rivers, from their sources to the ocean, reflects the culture and the civilization in the Indian Subcontinent. It is quite likely that the rate of mechanical erosion

J.D. Milliman and B.U. Haq (eds.), Sea-Level Rise and Coastal Subsidence, 151-168.

in particular has been influenced by local factors such as deforestation, shifting channels and tectonic events and land slides. Despite numerous popular accounts, actual scientific data on erosion and transport in the Himalayan rivers are scarce. To avoid generalization, a limited set of locally derived data are used in the present study and should be treated conservatively.

Description of the Basins

GANGES

The Ganges (Fig. 1A) originates at the Gangotri glaciers at an elevation of 7010 m and criss-crosses the Great and Lesser Himalayas over a distance of 250 km, descending into the plains at Rishikesh. The basin is bounded to the north by the Himalayas and to south by the sedimentary Vindyans (Precambrian). The western limits of the basin extend up to the semi-arid regions east of the Aravalli hills (Archean Lluartzite) whereas the eastern limits extend to the Bay of Bengal. The basin area of the river Ganges in India is about 860,000 km^2, extending between 74° to 89° E and 31° to 22° N. The Ganges Basin covers a wide variety of land forms, from the featureless plains in the central east-west belt (the Indo–Gangetic Alluvium) to the Himalayan ranges in the north. From source to the ocean, the river is about 2500 km long, with two distributaries, the Hoogly passing through Calcutta and the Padma passing through Bangladesh, where it merges with the Brahmaputra, forming one of the largest (Ganges-Brahmaputra) deltas in the world.

Two types of climate dominate the river basin : (a) Humid subtropical climate with dry winter (temperature range from 3°C in winter to 48°C in summer) and heavy rain fall in the east, progressively decreasing to the west, and (b) Mountain climate (heavy rain/snow with a temperature gradient of 0.6°C/100 m increase in elevation). 70% of the annual rainfall is delivered by the southwest monsoon (June to September).

The drainage network consists of the main Ganges River and a number of tributaries, such as the Yamuna, Ramganga, Ghaghra, Sone, Gandak and Kosi. Since the basin has varied climate, geology and physiography, a variety of soils have developed, both residual and alluvial. Distribution of various rock types in the Ganges River Basin are as follows: Recent Alluvium (57%), Deccan and Rajmahal traps (10%) (Tertiary), Granites, Gneisses and Charnokites (11%) (Tertiary), (Archean) Granites and Gneisses (5%), Cambrian and Upper Precambrian sedimentary rocks (16%).

BRAHMAPUTRA

The Brahmaputra rises (Fig. 1B) from a large glacier in the Tibetan Himalayas and traverses almost 2900 km to the sea. It flows for 1600 km in Tibet, 900 km in India and 400 km in Bangladesh. The Himalayan watershed consists of four topographic units. The lowermost ranges, the Sub-Himalayas (average elevation 1000 m), consist mainly of Tertiary sandstones and are characterized by many raised, young terraces (Gansser, 1964). The middle Himalayas (average elevation

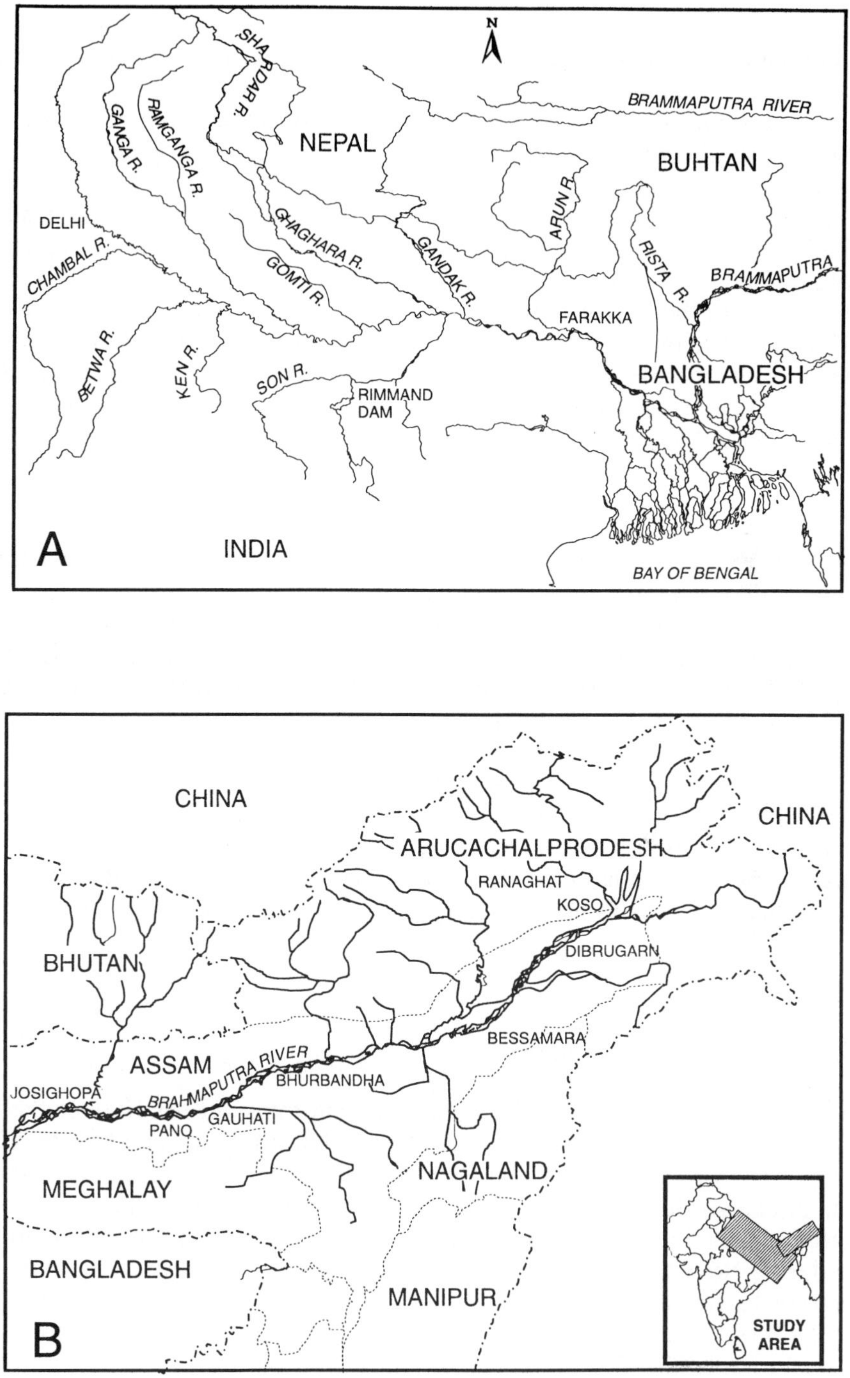

Fig. 1. (A) Ganges River basin; (B) Brahmaputra basin.

4000 m) are underlain by lower Gondwana (Paleozoic) shales, slates and phyllites overlain by a thick horizon of Tertiary basalts. The Greater Himalayas (average elevation 6000 m) consists primarily of younger granites and gneisses. Further north are the trans-Himalayas of Tibet (average elevation 4500 m), which are made up of sedimentary formations of Paleozoic to Eocene age (Wadia, 1975).

Except for the catchment of the Brahmaputra in Tibet, the drainage basin lies within the monsoonal regime of south and southeast Asia. The Himalayan sector annually receives 500 cm of rainfall. Monsoonal rains from June to September account for 60–70% of the annual rainfall. Overall in the basin, rainfall averages 230 cm/year. In the Indian sector of the river, temperature ranges from 10°C to 35°C with high humidity. Natural vegetation in the Brahmaputra Basin varies with altitude, from tropical evergreen to mixed deciduous forests. Soils in the Sub-Himalayan Tertiary sandstones are shallow and consist primarily of sand admixed with cobbles and boulders. Alluvial soils in the basin are Recent in origin.

DISCUSSION

There are several estimates of the sediment load of the Ganges–Brahmaputra system. The annual sediment discharge of the Ganges at Calcutta is estimated to be 660×10^6 tonnes (Abbas and Subramanian, 1984). For the Brahmaputra at Pandu, the annual sediment discharge is 402×10^6 tonnes (Goswami, 1985). Singh (1988) reported a similar annual value at 403×10^6 tonnes for the Ganges, and Subramanian (1987) reported an annual value of 710×10^6 tonnes for the Brahmaputra. Such wide variations reflect problems with the data base, perhaps largely a result of seasonal and interannual variations. Various factors involved in the sediment transport and its distribution are discussed below for the two Himalayan rivers and wherever possible, examples are cited from the available data base. Since the bulk of the data were collected during different seasons and from different sources (a number of agencies, at the states as well as at federal level), it is not possible to ascertain the quality of the data base. Discussion in this chapter suffers from these limitations.

Sediment and Water Discharge

The relation between the river flow and sediment transport has been studied by many workers (Leopold et al., 1964; Holland, 1978; Milliman and Meade, 1983; Probst et al., 1986; Milliman and Syvitski, 1992). In the Ganges, the average water discharge during the monsoon accounts for 80–90% of the total annual discharge (Table 1). Peak discharge (19 775 m^3/sec) occurred in July, and minimum flow during November. While these data represent a single year, over a six-year period the discharge and corresponding sediment load show considerable variation. Dasgupta (1984) gave a low mean water flow of 997 m^3/sec for April and 33 823 m^3/sec for September for the Ganges at Patna over the period 1965–81. Figure 2 shows monthly variations in discharge over a 5 years period for the Ganges at Patna. While the maxima and minima follow a similar trend, there are time lags from year to year, particularly for the minima.

TABLE 1
Monthly variation in discharge for the Ganges at Patna for 1986.

Month m^3/sec	Discharge 10^6 m^3	Monthly water yield 10^6 tonnes	Monthly sediment yield
January	1430	3830	0.33
February	1619	3916	0.36
March	1443	3865	0.27
April	1020	2644	0.16
May	1173	3142	1.92
June	2968	7693	2.44
July	19775	52965	40.3
August	14123	37827	61.6
September	14265	36975	48.7
October	5902	15808	13.2
November	1002	2597	1.70
December	2283	6115	1.40
		177377	172.4

TABLE 2
Variation of sediment load at Patna (based on daily observations).

Year	Sediment transported	Date	mg/l	Tonnes/day	Maximum/minimum
1980	Maximum	17 August	2472	12.4×10^6	8551
	Minimum	15 March	11	1450	
1981	Maximum	26 August	1416	3.4×10^6	3003
	Minimum	13 April	12	1132	
1982	Maximum	3 September	2275	12×10^6	10025
	Minimum	27 January	14	1197	
1983	Maximum	14 September	1693	6.6×10^6	7833
	Minimum	30 March	14	842	
1984	Maximum	12 September	1854	7.6×10^6	5998
	Minimum	29 May	30	1267	
1985	Maximum	23 August	1587	4.6×10^6	6855
	Minimum	14 May	12	671	
1986	Maximum	20 September	1287	2.4×10^6	503
	Minimum	25 March	58	4767	

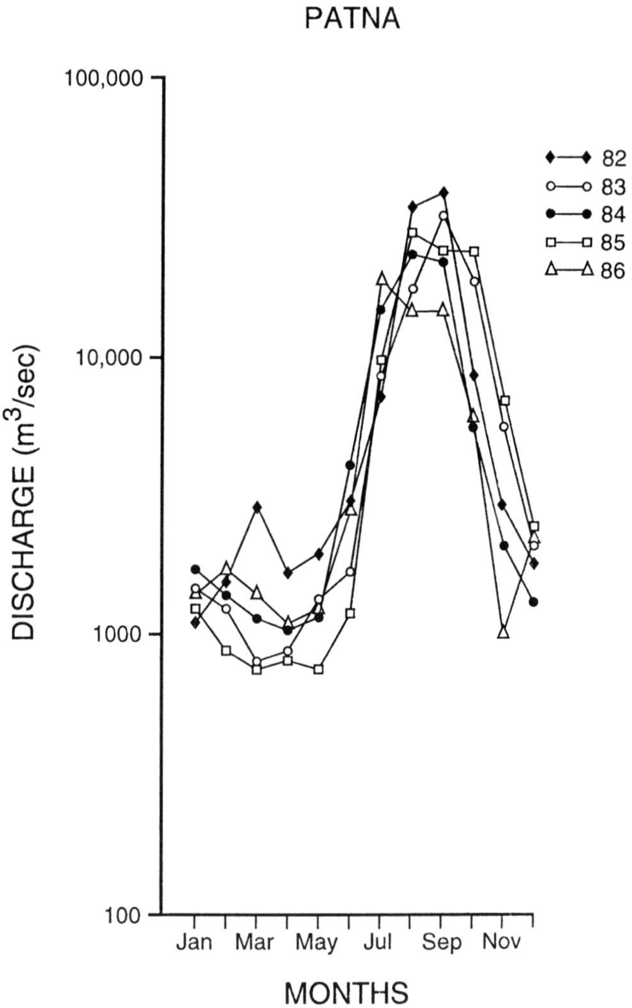

Fig. 2. Monthly and annual variations in the discharge of the Ganges at Patna. Data from Central Water Commission, Govt. of India.

Table 2 shows annual mean sediment load of the Ganges at Patna over a six-year period based on daily observations. The table shows only extreme values in the river; predictably, the maximum for any year occurs during the monsoon period, while the drier summer months show minimum values. Because of variations in the monsoonal characteristics from year to year, the maximum and minimum sediment loads may be supplied over a wider time span in this region. Figure 3A shows the spatial and seasonal variation of the sediment load at Kosi, one of the major tributaries of the Ganges. During the winter the sediment load is uniform whereas in the monsoon, the load increases in the flow direction towards its junction with the Ganges. Observations over a three years period at a single (Fig. 3B) show a

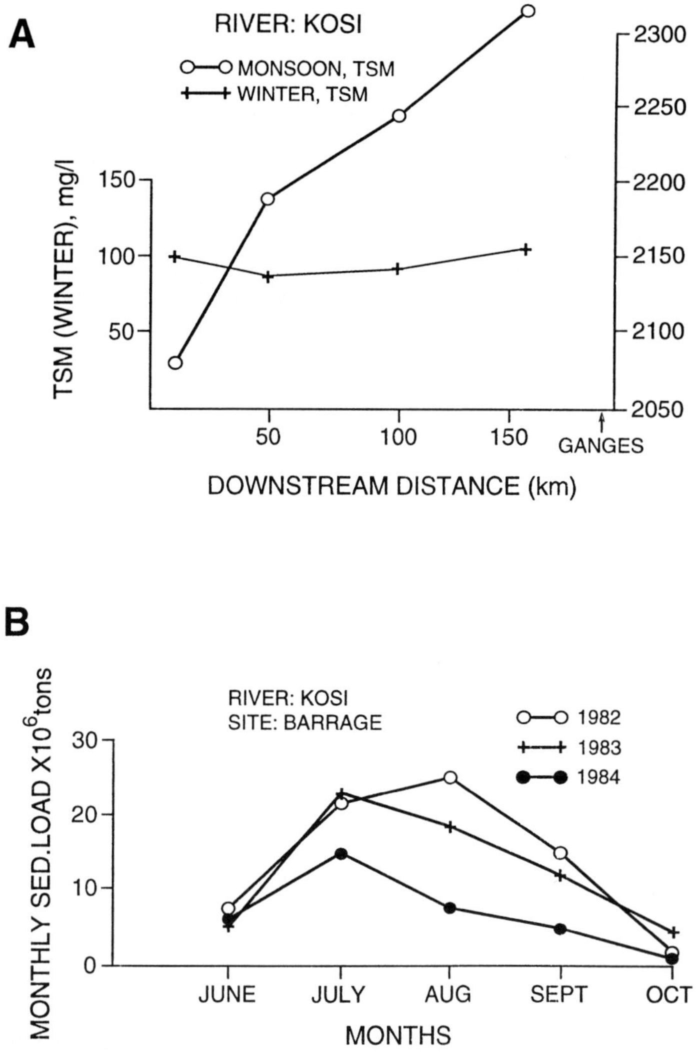

Fig. 3. Seasonal (A) and annual (B) variations in the sediment load of the Kosi River at the barrage site (200 km north of junction with the Ganges. Data from Public Works Dept., Bihar State Govt., India.

drop in the annual sediment load, even during the monsoon period, from 82 to 1984.

The highest level of sediment load recorded at Patna on a single day was in 1986 at 2.4×10^6 tonnes on 20 September (see Table 2). This figure is higher than the cumulative sediment transport in four months (December to March) of the same year (2.2×10^6 tonnes; see Table 1). Similarly, the highest sediment transport on a single day between 1980 and 1986 occurred on 17th August 1980, when 12.4×10^6 tonnes of sediments were transported, equal to all the sediment transported in eight months in the same year. These examples clearly indicate extreme variations in daily, seasonal and yearly sediment loads for the Ganges, primarily in response to the corresponding fluctuations in water discharge.

For Brahmaputra, annual average water discharge at Pandu is 51,156 m^3/sec (Goswami, 1985). Maximum discharge is observed for the monsoon periods from May to October, accounting for 95% of the mean annual flow at Pandu (Goswami, 1985). Thus, non-monsoonal seasons account for only 5% of the annual flow. Such summer-monsoon peak flows occur at least once every 2 years. Figure 4A shows the high fluctuations in the daily discharge of Brahmaputra at various locations for the year 1976, while Figure 4B shows the daily sediment concentration against discharge during the flood period at Pandu in 1960. Graphical representation by Goswami (1985) of the daily sediment and water discharge at Pandu for July 12–31, 1977 showed hysterisis between these two parameters. Similar to the Ganges, single day peak sediment load of the Brahmaputra may be much higher than the cumulative sediment load over much longer periods, as much as a few months (Fig. 4B).

Thus moderate events of relatively frequent occurrence (with variations in water discharge) are responsible for the transport of the bulk of the sediments in the Himalayan rivers.

Basin Lithology

To understand the extent of basin geological influence on the sediment transport of these rivers the lithology is catagorised into Precambrian or younger granites and other hard rocks, Mesozoic Deccan Traps (basaltic), Permo-Carboniferrous Gondwana sediments, Tertiary sandstone and Recent alluvium.

In the upper reaches, the main channel of the Ganges flows through Siwaliks (Tertiary) sediment with wide proportion of igneous intrusives. The middle and lower reaches are in alluvium, whose mean thickness in the Indo-Gangetic plain is 4 km (Wadia, 1975). Most of the Himalayan tributaries of the Ganges pass through limestone and other sedimentary rocks, and alluvium downstream, which contribute significant sediment. For example, Ghaghra provides 125×10^6 tonnes/year, Gandak 33×10^6 tonnes and Kosi 73×10^6 tonnes/year. The Yamuna (largest tributary of the Ganges) partly drains Deccan Traps (basalts) and has a sediment load of 125×10^6 tonnes/year, comparable to that of the Ghaghra. Due to its long traverse in alluvial plains and the coarse nature of the Yamuna sediments (mostly silt), most of the sediments are deposited and very little actually reaches the main Ganges channel. The Sone, with the lowest mean (325 m) and a small discharge (3180×10^6 m^3/year) as compared to the Gandak (1383 m and $52\,200 \times 10^6$ m^3/year) has a sediment load of 51×10^6 tonnes/year compared to that of Gandak (25×10^6) tonnes/year. Some 80% of the Sone's drainage area is in the Vindyan (Precambrian) rocks, composed of shale, sandstone and limestone, whereas in the Gandak there are considerable proportions of younger igneous intrusives.

The upper and middle reaches of the Brahmaputra's main channel and tributaries drain through granite and gneiss, Gondwalia shale, slate, phyllite and basalts,

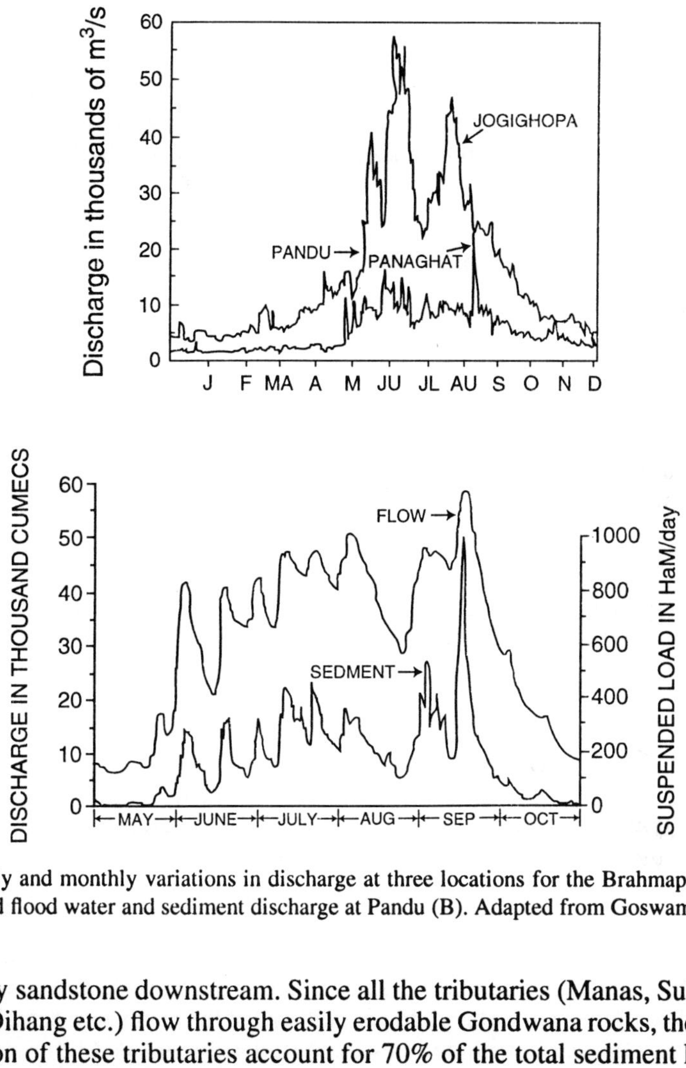

Fig. 4. Daily and monthly variations in discharge at three locations for the Brahmaputra River in India (A) and flood water and sediment discharge at Pandu (B). Adapted from Goswami (1985).

and tertiary sandstone downstream. Since all the tributaries (Manas, Subansiri, Jiabharali, Dihang etc.) flow through easily erodable Gondwana rocks, the sediment contribution of these tributaries account for 70% of the total sediment load of the Brahmaputra (Goswami, 1985). It is evident from the above observations that rock types and their distribution in the basin area are important factors in controlling the sediment load of both Himalayan rivers.

The presently estimated erosion rate for the Himalayan rivers is about 15.7 cm per 100 years (Goswami, 1985). The very low erosion rate for the Ganges (2 cm/100 years) reflects deposition of alluvium. The base of the 4-km thick Recent alluvium reaches the Mio-Pliocene boundary (Wadia, 1975). In the Indian part, the Brahmaputra has a soft-rock lithology whereas in the Tibetian part it flows through predominantly younger crystalline rocks and intrusives. As a result, the sediment yield of the Brahmaputra increases by 240% as it flows from Tibet to

India (Goswami, 1985). Thus, the denudation late for Brahmaputra in India is also very high (41 mm/100 years; Subramanian, 1978).

Compared to the erosion rate of 6 mm/100 years for North America (Judson, 1964) and a global average of 3 mm/100 years (Garrels and MacKenzie, 1971), the erodable sedimentary lithology of the Himalayas, coupled with a high degree of tectonic activity in the catchment area, makes the Ganges-Brahmaputra system one of the most unstable regions in the world, with an average denudation rate of 16 mm/100 years.

Natural (Tectonic) and Human Impact on Sediment Load

Erosion has been critical and pronounced in areas where man has indiscriminately used land and tampered with the natural balance by reshaping, modifying and defacing the topography — such as faulty agricultural practices, construction of roads, canals and buildings, the mining of hillsides, and deforestation. According to Carson (1985), rainfall erosion in small watersheds (63×10^4 m^2) in western Nepal shows a high rate of soil loss in degraded scrub forest and abandoned benches due to overgrazing. Overgrazing in forested areas of the Himalayas in Utter Pradesh in India (Kumayon and Garhwal region) is 2.5 to 4.5 times higher than the carrying capacity of the forests ((Valdiya, 1987).

According to Zachar (1982), the global rate of erosion, including riverine and soil erosion, is between 0.342 mm/year and 0.407 mm/year, being highest in the developing countries (0.5–0.43 mm/year) and lowest in Europe (0.06 mm/year). The use and misuse of land by humans has considerably augmented the pace of erosion. The amount of fluvial sediments discharged to the sea before man colonized the land is estimated to be one-fifth of what it is today (Garrels and MacKenzie, 1971). The most recent estimates of global sediment yield (Milliman and Meade, 1983) indicates a global value of 150 tonnes/km^2/year in comparison to 380 tonnes/km^2/year for the Asian region and 600 tonnes/km^2/year for the Himalayan region, suggesting that present yields partially reflect the impact of enhanced human activities.

Goswami (1985) estimated that the sediment yield for Brahmaputra between Ranaghat and Bessamara increases by 110%, while the drainage area only increases by about 55%; moreover, water yield per unit area of the basin here actually declines by about 3%. Goswami attributed this increased sediment yield to recent seismic events that accumulated and released sediment in/from various storage sites in the basin, including river beds. He also attributed the high rate of denudation to the uplift of Himalayan catchment areas following recent earthquakes; similar observations were made previously by Raymahasay (1970) for Kosi and Gandak Rivers in the Ganges Basin.

Extensive flooding in eastern Utter Pradesh and Bihar in India, especially by Kosi River in the Ganges Basin, is primarily due to the reduced carrying capacity of the rivers that are chocked with sediment derived from an accelerated erosion in the Himalayan regions (Valdiya, 1987). In southeastern Nepal, the river beds Bhabhar-belt have been observed to be rising at the rate of 15–30 cm/year (Valdiya, 1987).

Dredging channels for shipping (for example to keep the port of Calcutta open) and use of river-bank materials for construction has helped increase the erosion rates. The Brahmaputra River has migrated 9 km west in the last 200 years, and the bank width has increased from 2.5 km to 10 km in the 144-km long stretch between Pasighat and Dibrugarh, mainly during falling flood stage when excess sediment is deposited as bars within the channels, causing change in local flow direction (Coleman, 1969). During the 1986 earthquake, the river bed rose by 3 m at Dibrugarh, forcing the river to change its path (Valdiya, 1987).

In the Himalayan region there are already 44,000 km of paved and dirt roads that cut through dense forests, thereby causing slides and debris flows that have affected the ecological balance in the area (Valdiya, 1987). The impact of these human activities on natural erosion rates in the Himalayas needs to be quantified.

Erosion vs Deposition

Not all the material eroded by the rivers is discharged to the oceans. Rather, a considerable amount of it is deposited in river channels and flood plains. A number of reservoirs behind dams have become repositories for river-borne sediments. The present late of accumulation of sediments in different man-made reservoirs is 2.5 to 18 ha-m/100 km^2/year (Gupta, 1975), which is 3 to 5 times greater than the rates stipulated at the project planning stage for various dams in India. For example, in the reservoir behind the Tihri dam on the Ganges, the deposition rate is 1146 ha-m/100 km^2/year (Pant et al., 1986). Due to silting up of the reservoirs, the Subcontinent is losing both the hydroelectric and irrigation potential of these large rivers.

In the main channel of the Ganges near Oatna where major tributaries join the river, there is an increase in the sedimentation rate from about 0.4 cm/year above Patna to about 0.9 cm/year at Farakka, a distance of about 400 km and one of the most intensively populated regions in the river basin (Singh, 1988). The accumulation of the sediments within river channels thus diminishes their sediment-carrying capacity. For example, the bed of one of the smaller tributaries of Brahmaputra, rising in Bhutan, has been aggrading at a rate of 0.5 to 1.35 m/year, while some parts of the main channel of Brahmaputra have risen by 4 m/year during the same period (Singh, 1982). On the other hand, in the Yamuna, a tributary of the Ganges, sediments accumulate around the urban areas of Delhi at the very high rate of 8 cm/year (Subramanian et al., 1985; Fig. 5A).

Thus, the rate of deposition of sediments in these river systems is not in equilibrium with the reported rates of erosion due to a variety of causes, both natural (tectonics, earthquakes) and anthropogenic (deforestation, dredging, urbanization). In the Brahmaputra Basin the sedimentation rate decreases downstream (1 cm/year to 0.2 cm/year) over a distance of 900 km from Dibrugarh to Dhubri (Fig. 5B). There is a corresponding increase in particulate sediment load (150 fold increase from 30 mg/l TSM to about 5000 mg/l TSM) over the same distance (Fig. 5C). Thus, while the Ganges alluvial deposition tends to accommodate much of the eroded sediment, the Brahmaputra erodes its sediments in India and deposits them in Bangladesh.

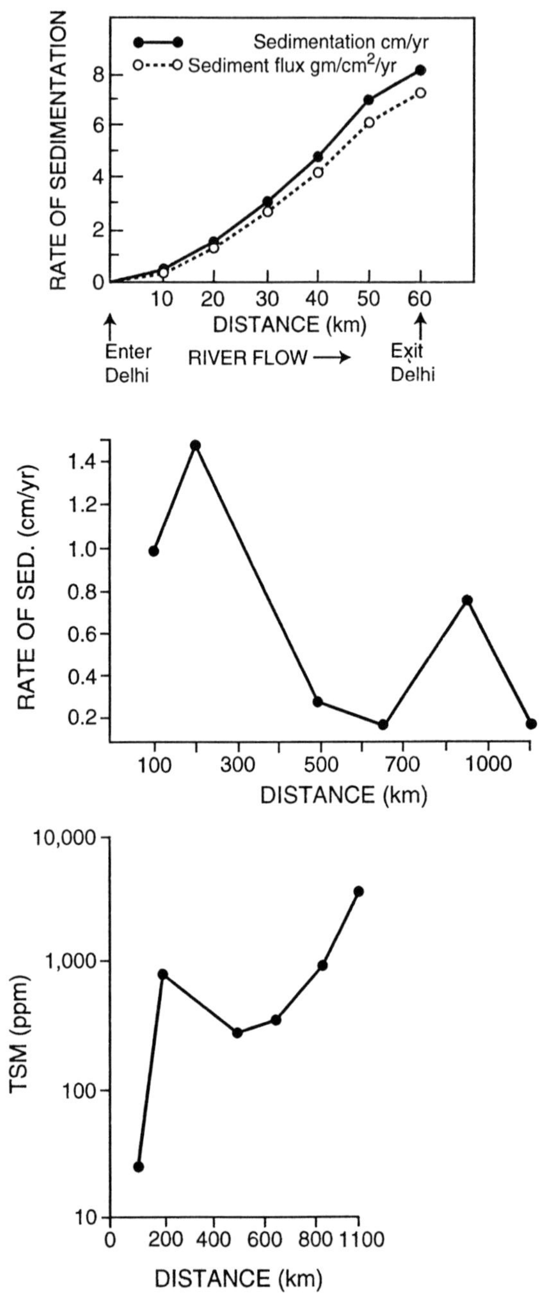

Fig. 5. Rate of sedimentation for the Yamuna (tributary of the Ganges) around Delhi (A) and for the Brahmaputra in India (B). Rates determined by $^{210}Pb/^{226}Pb$ technique. Downstream trend in suspended sediment load (total suspended matter = TSM) for the Brahmaputra in India (C). (A) adapted from Subramanian et al. (1985).

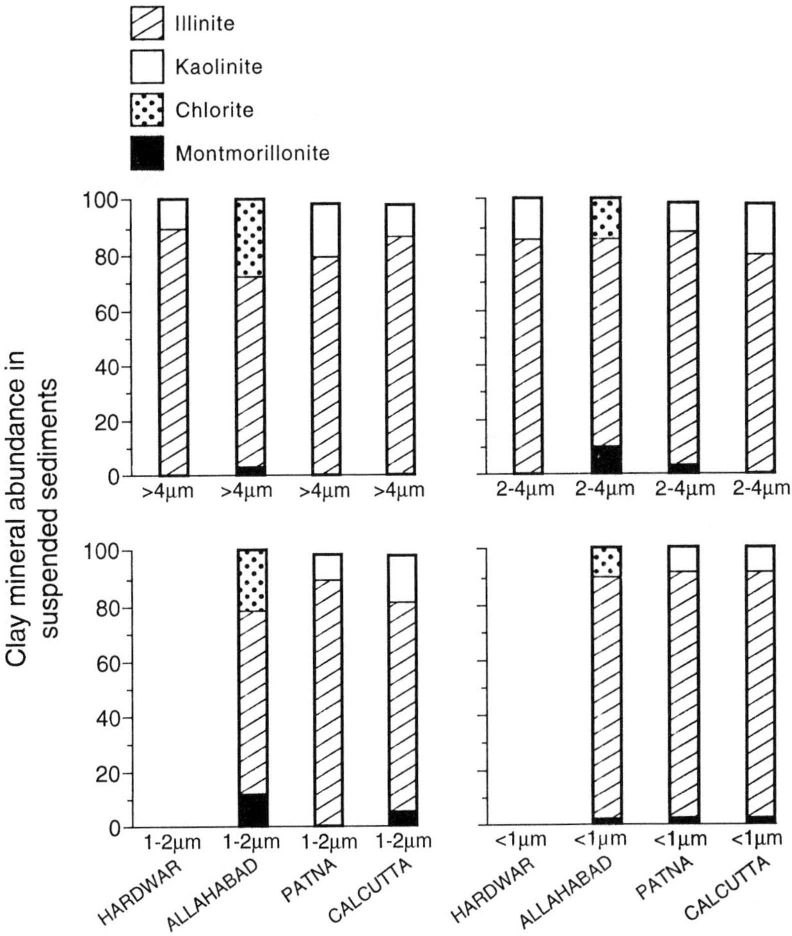

Fig. 6. Downstream variation in the mean grain size of bottom sediment in the Ganges River basin.

Nature of Sediment Load

Figure 6 illustrates the downstream profile of mean grain size of the riverine particulates as compared to those of the bed sediments in the Ganges Basin. Particulates are coarser upstream and mean size decreases downstream towards (Calcutta), where the mean size is nearly half the upstream value due to various tributary effects. Here the bed material is predominantly of sand.

Table 3 illustrates the role played by tributaries in regulating the mean size of the particulates for the River Ganges between Patna and Calcutta. The mean size of particulates vary from 39 mm (for the Yamuna) to that of 8.27 mm (for the Ghaghra). The relationship between total sediment load and the particulate size distribution for 3 locations in the Ganges near Patna is shown in Figure 7. The effect of the coarse-grained sediments in Gandak is actually felt on the particulates

TABLE 3
Mean size of Ganges sediment.

	Susp. load (10^6 tonnes/year)	$D\mu$ (mm)	$D\mu^x$ (mm)
Ganges at Patna	200	11.24	5.60
Ghagra	125	8.27	2.60
Sone	13	18.44	0.59
Gandak	31	34.35	2.64
Kosi	70	15.00	2.60
	439[a]		14 m[c]
Ganges at Calcutta	329[b]	12.1[d]	11.13[e]

[x] Sediment discharge weighted diameter

e = (c) (a-b)/a

d = observed value at Calcutta

of the Ganges only after mixing with the Kosi 200 km away. Before the junction with the tributaries, the sediments of the Ganges are relatively fine (medium to fine silt range) while after Kosi a large proportion of the sediment load is in the fine sand to coarse silt range. In general, silt-size sediments are common in the sediment load of the Ganges and its tributaries.

Figure 8 shows the size-mineralogical relation in the particulate load of the Ganges at four locations. At the foot hills (Haridwar) there is no <2 mm fraction and illite is the dominant clay mineral. While the mean size of the Gandak is coarsest, its clay mineral contents are also the highest (55%). There is a progressive increase in the clay mineral content from Patna to Kosi. Table 4 illustrates the variations in the mineral composition of particulate matter transported by Ganges and its tributaries. These mineralogical differences in the Ganges Basin reflect inherent differences in the size abundances and variable sorting, which gives rise to the observed mineral composition at any given location.

The predominant sizes of suspended sediments and bed material in the Brahmaputra at Pandu are 0.06 and 0.2 mm, respectively. Goswami (1985) reported five different clusters of particles, representing the various tributaries of the Brahmaputra. The observed ratio of medium sand, fine sand and silt (2:3:5) decreases from upstream (Ranaghat) to 1:2:3 downstream (Pandu). Goswami also observed that the peak sediment discharge depends strongly on the rate of sediment supply rather than on the hydraulics of stream flow, due to the influx of a large proportion of fine sediment during floods.

Bed material in the Brahmaptura is mostly fine sand to silt; the clay fraction generally represents less than 5% of the total sediment. Bank materials are mostly composed of varying amounts of fine sand and silt, with only occasional amounts of clay.

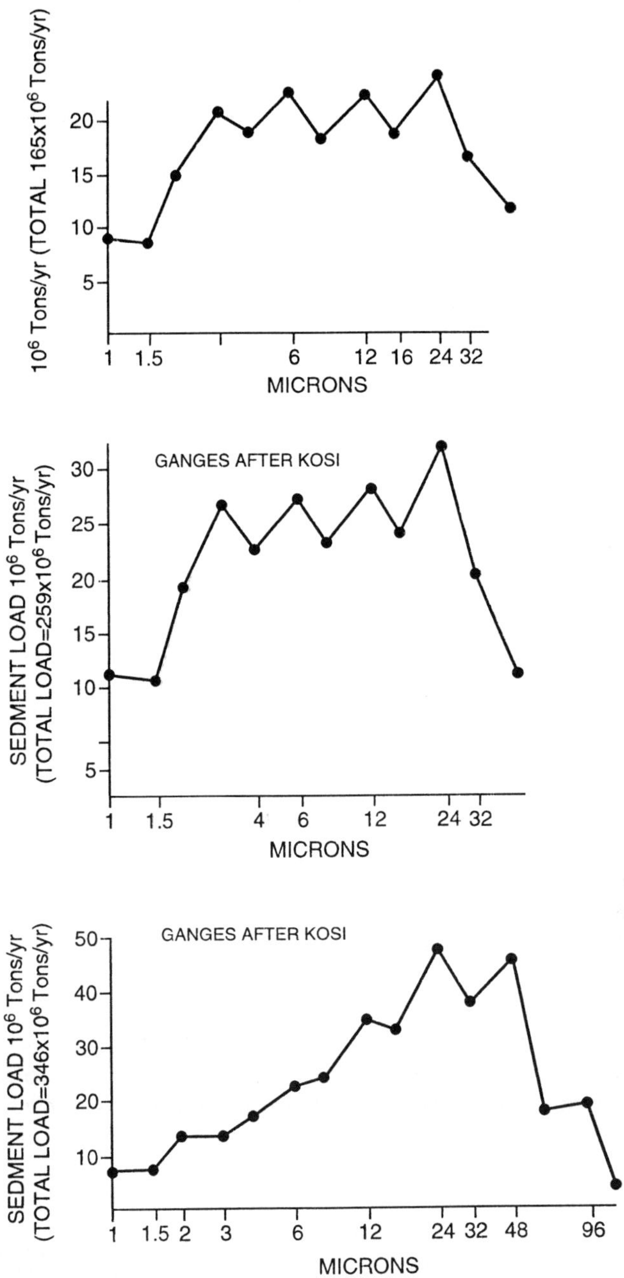

Fig. 7. Grain size distribution of the sediment load for the Ganges at three locations near Patna.

Most of the particulates in the Ganges are detrital, with carbonates dominating in in tributaries draining limestones. Among the clays, illite is most dominant, followed by kaolinite. The particulates of the Brahmaputra at Gauhati are sim-

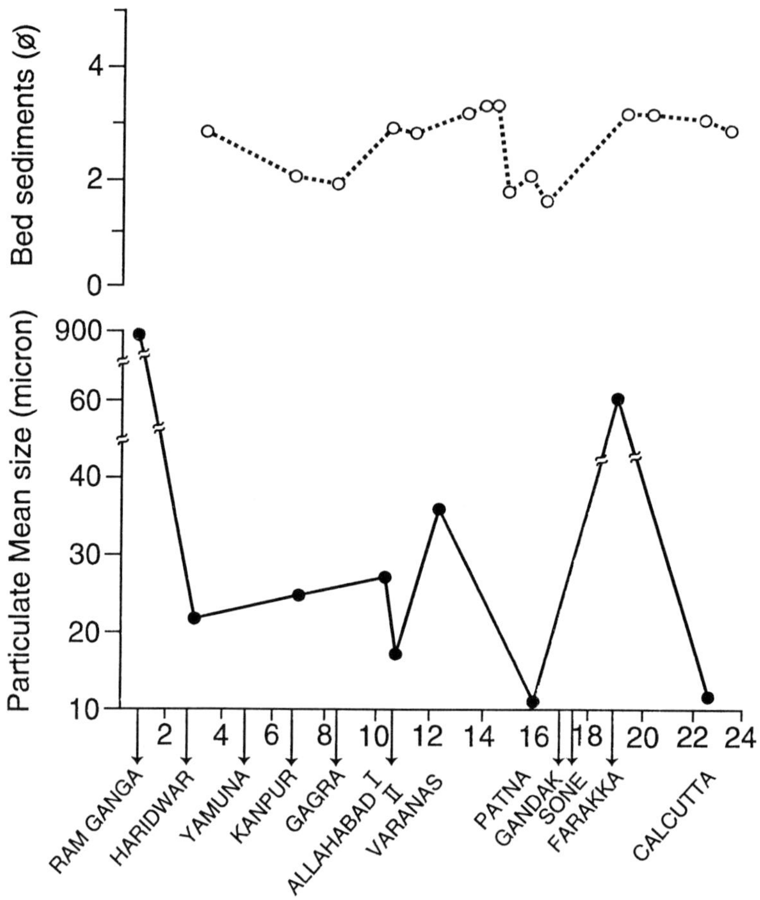

Fig. 8. Clay mineral composition of various grain size populations of suspended sediments in the Ganges River at four locations: Haridwar (upstream), Allahabad and Patna (midstream) and Calcutta (downstream).

ilar (Subramanian, 1980). Hence physical weathering appears to dominate over chemical weathering in the Himalayan drainage systems.

Conclusion

The sediment load of the Ganges-Brahmaputra River system represents an extreme rate of erosion compared to other drainage systems in the world. The depositional rates differ in the Ganges in that large areas are densely populated and urbanized, whereas most of the upstream and midstream parts of the Brahmaputra are either sparsely populated or densely vegetated. The respective erosion/depositional rates in these basins reflect this difference.

TABLE 4
Mineral composition of particulate matter transported by the Ganges

	(in percent)							
	Q	F	Car	I	Ch	K	Mon	Total Clays
Ganges								
at Patna	57	8	7	11	10	11	–	28
below Sone	57	9	10	10	7	7	–	25
below Gandak	52	7	10	8	12	12	–	31
above Kosi	50	6	6	18	12	12	–	38
below Kosi	44	7	1	34	4	4	–	48
at Farakka	54	9	3	20	6	6	–	34
Tributaries:								
Sone	10	20	40	–	–	21	–	21
Gandak	33	8	7	18	18	19	–	55
Kosi (left bank)	30	6	5	36	13	–	–	49
(right bank)	33	18	23	9	–	9	7	25
Ramganga	39	8	0.5	35	–	12	6	53
Yamuna	81	6	2.6	1	–	5	2	8
Gomti	49	9	0.5	32	–	9	–	41

Q=Quartz; F=Feldspar; Car=Carbonate; I=Illite; Ch=Chlorite;
K=Kaolinite; Mon=Montmorillonite

Diurnal, seasonal and annual variations in the discharge and sediment load in these rivers reflect the uncertainties in the existing data base. Even a decade of new observations may not lead to meaningful generalizations. The nature of the sediment load in these river systems points out their predominantly silty nature, which in turn reflects the overwhelming control of physical weathering in the river basins.

Our present knowledge about these rivers is grossly inadequate. Both rivers show evidence of human interferences in their natural erosional and depositional abilities. It is essential to study these rivers in detail if we are to understand the natural processes on the Bengal Delta before human activities inextricably alter the river basin and coastal area.

References

Abbas, N. and Subramanian, V., 1984. Erosion and Sediment transport in the Ganges river basin (India). *J. Hydrol.* 69, 173-182.

Carson, B., 1985. Erosion and sedimentation processes in the Nepalese Himalaya, ICIMOD Occasional Paper No. I,, ICIMOD, Kathmandu, 39 pp.

Coleman, J.M.,1969. Brahmaputra River: Channel processes and sedimentation. *Sediment. Geol.*, 3, 129-239.
Dasgupta, S.P., 1984. The Ganges basin, Part II. Central Board for the Prevention and Control Water Pollution, New Dehli. 204 pp.
Gansser, A., 1964. The Alps and the Himalayas. *Rep. 22nd Intern. Geol. Congr.*, 11, 387-399.
Garrels, R.M. and MacKenzie, F.T., 1971. *Evolution of Sedimentary Rocks*. Norton, New York, 367 pp.
Goswami, D.C., 1985. Brahmaputra River, Assam, India: Physiography, basin denudation, and channel aggradation. *Water Res. Res.* 21, 959-978.
Gupta, G.P., 1975. Sediment production-status resort on data collection and utilization. *Soil Conservation Digest*, 3, 10-21.
Holland, H.D., 1978. *The Chemistry of the Atmosphere and the Oceans*. John Wiley, New York, 351 pp.
Judson, S. and Ritter, D.F., 1964. Rates of regional denudation in the United States. *J. Geophys. Res.*, 69, 3395-3401.
Leopold, L.B., Wolman, M.G. and Miller, J.P. (eds), 1964. *Fluvial Processes in Geomorphology*. Freeman, San Francisco.
Milliman, J.D. and Meade, R.H., 1983. World-wide delivery of river sediment to the ocean. *J. Geology*, 91, 1-21.
Milliman, J.D. and Syvitski, J.P.M., 1992. Geomorphic/tectonic control of sediment discharge to the ocean: The importance of small mountainous rivers. *J. Geology*, 100, 525-544.
Pant, G., Narula, G.L. and Shome, S.K., 1980. Geotechnical Studies Relating to the Planning and Design of the Tehri Dam Project (Unpubl. Report to Geological Survey of India).
Probst, J.L., 1986. Dissolved and suspended matter transported by the Gironde River (France): mechanical and chemical erosion rates in a calcareous molasse basin. *Hydrol. Sci. J.*, 31, 61-79.
Singh, S.K., 1988. Nature of Chemical and Sediment Load in the Ganges River Between Sone and Kosi, PhD Thesis, Jawaharlal Nehru University, New Delhi, India. 187 pp.
Singh, S., 1982. Soil Conservation Problem, Approach and Progress in India, Soil and Water Conservation Division, Ministry of Agriculture, New Delhi, 40 pp.
Subramanian, V.,1978. Input by Indian rivers into the world oceans. *Proc. Ind. Acad. Sci.* 87A, 77-88.
Subramanian, V., 1980. Mineralogical input of suspended matter by Indian rivers to the adjacent areas of Indian Ocean. *Mar. Geol.*, 36, M29-M34.
Subramanian, V., 1987. Environmental geochemistry of Indian river basins - a review. *J. Geol. Soc. India*, 29, 205-220.
Subramanian, V., Sitasawad, R, and Joshi, L.U. (1985). Rate of sedimentation in the Yamuna river using Pb ^{210}- Ra ^{226} technique. *J. Radioanal. Chem.* 90, 271-276.
Valdiya, K.S., 1987. *Environmental Geology of the Himalayas*. Tata-McGraw Hill Pub., Delhi. 581 pp.
Wadia, D.N.,1975. *Geology of India*. Tata-McGraw Hill Pub.; Delhi. 506 pp.
Zachar, D., 1982. *Soil Erosion*. Elsevier, Amsterdam. 547 pp.

CHAPTER 8

Subsidence of the Ganges–Brahmaputra Delta of Bangladesh and Associated Drainage, Sedimentation and Salinity Problems

Mahmood Alam

ABSTRACT: The Ganges–Brahmaputra Delta is one of the most densely populated areas of the world. The delta occupies most of the Bengal Basin and is slowly subsiding as a result of isostatic adjustment of the crust due to rise of the Himalayas and dewatering of the Proto-Bengal Fan sediments which is now buried under thick Mio-Pliocene-Pleistocene deltaic sediments. Well-log data from northwest of Dhaka indicates that at least a part of the basin is subsiding at a rate of 2.2 cm/year. Three areas of the basin — the Hatiya Trough, Faridpur Trough and Sylhet Trough — may be subsiding at similar or higher rates. Engineering projects that do not consider the subsidence component in planning and designing may produce results detrimental to the environment.

There has been very little net progradation of the Ganges–Brahmaputra Delta in the last 200 years. The early and modern Meghna Delta, on the other hand, is changing rapidly, with new islands appearing and major channel shifting every few years. East of Tentulia River, there has been a general seaward growth of the coastline. The Noakhali coastline eroded about 8 km between 1789 and 1945, whereas it accreted about 20 km between 1959 and 1982. It is not clear if the latter seaward growth was due to increased soil erosion in the Himalayas, following deforestation or due to changes in estuarine circulation.

The subsidence and the relative sea-level rise could cause serious drainage and sedimentation problems in the Ganges–Brahmaputra Delta. With higher sea level, more areas will be affected by cyclonic surge; inland fresh water lakes, ponds and aquifers are likely to be affected by saline and brackish water intrusion. The present limit of tidal influence is expected to extend further north. Expected sea-level rise will cause soil salinity, as well as surface water and ground water salinity for a large part of the coastal area. The above conditions, together with lack of dry-season stream flow, may cause serious ecological and economic problems for the country.

J.D. Milliman and B.U. Haq (eds.), Sea-Level Rise and Coastal Subsidence, 169-192.

Introduction

The Ganges–Brahmaputra Delta lies on the eastern side of Indian sub-continent and covers most of Bangladesh and West Bengal of India. Bangladesh is a densely populated county with an estimated 115 million people and an area of 144,000 km^2 of which about 65% is composed of ancient and modern deltaic plain, 10% is the alluvial plain of modern Ganges–Brahmaputra River systems and 25% is hilly terrain, mostly in the Chittagong Hill Tracts (Fig. 1). The deltaic plain, which includes most of the agricultural land of Bangladesh, ranges from sea-level to about 20 m above sea level. The modern alluvial plain extends from the northern side of Bangladesh towards south and follows the river courses as a narrow belt to the Meghna Estuary. The hilly terrain of the Chittagong Hill Tracts ranges in altitude from about 15 m at the margin of the deltaic plain to over 460 m above sea level. The narrow, sinuous hill-ranges separate broad, flat-bottomed valleys, trending north-north-westerly from the Chittagong District into Tripura Sate of India. One of the significant topographic features of Bangladesh is the northern extension of the central lowland, at altitude less than 3 m above sea level, the land surface rise to an altitude of 1,525 m in less than 30 km. Within the deltaic plain (Fig. 1), three areas have been identified as Pleistocene Terraces: The Barind in the northwest, the Madhhupur Jungle north of Dhaka, and the Comilla Terrace in the southeast.

Geotectonics

The alluvial cover of the Ganges–Brahmaputra Delta obscures a complex and interesting geotectonic area. The Ganges–Brahmaputra Delta occupies most of the Bengal Basin. To the south the basin opens to the Bay of Bengal, where it passes into the Bengal Deep-Sea Fan. Figure 2 shows the interpreted depth of the basement rocks in the Bengal Basin and Figure 3 is the structural settings of the basin. Shield areas are exposed on two sides of the Bengal Basin. The Peninsular Shield, exposed west of the basin, disappears below a blanket of alluvium near the boundary of West Bengal, to reappear on the northern edge of the basin in the Shillong Plateau (Assam).

The western and northwestern parts of the Bengal Basin are underlain by "the shelf", which has a northeast-southwest trend along which the basement complex slopes steeply downward to form a hinge zone (Fig. 3). The Eocene Sylhet limestone, which dips very gently southeast is the characteristic formation of the shelf. The coastal plain consists of a hinge zone and a large elongated trough known as the Bengal Foredeep. According to gravity surveys and drilling data reported by Bakhtine (1966), Guha (1978) and Khan (1980), the Bengal Foredeep can be further sub-divided into several zones; (1) the Faridpur Trough; (2) the Barisal High; (3) the Hatiya Trough; (4) the Sylhet Trough and (5) the Madhupur High. These features are shown in Figure 3.

The eastern side of the Bengal Basin is bordered by a Mobile Belt (Orogenic Belt) that extends north-south as part of the Indo-Burmese Mobile Belt (Alam, 1972). Since the Cretaceous, sedimentation in the Bengal Basin has been con-

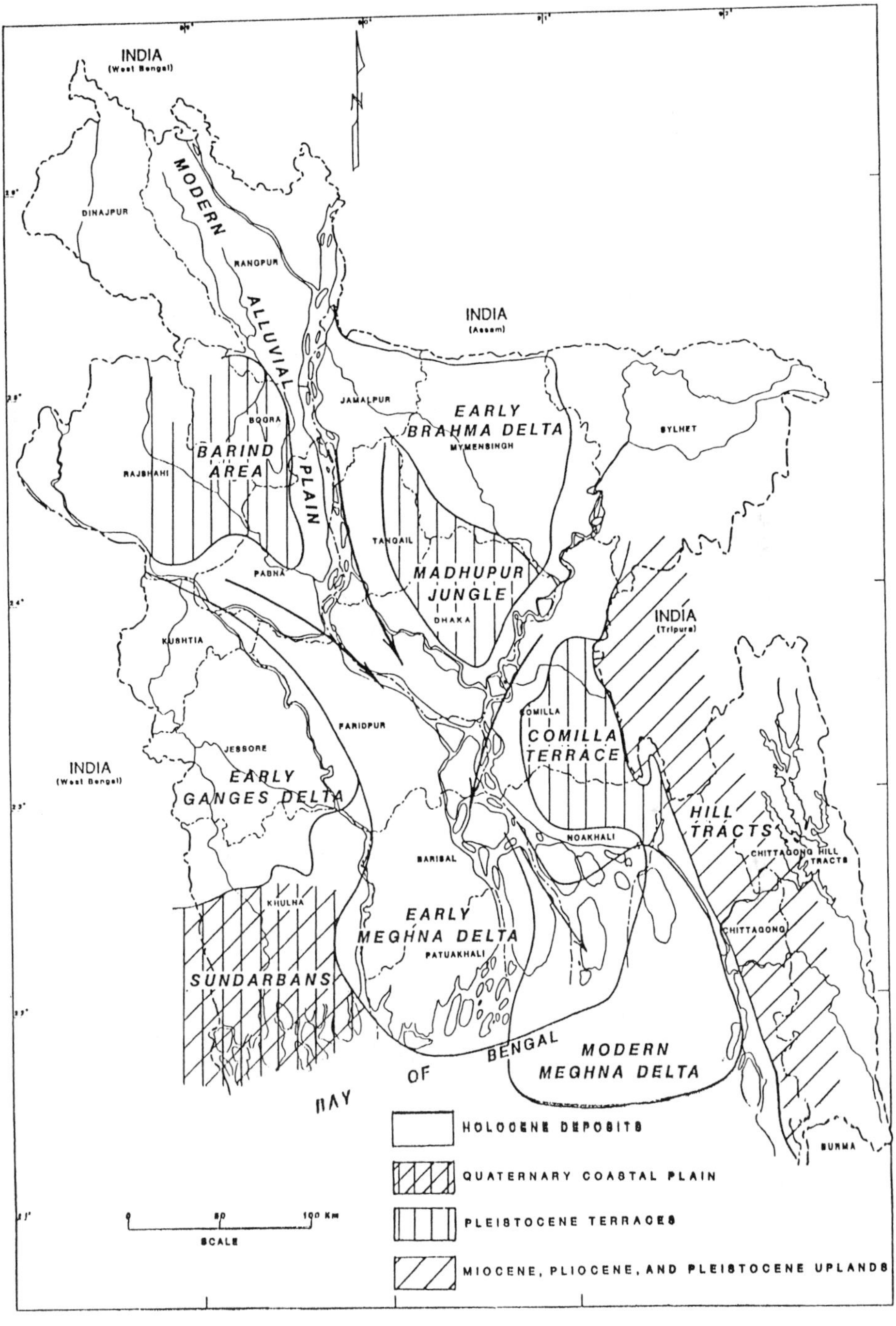

Fig. 1. Physiography of Quaternary deltas of Bangladesh (modified from UNDP/UNDTCD Ground Water Survey Project BGD/74/009 and Jones, 1985).

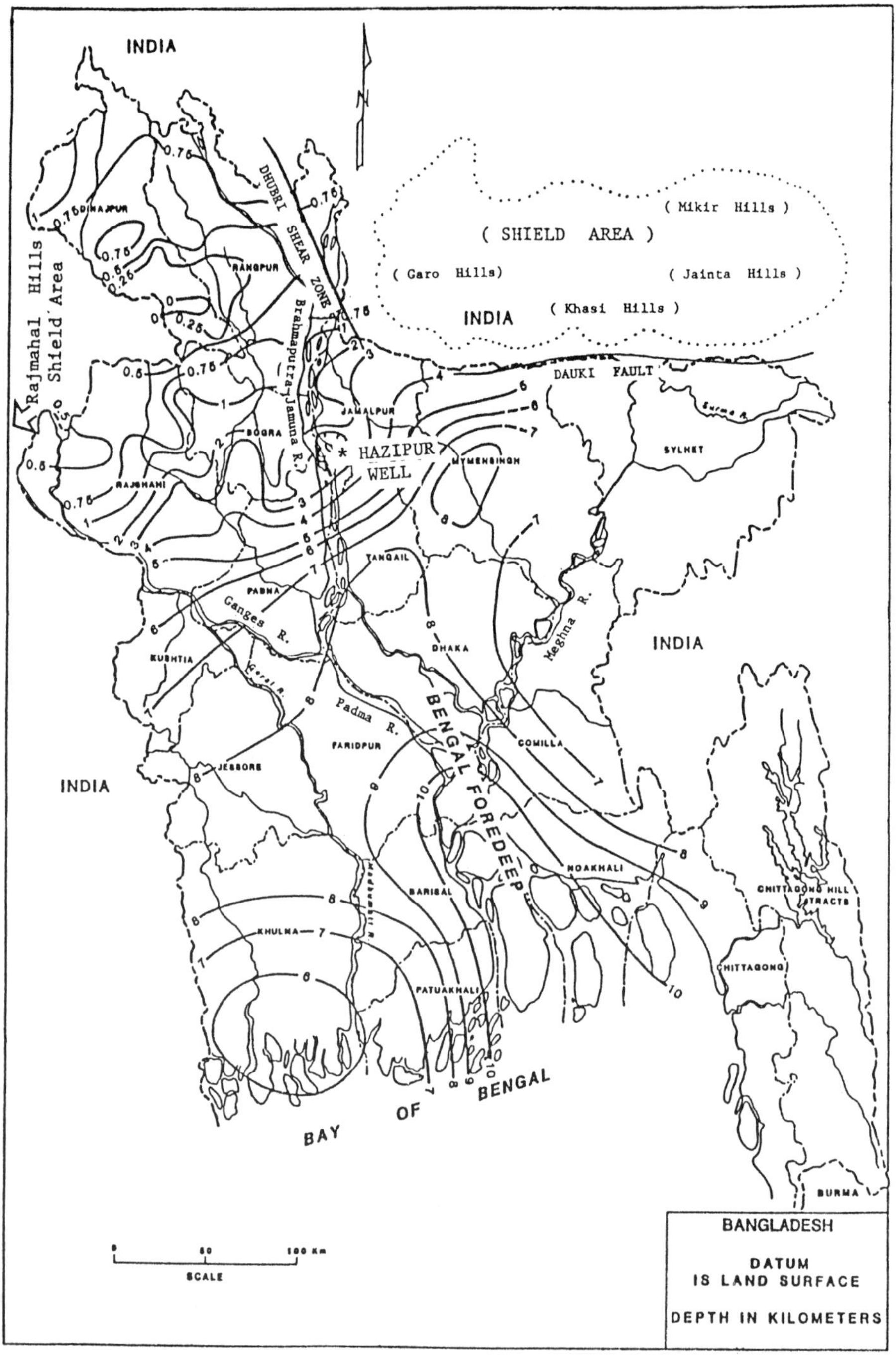

Fig. 2. Depth to basement in the Bengal Basin of Bangladesh (Data from aeromagnetic survey, Overseas Development Administration London, 1980, modified by Jones 1985).

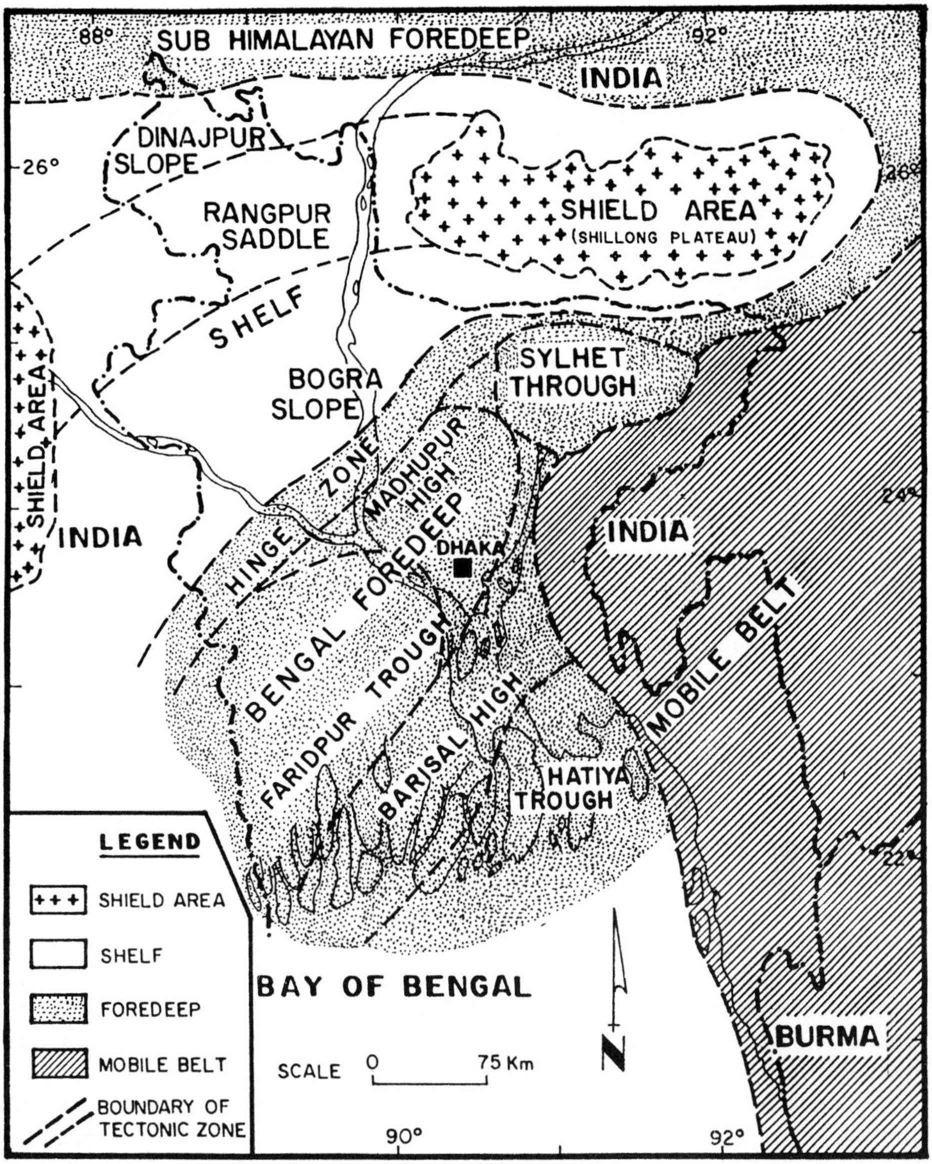

Fig. 3. Sketch map showing the structural setting of the Bengal Basin based on Alam, 1972, & 1989; Bakhtine, 1966; Guha, 1978; Zaher and Rahman, 1980). Five major structural features are shown: 1. Shield Area; 2. Shelf; 3. Bengal Foredeep (Deep Basin); 4. Sub-Himalayan Foredeep; 5. Mobile Belt. The Shelf in Bangladesh shows three structural zone: (a) Dinajpur Slope; (b) Rangpur Saddle; and (c) Bogra Slope. The Bengal Foredeep (Deep Basin) shows six major structural zones; (a) Faridpur Trough; (b) Hatiya Trough; (c) Sylhet Trough; (d) Barisal High; (d) Madhupur High; and (f) a transitional hinge zone.

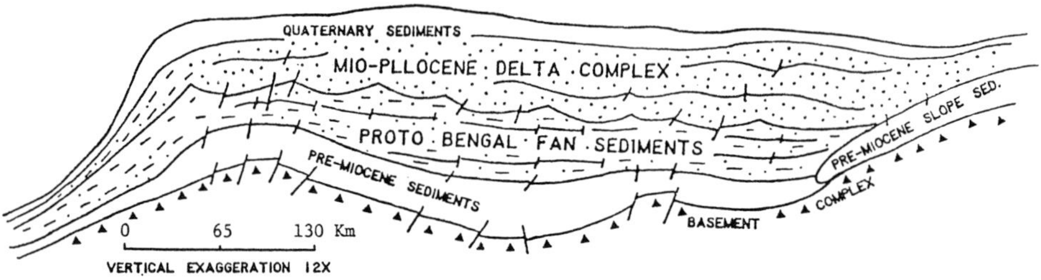

Fig. 4. Hypothetical cross-section of Bengal Basin from Sylhet to Khulna, showing progradation of Mio-Pliocene delta complex over the Proto-Bengal Fan.

trolled by the movement of the Indian Plate, by the collision pattern of the Indian Plate with the Burmese and Tibetan Plates, and by uplift and erosion of the Himalayas and Indo-Burmese ranges. Geochronologic and sedimentologic data indicate three prominent phases of uplift-denudation in the Himalayas: early Miocene (21–17 Ma); late Miocene (11–7 Ma) and Quaternary. The early Miocene phase reflects intense tectonic activities and a rapid pulse of uplift and denudation of Himalayas, development of Siwalik basin, and increase in clastic flux to Bengal Basin. The late Miocene seems to have encompassed initiation of the Main Boundary Thrust (located at the foot hill of Himalayas, see Fig. 5), another major pulse of clastic sediment to the Siwalik Ranges as well as to the Indus and Bengal Fans, and was coeval with onset of monsoon seasonality (Sorkhabi and Stump, 1993).

Depositional Sequence

Evolution of the Bengal Basin started in the Permo-Carboniferous with sedimentation in the faulted Gondwana basins. There were fracturing and block-faulting in the Permo-Carboniferous, several depressions formed in the basement. Mostly facies were continental (Gondwana sediments with coal), and some of these faulted Gondwana Basins can be traced below the sediments cover of the northern Bangladesh. The breakup of the Gondwandland in the Cretaceous Period led to the eventual separation of Peninsular India from the southern continents, with counter-clockwise rotation and northward drift, permitting a marine transgression and sedimentation in the Bengal Geosyncline.

Sedimentation in the Bengal Basin can be ascribed to two phases: 1) In the Cretaceous there was a major marine transgression, with possible sea-level oscillations, although these cannot be correlated with the Haq et al. (1987) sea-level curve, implying the importance of local tectonics that have obscured the the eustatic signal. There is no exposure of Cretaceous sediments in Bangladesh or in West Bengal, but limited drill hole data from West Bengal (Sengupta, 1966); Banerji, 1984) and North Bengal (Zaher and Rahman, 1980) suggest that most of the sediments on the stable shelf were deposited in fluvial, tidal flats, deltaic and lagoonal environments. Major sources of sediments were from the west and northwest. In Foredeep and Mobile Belt sedimentation possibly took place in a

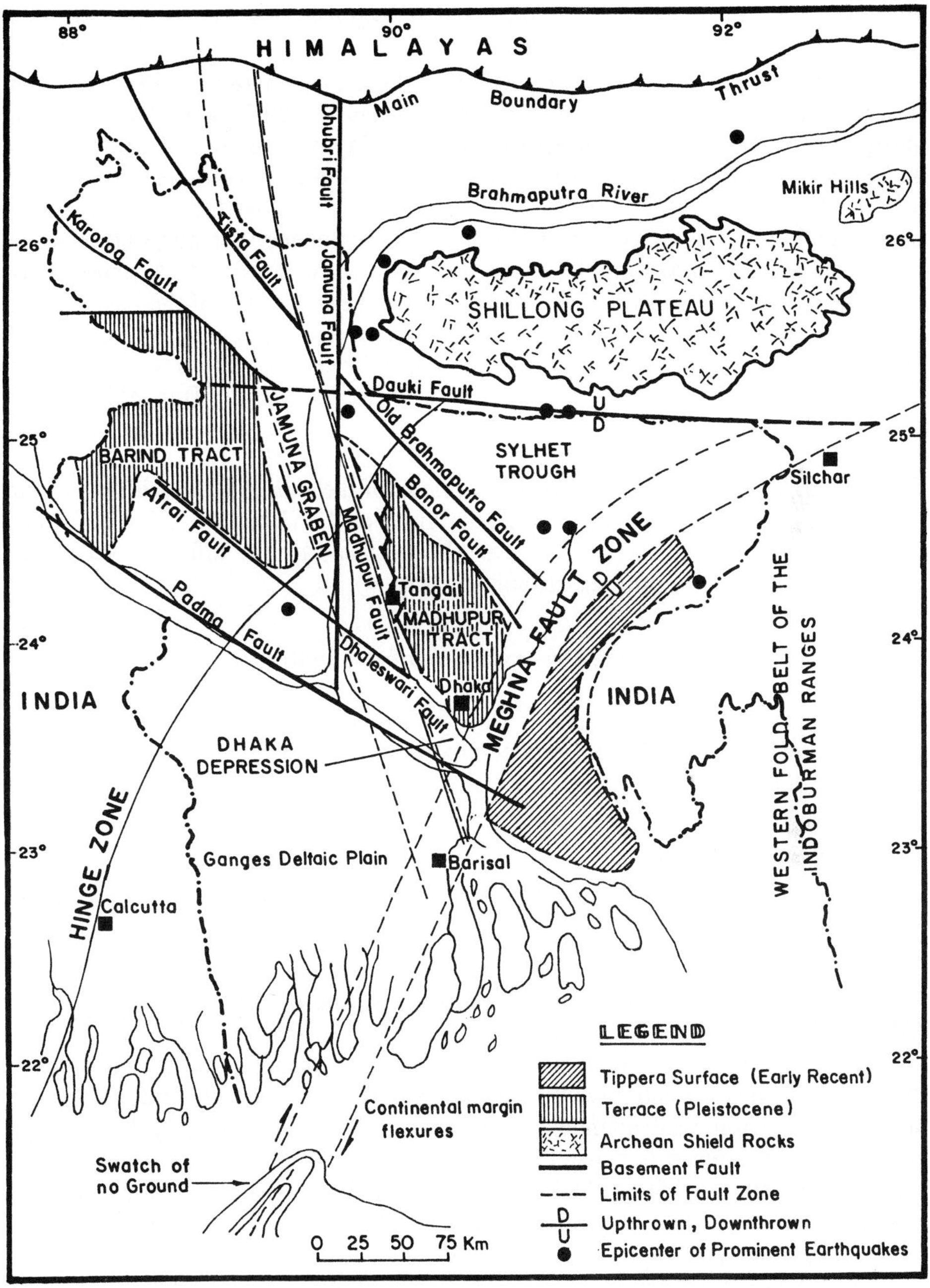

Fig. 5. Neotectonic map of Bangladesh, showing active faults and earth quake epicentres. (Modified from Khandoker, 1987).

marine environment, and turbidites probably played an important role (Graham et al., 1975). The Eocene was a period of maximum marine transgression - clastic sediment input on the shelf was dramatically reduced and limestone deposition was extensive.

2) In the late Oligocene to early Miocene there was a major marine regression in the Bengal Basin exposing most of the shelf. During this period, the basin was subjected to the impact of the rise of the Himalayas (Gansser, 1983, Sorkhabi and Stump, 1993), but the major source of clastic sediments was still to the west and northwest. However, during this period eastern and northeastern source of clastic sediment was activated, and by the mid-Miocene, the huge amounts of clastic sediments were funnelled into the basin from northeast. A major Mio-Pliocene delta complex started to build in the northeastern side of the basin. In the shallower parts of the basin, sedimentation was in fluvial, deltaic and open shelf environments, while in deeper parts of the basin the sedimentation was controlled by turbidites. The Proto-Bengal Fan, similar to the present day Bengal Deep Sea Fan (Curry and Moore, 1974; Graham et al., 1975) possibly played an important role in sedimentation in the Bengal Foredeep. Since the Miocene time, the Proto-Bengal Fan has been deformed by the collision of the Indian Plate with the Asian mainland and influence by the tectonic activities of the Himalayas. The deformed sediments of the Proto-Bengal Fan have been buried by the influx of sediments eroded from the newly uplifted Himalayas (Fig. 4).

Subsidence of the Basin and Progradation of the Delta

Although the entire Bengal Basin is slowly subsiding, the more rapid subsidence has taken place in the Bengal Foredeep, particularly in the Sylhet, Faridpur, and Hatiya (Hatia) Troughs.

TECTONIC SUBSIDENCE

The subsidence of the Bengal Basin is generally considered to be of tectonic origin and can be attributed to two major factors. One is related with the isostatic adjustment of the crust (sediment load and rise of Himalayas), and the other is related to dewatering and compaction of shale and mud of Proto-Bengal Fan (Alam, 1989).

Tectonic subsidence usually occurs over a large area and, unlike the subsidence due to dewatering, is relatively uniform and usually occurs at a slow rate. The areas subjected to tectonic subsidence are generally bounded by active faults or hinge zones. The subsidence is caused by the post-depositional downwarping due to relative movements of fault blocks or the crustal plate. Crustal loading by deltaic sedimentation also augments tectonic subsidence (particularly in coastal areas) by depressing underlying material (Costa and Baker, 1981).

There is not enough evidence to suggest that compaction of Quaternary sediments play a major role in subsidence of the basin. Most of these sediments were deposited, dewatered and pre-compacted during the Pleistocene lower sea level

(Jones, 1986). However, compaction of Holocene peat deposits may cause subsidence in certain areas in Bangladesh. Most studies (e.g Salahuddin, 1991; WASA, 1991) suggest, however, that the extraction of ground water may not be a major contributor to subsidence in Bangladesh.

EVIDENCE OF SUBSIDENCE

There are many indirect indications of subsidence in the Bengal Basin. Generally, very old buildings in Bangladesh show that in many buildings the previously above-ground-level floors are now a few feet below the land surface. Evidence of such subsidence can be seen in many old buildings in the old part of Dhaka city and other places.

Well-preserved wood logs and vegetable material can be found buried a few to tens of cm beneath the present ground level in many parts of the Bengal Basin. Their presence indicates that the basin is still receiving sediment and that burial is rapid. Huge amounts of sediments bought into the basin by the Ganges and Brahmaputra Rivers have not significantly increased the land areas of the delta however (see below), suggesting that at least a part of this river-borne sediment is deposited in the basin and compensates for the subsidence.

Subsidence is also indicated by neotectonic activity. Figure 5 shows presently active faults and epicenters of recent earthquakes in Bangladesh. Evidence of neotectonic activities is also documented by uplift in the Eastern Folded Belt (Hills of Chittagong and Sylhet) and subsidence in the basin is indicated by topographic depressions, earthquakes and changes in stream course, especially in the Brahmaputra River.

MEASUREMENT OF SUBSIDENCE

There are no quantified data on the rate of subsidence in the Bengal Basin. There is no station in Bangladesh that actually measures subsidence in the field (as in Bangkok) or gathers long-term tide level data. Rates of uplift and subsidence also can be determined by accurate geodetic and geomorphic surveying, but such techniques have not been employed in Bangladesh.

Estimated values of subsidence given by MPO (1985) range from 0.6 mm/year to 5.5 mm/year, and may exceed 20 mm/year in the Surma Basin. Estimates given by Morgan and McIntire (1959) for the vicinity of Calcutta, Khulna and the Sundarbans indicate an average subsidence of the Ganges deltaic plain in the order of about 1 to 2 mm/year.

Hoque et al. (1982) estimate the rate of subsidence in Dhaka Depression (located between the Madhupur Tract and the confluence of the Padma and Meghna Rivers; see Fig. 5) at 1.88 mm/year. According to radiocarbon dates obtained by Brammer(1971), M.K. Alam (1988), Monsur (1990) and WASA (1991), the average rate of subsidence in Dhaka City area is about 0.65 mm/year. The rate of subsidence in Dholai Khal area is estimated to be 0.75 mm/year, 0.57 mm/year in Gulshan area, 0.75 mm/year in Kalibari pond area, 0.52 mm/year in Dakshingoan area, 0.96 mm/year in Mirpur area and 0.43 mm/year in Rampura area.

TABLE 1
Radiocarbon dates from the Ganges–Brahmaputra Delta. Most data from Umitsu (1993).

Site	Age (Yr BP)	Code No	Material	Depth (Altitude In meter)	Reference
Sankrail, Howrah, Calcutta, India	2,615 ± 100	TF-850	Peat	1.37	Vishnu-Mittre and Gupta (1970)
	4,925 ± 100	?	Wood	1.52	
	4,075 ± 100	TF-851	Peat	1.82	
	4,720 ± 135	TF-855	Peat	3.04	
	5,440 ± 115	?	Wood	4.87	
	5,810 ± 120	TF-856	Peat	6.25	
Sirajganj, Rangpur, Bangladesh	28,320 ± 1585	GaK-5158	Wood	112 (-101)	Umitsu (1985)
Kanailhat, Sylhet, Bangladesh	4.180 ± 120	GaK-11955	Peat	8.55	Umitsu (1985)
Dewlatpur, Khulna, Bangladesh	3,230 ± 110	GaK-12952	Peat	5 (-3)	Umitsu (1987)
	6,490 ± 100	NUTA-342	Wood	16 (-14)	
	6,880 ± 130	GaK-12953	Wood	13 (-11)	
	7,640 ± 100	NUTA-343	Wood	27 (-25)	
	8,890 ± 150	NUTA-344	Plant Fragment	34 (-32)	
	10,190 ± 210	GaK-1254	Wood	30 (-28)	
	12,320 ± 240	NUTA-345	Plant Fragment	48 (-46)	
	7,060 ± 120	NUTA-1342	Shell	20 (-18)	Umitsu (1993)
	8,910 ± 150	NUTA-1345	Shell	35 (-32)	

Source of Data: Umitsu, 1993.

The rate of subsidence can also be measured indirectly from well logs. There are numerous water wells, but records do not show the stratigraphy or the age of the sediment. Although there are a number of oil and gas exploration wells, most are located in the Mobile Belt or on the shelf, and almost no stratigraphic analyses have been carried out on the Quaternary section of these wells. However, at the Hazipur well (Fig. 2), the Holocene-Pleistocene boundary is shown at 244 m. Taking the base of Holocene as 10,000 B.P., the rate of subsidence in the area around Hazipur would be about 2.2 cm/year. The Hazipur well is not located in the middle of any known trough, and, consequently may not represent the highest rate of subsidence in the basin. Clearly, more basin-wide data on Holocene-Pleistocene boundary and C^{14} dates are needed.

Radiocarbon dates (Table 1) obtained by Umitsu (1993) suggest that the coastal areas of Bangladesh (Khulna area) are subsiding at a rate of about 4 mm/year. His data also suggest a high rate of subsidence in the shelf (3.9 mm/year in Sirajganj, Rangpur), and in Sylhet Basin (2.04 mm/year in Kanailghat, Sylhet). The high rate of subsidence in Sirajganj may indicate influence of local faulting. Vishnu-

TABLE 2
Annual suspended sediment discharge at selected locations in the Ganges–Brahmaputra Delta.

River/Location	Annual Suspended Sediment Discharge (Million metric tons)
Brahmaputra River at Bahadurabad	387
Ganges River at Hardinge Bridge *	212
Padma River at Baruria *	563
Old Brahmaputra River at Mymensingh	2
Dhaleswari River at Jagir	40
Gorai River at Gorai Rly. Bridge *	30
Gorai River at Kumarkhali *	43

* Post-Farakka Discharge Conditions. (Source: Master Plan Organization, Bangladesh 1986). For the locations and water discharge through these rivers, please see Figure 7.

Mittre and Gupta's (1970) data presented by Umitsu (1993) suggest that the rate of subsidence around Calcutta varies between 5 and 10 mm/year.

SEDIMENTATION AND SUBSIDENCE

Although there are no accurate estimates about the sediment discharge to the Ganges–Brahmaputra Delta, it appears that a large portion of the sediment load entering the delta may be deposited subaerially, offsetting the basin's subsidence. A comparison of old and new coastal charts shows that delta has prograded only slightly in the past 200 years even though more than 200 km^3 of river-borne sediment has been introduced to the delta. Assuming that about 30% of the basin (43,200 km^2) is subsiding at the rate of subsidence at Hazipur, and that 14 tons of sediment are required per hectare to balance 1 mm/year subsidence, about 1.33 billion tons of sediments are required to balance the subsidence of Bangladesh part of the basin. Milliman and Meade (1983) estimate that about 1.67 billion tons of suspended sediments are discharged annually through the Ganges–Brahmaputra rivers of Bangladesh, whereas the Bangladesh Water Development Board estimates 1.27 billion tons/year suspended sediments discharge through six major rivers of Bangladesh (Table 2). Neither of these estimates considers the bed load sediments, which may account for as much as 50 percent of the total sediment load (J.M. Coleman, in Curray and Moore, 1971). This, together with the fact that at least a part of suspended sediments and possibly some sand are lost to the Bay of Bengal, suggests an imbalance in sediment budget estimates. A critical examination of sediment discharge data is required for the Ganges–Brahmaputra delta for meaningful sediment budget estimates.

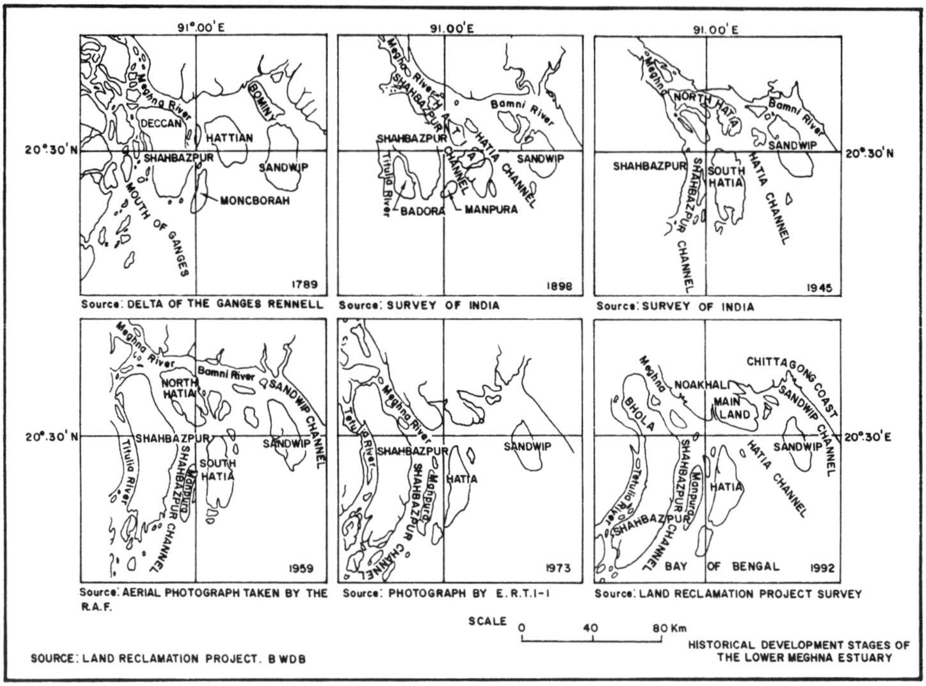

Fig. 6. Historical development of Lower Ganges–Brahmaputra Delta as complied by the Bangladesh Water Development Board from the Land Reclamation Project.

The ancient and modern Meghna Delta (Fig. 1) is changing very rapidly, with new islands appearing and major channel shifting within a few years time (Fig. 6). Channel system of the Lower Meghna estuary underwent a complete change between 1789 to 1982. Eastward of Tentulia River, there has been a general seaward growth of the coast line, especially in the region of Bhola and Hatiya (Hatia) Islands and on the mainland near Noakhali. Between 1959 and 1982, Noakhali accreted about 20 km seaward. However, the same area showed about 8 km landward erosion between 1789 and 1945. Master Plan Organization of Bangladesh (MPO, 1986) estimates that there has been a net accretion of about 110 km^2 south of Bhola/Shabazpur, Manpura and Hatiya Islands from 1972 to 1979. It is not clear if the accelerated seaward growth of the Noakhali area between 1959 and 1982 (Fig. 6) is due to increased soil erosion in the Himalayas because of deforestation or due to change in estuarine circulation. The historical evidence from the Ganges–Brahmaputra Delta, however, suggests the land accretion is caused by channel migration, and accreted land is offset by erosion elsewhere.

Floods and Drainage Problems

Due to the monsoonal climate, a major problem faced by Bangladesh is that there is too much water in the wet season and too little in the dry season. In the dry season water shortage problem is also accompanied by increased salinity in a large part of the coastal area.

In Bangladesh four main types of floods occur (World Bank, 1989). These are:

1. **Monsoonal flooding from the major rivers**. Overbank flood spills of the major rivers and their distributaries cause extensive floods in Bangladesh. With the onset of the monsoon, all the major rivers start rising, generally cresting slowly over several weeks.

2. **Local flooding due to heavy and intense rainfall**. Local flooding is caused by very heavy monsoon rainfall within Bangladesh which causes runoff volumes in excess of local drainage capacity, especially when the main rivers are already at high levels.

3. **Flash floods in the eastern and northern rivers**. These are characterized by a sharp rise followed by a relatively rapid recession. High flood velocities often damage crops and properties.

4. **Floods due to storm surges in the coastal area**. The coastal areas of Bangladesh consist of large estuarine channels, extensive tidal flats and low-lying islands. Storm surges generated by tropical cyclones cause widespread damage to life and property. Tropical cyclones are most likely to occur during the pre- and post-monsoon periods (April–May and October–November, respectively), and are not known to coincide with monsoonal flood peaks.

Flood control is among the highest priority for Bangladesh. The 1987 and 1988 floods were catastrophic events which led to widespread destruction and loss of life. But they differed in nature and origin.

The main 1987 flood resulted from heavy rainfall from July to September over northwest Bangladesh and the area of West Bengal immediately to the north. The heavy local precipitation caused severe flooding in many minor rivers in the northwest region. These effects were further aggravated by the highest flood peak ever recorded on the Ganges, and exceptionally high flood on the Teesta, and (despite a Brahmaputra flood only slightly above average) breaches in the right embankment of the Brahmaputra.

The August 1988 floods, in contrast, were caused by intense rainfall for ten days over northern and northeastern Bangladesh, India, Nepal and Bhutan. The most intense local concentrations were in Assam, Meghalaya (India), Bhutan and Arunchal Pradesh (India). The flood peak on the Brahmaputra was the highest ever recorded. Those on the Ganges was also high, but more significantly, the two peaks coincided, with devastating effects downstream of the Brahmaputra–Ganges confluence. Very large areas along the Brahmaputra, Ganges and Padma were flooded and Dhaka was seriously affected. There was also an exceptionally high flood in the Meghna which added to flood congestion in the Lower Meghna.

The Bangladesh government estimated the damage in 1987 at around $3 billion, and the 1988 flood damage even higher. The 20-year average flood peak discharge through the Ganges–Brahmaputra River system is about 152,200 m^3/sec (Fig. 7);

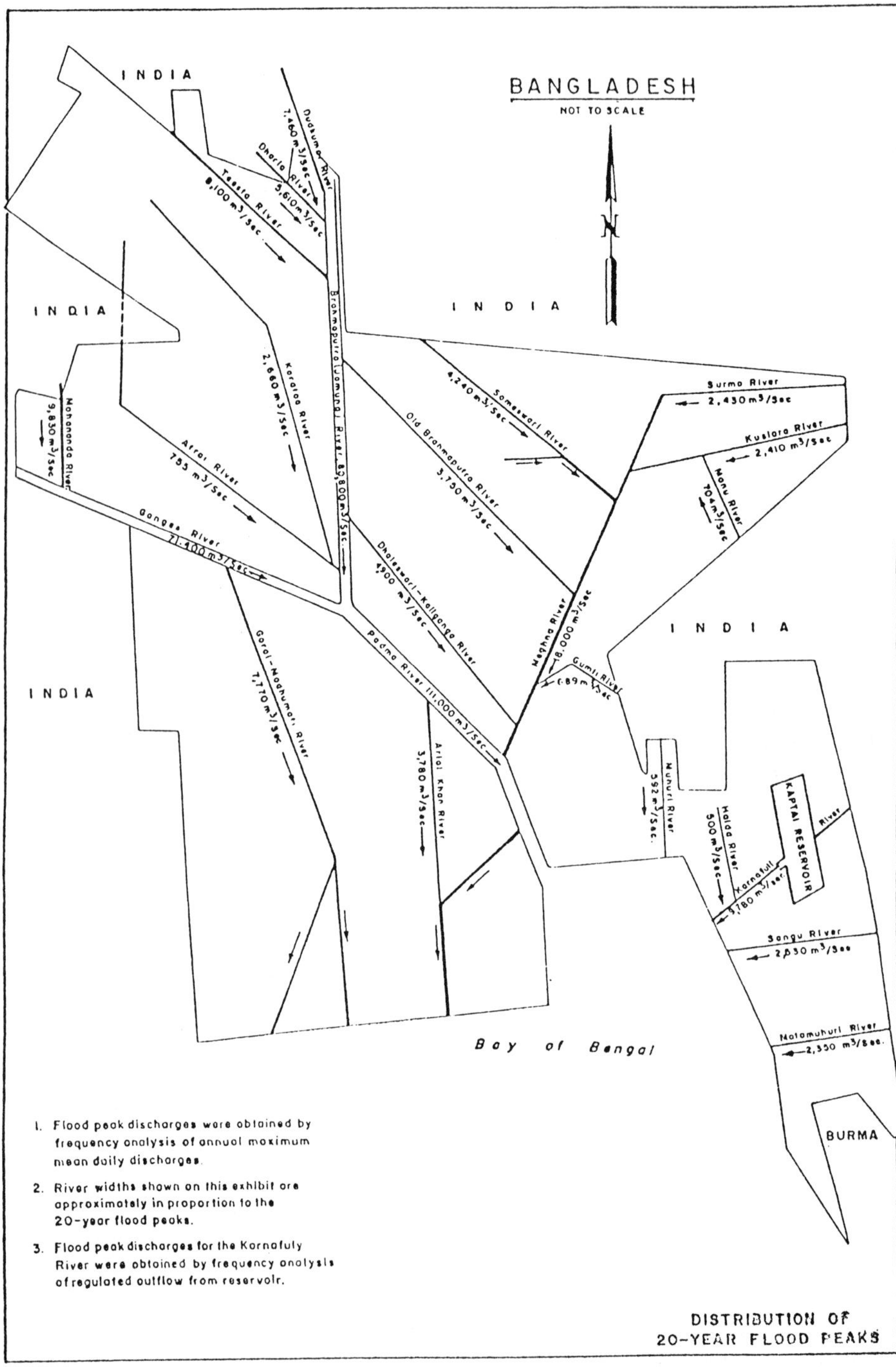

Fig. 7. Distribution of 20-year flood peaks in Bangladesh as compiled by Bangladesh Water Development Board. (Source: MPO, 1986)

of this the discharge through the Ganges at Harding is about 71,400 m^3/sec, and Brahmaputra at Bahadurabad is about 80,800 m^3/sec (MOP, 1986). The peak flood flow of 1988 flood at Bahadurabad on the Brahmaputra was estimated to be 98,300 m^3/sec, and at the Harding Bridge on the Ganges it was 72,300 m^3/sec (World Bank 1989). As a result of the high floods on both the Brahmaputra and Ganges, and the close synchronism of their peaks, the Padma flood peak at Baruria (132,000 m^3/sec) was the highest on record. The Meghna flood peak was the highest for more than 20 years. The 1988 flood peak on the Brahmaputra is estimated to have been of the order of a 100-year event, on the Padma and Meghna a 30-year event and on the Ganges a 40-year event.

There are several proposals to control flood or to reduce the impact of floods in Bangladesh (e.g., Rasid and Paul, 1987; Rogers et al., 1989; Solomon, 1992; World Bank, 1989). Most flood control proposals, however, concentrate on the Ganges River, although peak flood discharge through the Brahmaputra River is larger than that of the Ganges River. At present Bangladesh is pursuing a strategy of protecting human settlements and farmland from floods by constructing flood control works (mostly polders) at a rate of 130,000 ha/year. The benefits of this in terms of income and safety are fairly well documented, however, the longer-term environmental impacts have not been properly assessed.

There is a proposal to build a series of dams in Nepal to store the monsoonal runoff in order to reduce the impact of flood in Bangladesh. The stored water can then be released in the dry season to augment the dry season flow of the Ganges River. However, this involves the construction of a large number of costly and technically complex dams in the Himalayas. This also will involve complex riparian issues between Nepal, India and Bangladesh. The other proposal involves construction of a large number of deep artesian wells in the Terrai Plain of Nepal. Free-flowing wells would lower the water-table in the recharge areas, in this case, the Bhabar Zone of Nepal. Space created by lowering of water-table could be used to store monsoon rain water and reduce the impact of flood in downstream areas. The wells would operate only in the dry season and would help to augment dry season flow of the Ganges River. The deep aquifer option (artesian well option) deserves further attention and could be supported by a program to test its feasibility.

A study by Rogers et al. (1989) on eastern Himalayas suggested that sufficient space can be created in the aquifer during the dry season to hold rain water during the monsoon and reduce the impact of flooding. Another proposal that is receiving considerable attention is to build embankments along the Ganges and Brahmaputra Rivers. The embankment option, when accompanied by selective flooding at peak flood stage and occasional channel dredging, could offer a short- to medium-term solution to the floods in Bangladesh, although perhaps not in the longer term. However, traditional method of coping with floods by building villages on moulds, and agriculture based on use of receding or advancing flood water may still be the only viable option for many parts of the country.

The diversion of the Ganges River at Farakka (close to the Bangladesh–India border) through the Hoogly River to save Calcutta Port from silting, left the southwestern side of Bangladesh with serious problems of excess salinity and reduced sedimentation. It has disrupted agriculture, navigation, industry, fisheries, forestry and drinking water supply through an area of some 40,000 sq. km. Since the operation of the Farakka Barrage, salinity (during dry season) at the river port of Khulna, a city some 80 km inland, has increased by two-fold. The discharge of the Ganges River has been dramatically changed by man-made projects. Recorded monthly low flow of the Ganges River at Harding Bridge was 728 m^3/sec, (for the month of April, 1976) which is less than 1% of the peak discharge. Recorded monthly low flow for the Brahmaputra River at Bahadurabas is 3,355 m^3/sec, (February, 1973), which is about 4% of the peak flood discharge. Under natural condition, low monthly average of Ganges Rivers is unlikely to show such large differences. It is anticipated that water demand during the lean-flow times will increase in the coming years due to increasing population.

Effects of Recent Engineering Projects

In the Ganges–Brahmaputra Delta, sedimentation and drainage patterns are being significantly altered by coastal engineering projects, for instance by building embankments and polders. In may parts of the delta drainage channels have been filled and sluice gates clogged. Construction of the Bangladesh Coastal Embankment Project started in 1961. By 1984, about 626 km of channels had been silted up in Southwestern Region alone (Delta Development Cell, Bangladesh Water Development Board, 1984, cited in the Water Master Plan, 1986). In the Polder Projects, dikes designed to protect areas from floods have aggravated flooding and silting conditions on the delta.

A report by the chief and superintendent engineers of Bangladesh Water Development Board (Moshin-Uddin and Islam, 1982) states, "initially the primary objective (of polder construction) was the protection of agricultural lands from tidal inundations and flooding. But the solution of one set of problems has given rise to another, calling for immediate and careful consideration". They found that empoldering had let to containment of silt deposition within the channels and many of them had been silted up. Silting resulted in drainage sluices to became inoperative and caused serious drainage problems in the project areas.

Most often lands inside the polders are leached by rain during the monsoon, but during the dry seasons they are affected by capillary rise of brackish ground water, which leads to a reduction in yield of winter crops in the area protected by polders. In addition, uncompensated subsidence in the delta has resulted in wide-spread water logging. For example, in Polder 24 (Jessor District), about 50 percent of the land became non-productive by the development of beels (shallow lakes) and wetland areas on previously productive paddy fields. Similar problem can be seen in other polders of the area, especially polders older than 20 years that have begun to show serious water-logging and drainage problems.

Confinement of rivers also increases the siltation of stream channels because sediments once deposited on flood plains are now deposited in stream channels, resulting in a rise of the stream bed and stream level as well as the water table in the surrounding land, leading to formation of widespread wetlands in the flood plain. Other side effects include the more severe flooding of nearby unprotected areas. Inland open water fisheries also are disrupted by embankments or by siltation in water ways. The deposition of nutrient-rich sediments deposited during flooding are reduced, with long-term effects on soil fertility. Insects, salts and pollutants normally washed away by floods may build up in the protected areas. It appears that there is an urgent need to study the impact of coastal embankment project on the overall drainage of the basin. Designs should be modified so that the area protected by polders occasionally can be flooded to allow sediment accumulation in the protected area to compensate for subsidence.

Salinity

Increasing salinity is considered an important factor in limiting further agricultural and industrial development in the coastal area of Bangladesh. It limits irrigation potential and damages crops by flooding during high tides. There seems to be little potential for further large-scale development of surface water resources given the low relief and saline intrusion in the coastal areas.

Quality of urban and industrial water supply is also adversely affected by salinity. There is already a problem in obtaining acceptable water quality for large communities in the coastal areas. Some industries, such as pulp and paper, food processing (shrimp) and textile face serious water quality problems during the dry season. For example, Khulna News Print Mill has to import fresh water by tanker during the dry season to fulfill their needs.

Coastal groundwater aquifers are badly affected by salinity. Some of these are affected by seepage from saline surface water, others by vertical and lateral migration of saline groundwater. Salinity limits the use of groundwater in agriculture, industry and for drinking purpose. The shallow groundwater aquifers in the coastal area of Bangladesh have high salinity. At present water supply wells must penetrate 250 meters or more to find water of acceptable quality (MPO, 1986). Relatively fresh potable water supply for small communities can be developed using either very shallow gravel shrouded hand tube wells or very deep hand tube wells. However, there is concern that high capacity tube wells will cause local upwelling of saline groundwater over and above the regional saline-fresh water interface.

Present Coastal Soil Conditions

About 0.850 Mha of lands in Bangladesh have been affected by soil salinity (Karim, 1990). The soil salinity generally refers to condition in the dry season. Heavy rainfall and freshwater flooding during the monsoon season generally maintain non-saline soil conditions, except where there is continuous flooding by saline water.

During the dry season, capillary movement of moisture to the soil surface concentrates salt in a layer at or near the surface. This localized concentration of salt is frequently sufficient to limit crop cultivation until there has been sufficient pre-monsoon or monsoon rainfall to dilute or leach the salt. Capillary upward movement of moisture is significant in the young Meghna estuarine floodplain and

Ganges tidal floodplain. Most of the top soils in these areas turns saline during the dry season.

New char (islands) land formed at the mouth of the lower Meghna is saline due to residual salts present in sediments, but the salt contained in the pore fluid is less than sea water as it forms during the monsoon when freshwater flushing is dominant. Subsequent leaching by monsoonal rainfall quickly reduces soil salinity which has a seasonal peak at the end of dry season (Smedema, 1984). Periodic saltwater flooding during exceptionally high tides and cyclonic storm surges affect these areas. Hoq (1989) also found a similar pattern in other coastal areas of Bangladesh.

STREAM SALINITY

Water salinities greater than 2,000 micro-mhos in the dry season affect 19,574 km^2 of lands in southwest, south-central and southeastern parts of Bangladesh. In addition about 3,800 km^2 within the polders of the Coastal Embankment Projects in the southwest and south-central parts are affected by similar level of streamflow salinity (MPO, 1986).

Streamflow salinity depends on several variables such as freshwater discharge, geometry and tide level of the basin. Freshwater discharge plays the dominant role in limiting salinity intrusion. Any reduction in the freshwater flow due to increased withdrawal for irrigation or for other purposes upstream, or a reduction in discharge due to low rain fall, causes an increase in the penetration of saline water. The saline front (defined by the 500 micro-mhos/cm iso-saline) has penetrated in the Nabaganga River as far north as Magura, 241 km from the coast. Similarly, a saline front of about 6,000 micro-mhos/cm has penetrated 173 km from the sea in the Atharobanka River to the vicinity of the confluence of the Madhumati River.

Due to reduced freshwater discharge from the Ganges in recent years following the diversion at Farraka, more areas are being affected by streamflow salinity. However, there are no quantitative data to indicate how much more lands are being affected by this increased salinity. The effects of increased salinity on the ecosystem of the Sunderbans have started to become evident through the demise of Sundri trees, retrogression of forest types, slowing of growth of forests and reduced productivity of the forest sites. Increased salinity is also affecting drinking and irrigation water supply for a larger area in southwest and south-central regions. The Gorai-Maadhumati Basin is directly affected by withdrawal of water from the Ganges River.

GROUNDWATER SALINITY

Groundwater is the primary potable water source for the coastal region. In many coastal areas of Bangladesh, groundwater in coastal aquifers is affected by salinity. There are also groundwater salinity problem in the Comilla Region. The position of the saline groundwater front is dictated by local seasonal recharge and the flow of regional groundwater from the north. If withdrawal exceeds recharge or the rate of through-flow is reduced, the imbalance in the flow system leads to saline encroachment. In the coastal areas, the fresh groundwater flow is spilt into two zones: shallow fresh groundwater lenses and deep freshwater. This is reflected by the fact that in the coastal area hand tube wells have to be drilled at shallow, 10 to 20 m, or to depths greater than 200 m.

The recharge area of deep aquifers are located inland away from the coast in the Joshore, Kushtia, Faridpur and Comilla, and may be even farther north. Reduced surface water flow or surface saline water intrusion in these areas will affect the dynamic balance between recharge, withdrawal and natural discharge. However, there are few comprehensive hydrogeological data available on these deep aquifers.

The shallow fresh groundwater above the saline groundwater aquifers is sustained by annual recharge and the depth to saline groundwater is established by the delicate balance between fresh and saline water heads. The Ghyben–Herzberg relationship shows that the balance is maintained by a ratio of 1:40, that is one meter depth of fresh water above sea level, pushes the freshwater salt water interfere to a depth of 40 below the sea level. Because the land surface seldom exceeds one to five meter in the coastal area, saline groundwater is generally found between 40 and 200 m below surface. The 1:40 relationship, if disturbed by a decrease in fresh water head or rise of sea level, may cause salinization of fresh water aquifers. For example, a seasonal lowering of the water table from +2 m elevation to +1 m elevation will cause saline water to rise from 80 m to 40 m below sea level. Similarly, rise of sea level by 1 m will cause the saline groundwater level to rise 40 m. Large scale groundwater development of coastal aquifers is not possible because it will upset the delicate balance. High capacity irrigation will cause localized upwelling of saline groundwater above the regional fresh and saline water interface. However, there are areas in the coastal zone where relatively fresh potable water supplies can be developed using either very shallow tube wells or very deep tube wells.

The groundwater salinity in the Comilla-Brahmanbria area may have been caused by saline groundwater upwelling in response to withdrawal of water for irrigation. The source of saline water may be deep seated Miocene rocks of the Bengal Basin.

Sea-Level Rise

There have been several studies on possible impact of sea-level rise on coastal Bangladesh, including Resource Analysis (1993), Milliman et al. (1989), Muhtab (1989), and Ali and Hoq (1990). Unfortunately, there are no quantified data from Bangladesh coast regarding sea level. However, considering the overall basin subsidence and average global sea level rise (Raper, this volume) a part of the Bangladesh coast line may be experiencing 2 to 3 cm/year relative sea-level rise. The immediate effect of the sea-level rise would be felt on water quality and drainage of the basin. The long-term effect would result in inundation of low lying areas by shallow seas in the coastal areas. A large part of the coastal zone will be subjected to erosion, this will include coastal embankment and infrastructure.

Some of the inland depressions in the Faridpur and Sylhet Troughs could experience expansion of standing body of fresh and brackish water (beels; shallow lakes) due to rise of water table. Another effect of the rise of sea level would be felt in the form of silting of coastal channels as the nearshore sand prism moves landward to adjust to the new hydrodynamic conditions of the continental shelf. This may result in increased intensity of floods in the coastal areas as well as in the inland. The salt water intrusion zone also would expand, affecting dry season water quality farther inland (Fig. 8). The situation is expected to be aggravated by water resources development in the Ganges–Brahmaputra River Basin. Because of the funnel shaped nature of the Bay of Bengal, each centimeter rise of sea level would result in further intensification of storm-generated tidal waves and the impact would be felt further inland. The present limit of tide (Fig. 8) is also expected to expand, which may result in further reduction of discharge of flood water to the Bay of Bengal.

In short, with changes in climate and associated sea-level rise, coastal areas could be affect in several ways:

1. Sea-level rise will cause a change in the pattern of inundation by saline and brackish water. A large area will be subjected to coastal erosion by waves, tides and coastal currents.

2. Recharge to groundwater will be affected and freshwater aquifers will be contaminated by saline water.

3. Change in major river discharges will result in a change in saline water intrusion into these rivers.

4. Change in precipitation-evapotranspiration pattern will cause a change in the dilution of the salt in the root zone during the dry season.

5. With higher sea level more areas are likely to be affected by storm surges, and fresh water lakes, ponds and aquifers will be affected by saline and brackish water. Sea-level rise will cause an increased saline upward seepage and will raise the saltwater-freshwater interface in coastal aquifers. With higher sea level more areas will be affected by soil salinity.

Muhtub (1989) suggested that sea-level rise will cause salinization of the drinking water supply of Dhaka City. Broadus et al. (1986) suggested that approximately 9% population if Bangladesh would be directly affected by a 1 m rise of sea level and 27% of the total population would be affected by 3 m rise. The densely populated areas around the southwestern city of Khulna, where population density exceeds 2,900 per sq. km., stands barely 1 m above sea level (Fig. 7). Similarly, the heavily populated environs of Meghna River would clearly be adversely impacted by a 3 m rise of relative sea level. About 11% of the agricultural GDP is contained within the area to be affected by a 1-m rise in relative sea level (Milliman et al., 1989).

Conclusions

Most of the Bengal Basin is subsiding, due largely to tectonic subsidence and isostatic readjustment of the crust. There is no evidence to suggest that the extraction

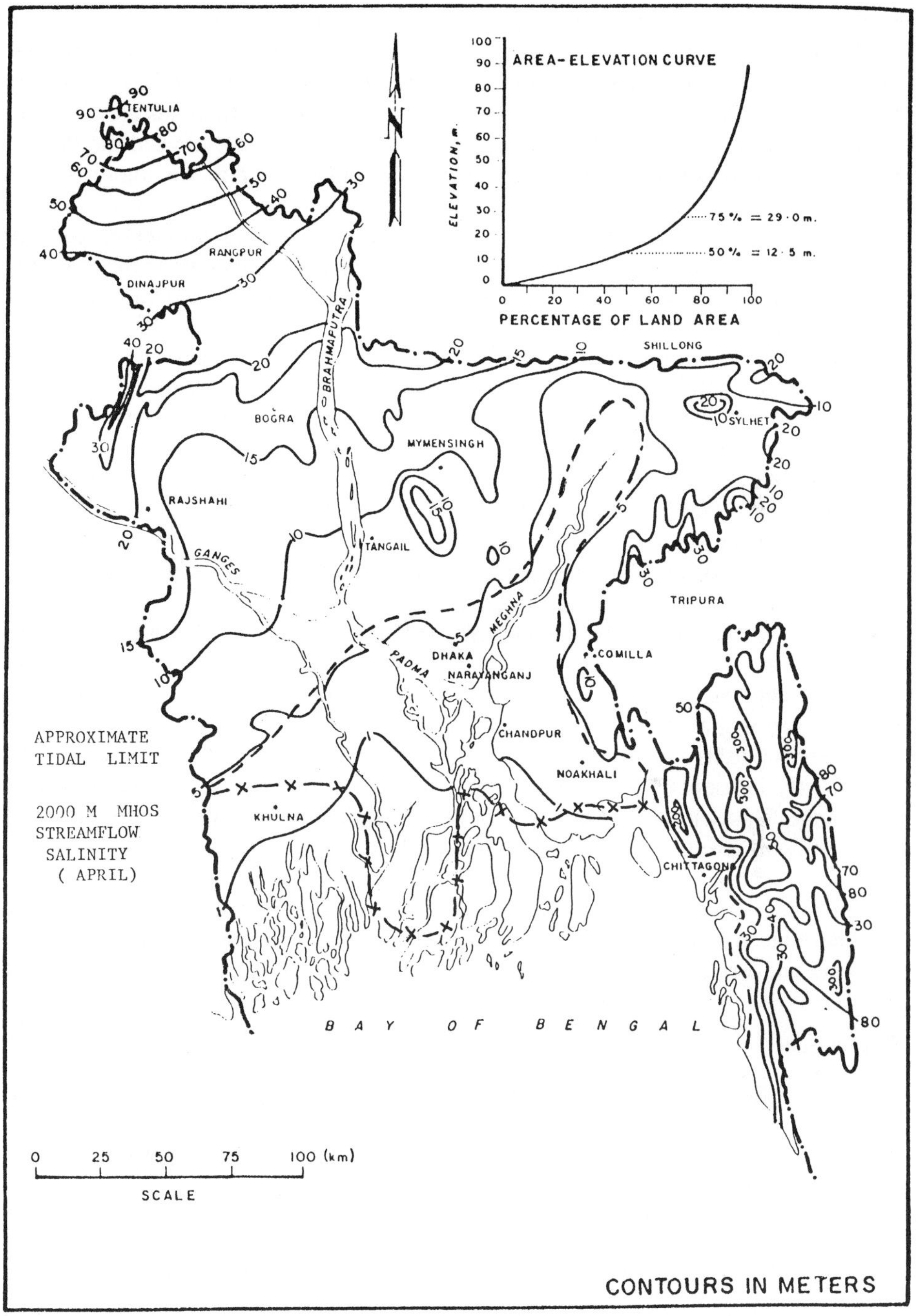

Fig. 8. Generalized relief map showing contours (in meters) of approximate tidal limit and 2,000 micro-mhos/cm stream-flow salinity (April). Source of data: Master Plan Organization, Bangladesh.

of groundwater has played a major role in the subsidence of the basin. However, compaction of peat deposits may cause increased local subsidence.

Many coastal embankment and polder projects in Bangladesh did not consider the subsidence component of the basin, and thus altered the delicate balance between sedimentation and subsidence. The altered coastal processes which affect tidal flow, fluvial flow, coastal circulation, cyclic pattern of deltaic sedimentation, have resulted in serious water logging and drainage problem for many project areas.

Proposed water resource projects o the Ganges and Brahmaputra River basin (in India and Nepal) would also alter sediment-water discharge patterns in the drainage basin areas, and may influence the delicate physical and ecological balance of coastal areas of Bangladesh.

The rate of subsidence varies considerably and depends on local geological conditions. Around Dhaka City the average rate of subsidence is about 0.65 mm/year, whereas in the Sylhet Trough the rate may exceed 20 mm/year and at Hatiya Trough it may over 30 mm/year. Quantitative data on the rate of subsidence obtained by instrumental measurements or geodetic surveys are lacking in Bangladesh.

Coastal areas undergoing rapid subsidence will be impacted more severely by the sea-level rise. In some of these areas relative sea-level rise will be much higher than the global average. These areas are likely to undergo more rapid erosion than the shoreline not being affected by subsidence.

Sea-level rise cannot be quantified for Bangladesh coast. However, due to subsiding nature of the basin and dense population of the area, the effect could be severe. Recent accelerated sea-ward growth of the Meghna Delta possibly reflects increased erosional rates in the Himalayas.

Salinity patterns also will be affected. Stream water salinity and soil salinity are directly influenced by stream flow, gradient of land, coastal circulation and tidal range. Upstream water resource development is reducing the stream flow and is adversly affecting a large area by increased salinity.

Sea-level rise is also expected to change the sedimentation patterns of the coastal areas of Bangladesh and indirectly affect the rate of subsidence in the area.

Sea-level rise and accompanying subsidence may cause serious problems for drinking and irrigation water supply over large areas.

References

Alam, M., 1972, Tectonic Classification of Bengal Basin. *Geol. Soc. Am. Bull.*, 83(2), 519-522

Alam, M., 1989, Geology and Depositional History of Cenozoic Sediments of the Bengal Basin of Bangladesh. *Paleogeogr., Paleoecol., Paleoclimat.*

Ali, S.I. and Hoq, S.,1990. International Sea-Level Rise: National Assessment of Effects and Possible Responses for Bangladesh. In: National Assessments of Sea-Level Rise Vulnerability, S.P. Leatherman (ed.). College Park, Center for Global Change, University of Maryland.

Bakthtine, M. I., 1986, Major Tectonic Features of Pakistan. Eastern Province. Science and Industries Pakistan 4(2), 89-100.

Banerji, B.K., 1984, Post-Eocene Biofacies, Paleoenvironments and Paleogeography of the Bengal Basin, India. *Palaeo., Palaeo., Palaeo.*, 45, 49-74.

Broadus, J., Milliman, J., Edwards, S., Aubrey, D., and Gable, F., 1986. Rising Sea Level and Damming of Rivers: Possible Effects in Egypt and Bangladesh In Effects of Changes in Stratosphereic Ozone and Global Climate, Vol 4, Sea-level Rise P 165-189, UNEP and USEPA.

Coleman, J. M., 1981. *Deltas Processes of Deposition and Model for Exploration.* Burgess Publishing Co., Minneapolis, USA, 125 p.

Costa, J. E., and Baker, V. R., 1981. *Surficial Geology Building with the Earth*, John Wiley and Sons, New York.

Curray, J. R., and Moore, G. D., 1971 Growth of Bengal Deep Sea Fan and Denudation of the Himalayas. *Geol. Soc. America. Bull.*, 82, 563-572.

Curray, J. R., and Moore, G. D., 1974. *Sedimentary and Tectonic Processes in the The Geology of Continental Margin*, Springer-Verlag, New York, p. 617-627.

Gansser, A., 1983. *Geology of the Bhutan Himalaya.* Birkhuse-Verlag, Basel, 181 pp.

Graham, S. A., Dickinson, W. R., and Ingersoll, R. V., 1975. Himalyan-Bengal Model for Flysch Dispersal in the Appalachian-Ouchita System. *Geol. Soc. Am. Bull.*, 86, 273-286.

Guha D. K., 1978. Tectonic Framework and Oil and Gas Prospects of Bangladesh. Bangladesh Geol. Soc. 4th Annual Conf. Proc. Dhaka, p. 65-76.

Haq, B.U. et al., 1987. Chronology of fluctuating sea level since the Triassic. *Science*, 235, 1156-1166.

Hoq, S., 1989. Land Reclamation from the Sea in Bangladesh. Coastal Zone'89, Proceedings of the Sixth Symposium on Coastal and Ocean Management, Washington, DC: American Society of Civil Engineers. p. 1602-1608.

Hoque, Monirul, Jabbar, M.A., Khan, Shamsuddin and Islam, Shaheedul, 1985. Land Accretion and land Reclamation of The Coastal Areas of Bangladesh in Development of Coastal Areas of Bangladesh, Bangladesh Planning Commission's Report , p 17 -23.

Huh, O. K., Ali, A., and Quadir, D. A., 1985. Use of NOAA Satellite AVHRR Data to Monitor River Flood Hydrology in Bangladesh, Project Sponsored by UNDP and FAO, Under the Applied Remote Sensing Project (Phase II) of SPARRSO (Bangladesh), Published by Coastal Studies Institute, Louisiana State Institute, Baton Rouge, Louisiana, USA.

Jones, P. H., 1985. Geology and Groundwater resources of Bangladesh. P.H. Jones Hydrogeology Inc., Baton Rouge, Louisiana, USA, November, 1985.

Karim, Z, S. G. Hussain and M. Ahmed, 1990. Salinity Problems and Crop Intensification in The Coastal Region of Bangladesh, Soil and Irrigation Division, Bangladesh Agriculture Research Council.

Khan, M. A. Maroff, 1980. A Brief Account of Geology and Hydrocarbon Exploration in Bangladesh. Offshore South Asia Conf. February 1980, SEPEX Session, 6pp.

Khandoker, R. A., 1987. Origin of Elevated Barind-Madhupur Areas, Bengal Basin: Results of Neotectonic Activities. *Bangladesh Jour. Geol.,* 6, 1-9.

Master Plan Organization (MPO), Bangladesh, 1986. National Water Plan. December 1986, Three Volumes. Ministry of Irrigation Water Development and Flood Control, Government of Bangladesh, Dhaka.

MPO, 1985. Geology of Bangladesh, MPO Technical Report No 4. Master Plan Organization. Ministry of Irrigation, Water Development and Flood Control, Dhaka, Bangladesh.

Milliman,J.D., Broadus, J.M. and Gable, F., 1989. Environmental and Economic Implications of Rising Sea Level and Subsiding Deltas: the Nile and Bengal Examples. *Ambio*, 18, 340-345.

Milliman, John D., and Meade, Robert H., 1983. World-Wide Delivery of River Sediments to the Oceans. *J. Geology*, 91, 1-21.

Monsur, M.H., 1990. Stratigraphical and Palaeomagnetical Studies of some Quaternary Deposits of the Bengal Basin, Bangladesh, PhD Thesis (unpub.), Vrije University, Brussels, Belgium.

Morgan, J.P., and McIntire, W.G., 1959. Quaternary Geology of the Bengal Basin, East Pakistan and India, *Geol Soc.. Amer. Bull.*, 70, 319-342.

Moshin-Uddin, Md., and Islam, S., 1982. Polder Development in Bangladesh Paper I: Past and Present Development. In: *Polders of the World* International Institute for Land Reclamation, Netherlands, p. 288-295.

Muhtab, F.U.,1989. Effect of Climate Change and Sea-level Rise on Bangladesh. Report prepared for the Commonwealth Expert Group on Climate Change and Sea-level Rise. Commonwealth Secretariat, London.

Nishat, A, and Hoque, M., 1985. Sedimentation in Coastal Areas of Bangladesh. *Journal of the Institution of Engineers, Bangladesh*, 13(2-3), 41-49.

Rasid H. and Paul, K.B.,1987. Flood Problems in Bangladesh: Is There an Indigenous Solution? *Environmental Management*, 11, 155-173.

Resources Analysis 1993. Assessment of the Vulnerability of Coastal Areas to Climatic Change and Sea-level Rise; A Pilot Study of Bangladesh, Final Report 2 Vols. Prepared for Intergovernmental Panel on Climate Change (IPCC), by Resources Analysis, Bangladesh Centre for Advanced Studies and Approtech, Financed by Government of Bangladesh, Ministry of Environment and Government of The Netherlands, Ministry of Foreign Affairs.

Rogers, P., Lydon, P. and Seckler, D., 1989. Eastern Water Study: Strategies to Manage Flood and Drought in the Ganges–Brahmaputra Basin. Prepared for US AID by the Irrigation Support Project for Asia and Far East, Washington D.C., 83 pp.

Salahuddin, M. M., 1991. Subsidence in Dhaka City. Unpublished Thesis, University of Dhaka, in Partial Fulfillment of the Requirement for the Degree of Master of Science in Geology.

Sengupta, S., 1966. Geological and geophysical studies in Western part of Bengal Basin, India. *Am. Assoc. Petrol. Geol. Bulll.*, 50, 1001-1018.

Smedema, L. K., 1984. Land Reclamation on Chor Baggar Dona - Drainage and Salinity Investigation 1983-1984, BWDB Land Reclamation Project, Technical Report No. 21.

Solomon, S.,1992. Floods in Bangladesh: Is There a Solution? *Ecodecision*, 6, 83-85.

Sorkhabi, R. B., and Stump, Edmund, 1993. Rise of the Himalaya: A Geochronologic Approach, *GSA Today*, 3, 85, 88-90.

Umitsu, M., 1985. Natural Levees and Landform Evolution in the Bengal Lowland. *Geogr. Rev. Japan*, 58, 149-164.

Umitsu, M.O., 1993. Late Quaternary Sedimentary Environments and landforms in Ganges Delta. *Sediment. Geology*, 83, 177-186.

Vishnu-Mittre and Gupta, H. P., 1970. Pollen Analytical Study of Quaternary Deposits in Bengal Basin., 19, 297-306.

WASA, 1991. Dhaka Regional Ground Water and Subsidence Model, Dhaka Water and Sewerage Authority, Dhaka, Bangladesh.

World Bank, 1989. Bangladesh Action Plan for Flood Control, Document of the World Bank, Asia Region Country Department I, World Bank, Washington. D.C., 91 pp.

Zaher, M. A., and Rahman, A. 1980. Prospects and Investigation for Minerals in the Northern Part of Bangladesh. Petroleum and Mineral Resources of Bangladesh, Seminar and Exhibition, Oct. 1980, Dhaka, p. 9-18.

CHAPTER 9

Tectonic, Eustatic and Isostatic Changes along the Indian Coast

K.R. Subrahmanya

ABSTRACT: The present shape of India is due to a combination of events that took place over geological time, including the breaking away of the Indian Plate from Australia and Antarctica, rifting and drifting between India and Madagascar along the west coast, and collision of the plate with the Asian continent. Following these tectonic events, eustastic changes have played a major role in modifying the east and west coasts of India. Another factor responsible for both past and present regional changes is isostatic movement, resulting in uplift and subsidence, especially in deltaic areas.

Introduction

The Indian peninsula acquired its present configuration in several stages. Before the breakup of Gondwanaland, India had a coastline to the north facing the Tethys Sea. With the breakup of India and Madagascar from Australia and Antarctica during the late Jurassic or early Cretaceous period (140–120 m.y. ago), an east coast was added to the coast line. This is illustrated by the occurrence of marine intercalations in the upper Gondwana sequences now outcropping at Sivaganga, north of the Gulf of Mannar. The event was followed by fully marine sedimentation during the Cretaceous. The Cretaceous sequence at Tamil Nadu reveals that from west to east stratigraphically younger horizons are found (Ramanathan, 1968). This may indicate any or all the three possibilities: 1) a gradual upward shift of the east coast resulting in withdrawal of the sea; 2) the lowering of the sea level; or 3) the expansion of land due to deltaic progradation.

In the middle Cretaceous, the western part of the peninsula, which consisted of India and Madagascar, started doming up along an axis parallel to the present west coast (Subramanian, 1987; Subrahmanya, 1987). The doming was succeeded by rifting around 70 m.y. ago. This was also the period of intense volcanic activity as exemplified by the occurrence of Deccan basalts in the Konkan region. Rifting gave way to drifting of Madagascar from India, resulting in a unique ocean basin structure. Naini and Talwani (1982) have suggested that the Arabian Sea consists

J.D. Milliman and B.U. Haq (eds.), Sea-Level Rise and Coastal Subsidence, 193-203.

of two tectonically distinct basins — an eastern basin with a transitional or rift stage crust and a western basin with oceanic crust. Underlying the Tertiary rocks on the eastern basin one finds Deccan basalts to the north and Precambrian rocks to the south (Closs et al., 1974).

The tectonic activity which gave rise to the east and west coasts during the Mesozoic also resulted in the disappearance of the northern coast of India due to the collision of the Indian Plate with the Asian mainland. The initial contact between India and Asia occurred in the late Paleocene, approximately 50 Ma ago (Curray and Moore, 1974), resulting in more or less the present Indian coastline. Subsequent changes to the coastline have occurred through a combination of factors that include sedimentation, sea-level changes, isostasy, marine erosion, and littoral transport. Figure 1 shows the nature of the coastline in different parts of India.

Quaternary Eustatic Changes

West Coast

Available data for the early Pleistocene sea level along the west coast indicate that around 220,000 years before present (BP) the sea level was higher by 55 to 60 m than at present (Verma and Mathur, 1979). By dating relict corals, Gupta (1972) and Gupta and Amin (1974) concluded that the sea levels on the Saurashtra coast stood higher by 2 to 6 m around 120,000 years BP. Data on the changes in sea level between 120,000 years and 11,000 years BP are not available, a period when sea level was substantially lower than at present. Agarwal et al. (in Merh, 1987) are of the opinion that the sea level was much lower 35,000 years BP and began to rise around 30,000 years ago.

The sediment distribution map of the Arabian Sea (Geological Survey of India, 1975) shows a linear band at water depths of 75 to 100 m. This sand body could represent an ancient beach and shallow marine environment, which would indicate a lowering of sea level all along the west coast.

Nair (1975) recognized several well-defined terraces off Ratnagiri at −65, −75, −85, −95 m. Radiocarbon dating of the samples from these terraces has shown their ages ranging from 9000 to 11,000 years BP. This finding agrees well with the global picture in which sea level was at its lowest (−100 to −150 m) between 15,000 and 18,000 years BP following the last glacial maximum. Therefore, the terraces and the sand body must have formed when the sea was transgressing. During this transgressive phase, sea level rose rapidly at about 1 cm per year until 6000 years ago (Hashimi and Nair, 1988). There is some evidence to suggest that the sea level stood slightly higher than at present some 6000 years ago. For example, relict corals exposed along the Saurashtra coast (Gupta, 1972; Gupta and Amin, 1974; Somayajulu et al., 1985), oyster beds in Goa coast, as well as data presented by Verma and Mathur (1979). These deposits point to a sea level ca. 2 m above persent some 6000 years ago.

The sea appears to have switched from a transgressive to a regressive phase 6000 to 3500 years BP. A marine archaeological survey conducted off the Saurashtra coast by Rao (1987) revealed the existence of a submerged port city of Dwaraka

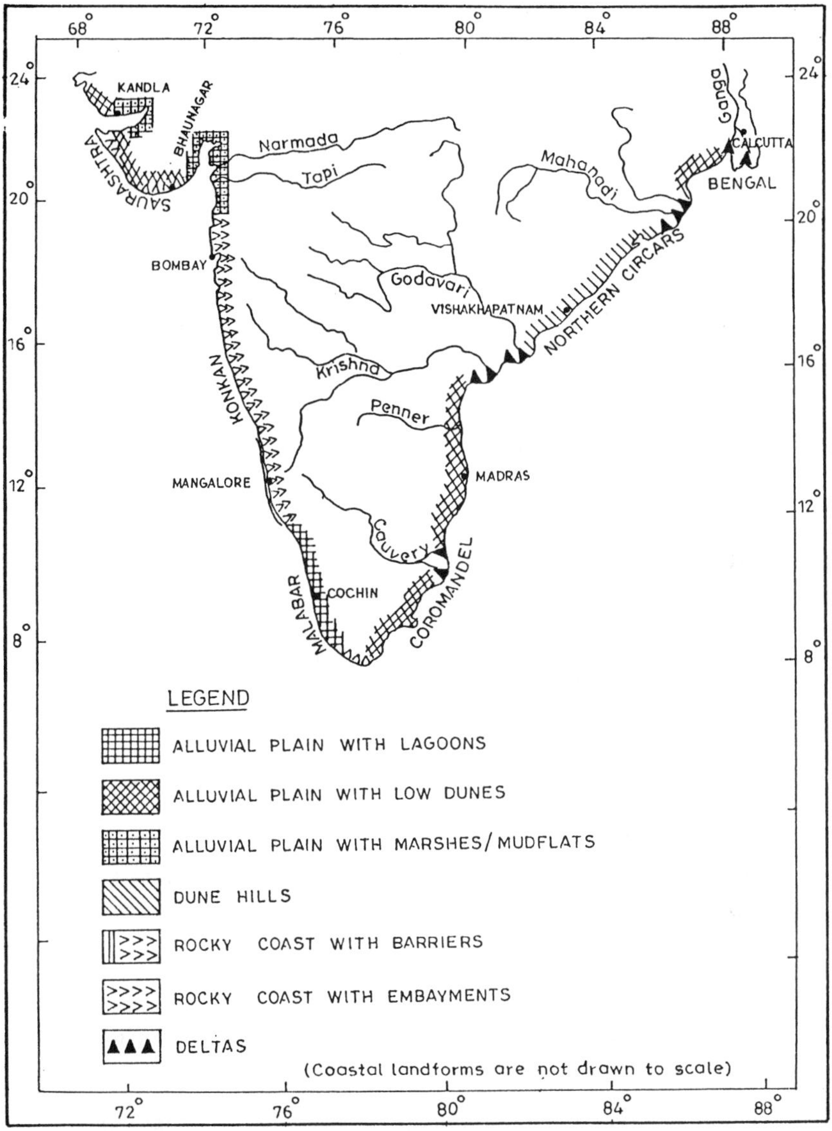

Fig. 1. Nature of coastline in different parts of India (Modified from Coastal Landforms of India, GSI, 1972).

at a depth of 5 m. This city, which is considered to be part of late Harappan culture, was inhabited up to 1400 B.C. (3400 years BP), suggesting a rise of 5 to 6 m in sea level in this area during the last 3300 years (Rao, 1987). However, Agarwal et al. (in Merh, 1987) argue that there was no fall after 6000 years BP, and that, on the contrary, there was an uplift of the margin of 1 to 6 m between 6000 and 2000 years BP (Fig. 2).

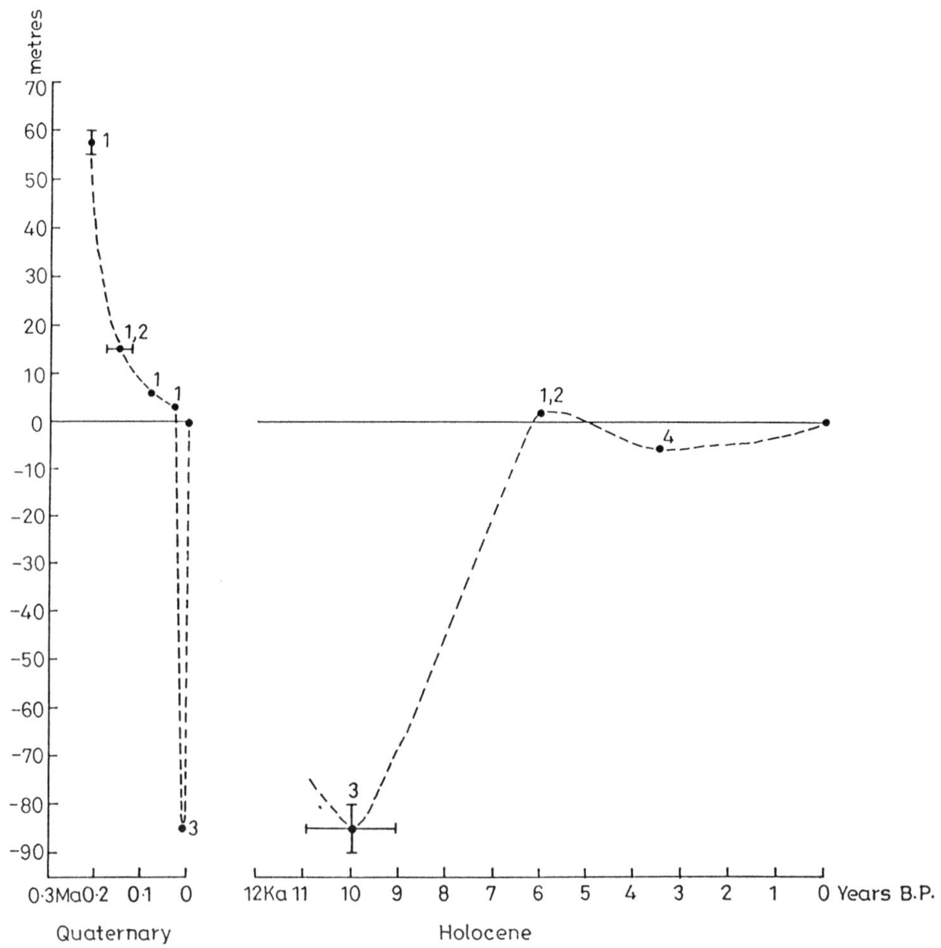

Fig. 2. Sea-level changes along the west coast of India during Quaternary and Holocene (Sources of data: 1. Verma and Mathur, 1979; 2. Somayajulu et al., 1985; 3. Nair, 1975; 4. Rao, 1987)

Because the data available are scanty and unevenly distributed, the validity of extending such results to the entire west coast is questionable. However, three lines of evidence suggest that the entire west coast may have acted as a stable block, with very little subsidence or uplift during the Quaternary: 1) the occurrence of linear sand bodies in the Arabian Sea along the coast at more or less uniform depths; 2) the presence of relict higher strand lines nearly at the same elevation in geographically-distant locations; and 3) the occurrence of off-shore terraces at uniform depths in several places.

Verma and Mathur (1979) are also of the opinion that the marine transgressions and regressions along the west coast are not due to tectonic rise and fall of the land, but are mainly of eustatic origin.

EAST COAST

In comparison with the west coast, east coast sea-level data are scanty. Venkataraman and Rangaraju (1968) noted the existence of Quaternary sediments in the Cauvery deltaic region of Tamil Nadu, and in the Kolanka-Masulipatnam region of Andhra Pradesh. This indicates that during the early part of the Pleistocene sea level was higher to be followed by a lowering of sea level. Based on the existence of oolitic sands at depths of 90 to 180 m off Visakhapatnam, Subba Rao (1964) concluded that these were relict deposits that formed during low seastand. Such deposits are scattered unevenly across the eastern shelf of India at the mouths of major rivers. Because similar deposits are found along the western shelf, one can assume that these deposits formed around 11,000 years BP. Prudhvi Raju and Vaidyanadhan (1981) have recorded a −23 m level in Visakhapatnam, which they consider a paleo sea-level that prevailed from 7000 to 8000 years BP. Subsequent to these low seastands, sea level rose to +2 to +5 m, as evidenced by paleo coral-reefs in the Rameshwaram peninsula. Again, as with the west coast (Saurashtra), these reefs appear to have been formed around 6000 years BP. Based on their investigation of subsurface Holocene sediments around Calcutta, Banerjee and Sen (1985) indicated relict mangrove forests dating back to 6000 to 7000 years BP. No definite indicators are available regarding subsequent fluctuations in sea level.

Holocene Uplift and Subsidence

WEST COAST

The nature of drainage around 13_N along the Mulki-Pulicat Lake axis (Fig. 3) indicates the presence of a major water divide separating the north and south flowing rivers. This divide extends along the coasts, both of which have Quaternary landforms (Subrahmanya, 1994). The coasts near Mulki along the west coast and near Pulicat Lake along the east coast prograde outwards, indicating coastal accretion (Ramasamy, 1989). There are other indications of uplift, such as elevated oyster reefs, beach ridges, highly fragmented lateritic terrain, and the shallowing of nearshore waters. Tide gauge data (discussed below) further demonstrates uplift.

Subsidence is seen to the north of 15° N in the form of cliffed coastline. It is difficult to assign a definite age to this feature.

EAST COAST

Mahmood Alam (this volume), has noticed that the entire Bengal basin is slowly subsiding due to isostatic adjustment and dehydration of sediments. Based on the log data from the Hazipur well, he suggests a rate of subsidence of about 2.2 cm/year. However, this figure may only have local significance since the evidence along the Indian coast indicates that the sea level 11,000 years BP was lower by 100 m; thus net Holocene subsidence would be 144 m, or 1.3 cm/year. This figure for subsidence agrees closely with the Holocene subsidence near Calcutta — 150 m (Ahmad, 1972b).

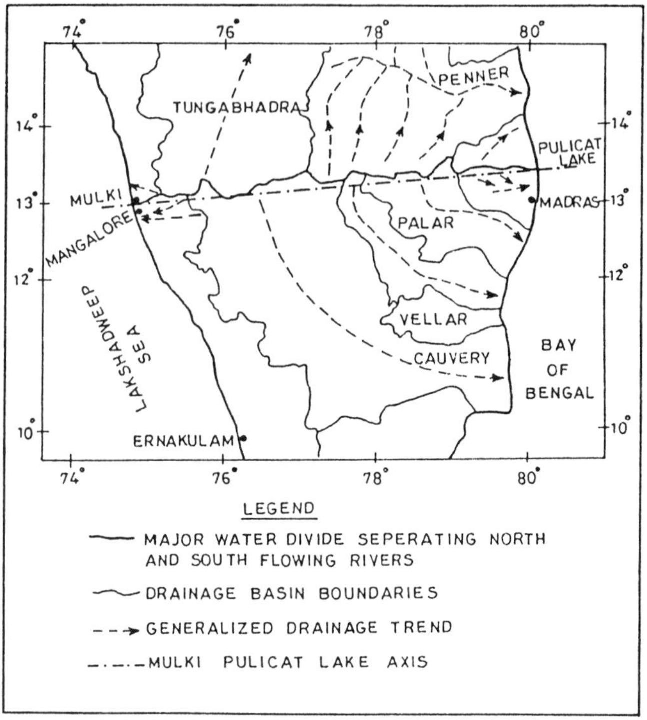

Fig. 3. Generalized drainage map of Southern peninsula. A major divide along the Mulki-Pulicat Lake axis separates the northward and southward flowing rivers. The divide is an active zone of uplift that causes river channel migration. The region also displays an anomalous gravitational field.

Venkataraman and Rangaraju (1968) presented evidence to show Mahabalipuram and the city of Pumpuhar in Tamil Nadu have undergone submergence whereas some areas along the Ramnad and Tirunelveli coasts have emerged recently. Similarly, they found evidence for subsidence of the Nellur-Gudur coast and the emergence of Godavari-Krishna deltaic area along the Andhra Pradesh coast. Other than these general observations, there are no specific data indicating the extent of uplift or submergence along the east coast. The picture in the east coast, however, may be highly complicated owing to the following factors:

- Almost all of the peninsula's discharge enters the Bay of Bengal, and thus the major rivers Cauvery, Krishna, Godavari, Mahanadi, Ganga, and Brahmaputra bring an enormous quantity of sediment that prograde seaward. These sediments are distributed by strong littoral currents, which are estimated to transport many millions of ton of sediment northward annually.
- Most of the deltaic regions may be undergoing active subsidence that will accelerate in the future. This is because the deltaic regions are rich agricultural lands that are being extensively cultivated. Since the outflow in the rivers has been reduced due to construction of upstream dams, ground water is being

mined at an increasing pace. In addition, all the deltaic areas and their offshore extensions, rich in oil and gas, to a large extent are being exploited, which is likely to increase in the near future.

These factors contribute to the subsidence of poorly consolidated sediments. Also, cyclonic storms, to which the east coast is normally subjected, may increase in frequency and cause extensive damage to the life and property along the east coast. Consequently, there is a real need to closely monitor the vertical movement along the east coast and take precautionary measures to minimize any adverse effects.

Most Recent Sea-Level Changes along the Indian Coast

Tide gauge data for the Indian stations collected by the No. 68 (Tidal) Party, Geodetic and Research Branch, and by the Survey of India, have been obtained by permission of the Surveyor General of India, through the Permanent Service for Mean Sea Level, Bidston Observatory, U.K. The data have been presented in Figures 4 and 5.

West Coast

The graphs for the west coast stations do not show any consistent pattern. Three stations, Kandla, Bombay, and Cochin show an increase in sea level, whereas two stations, Bhaunagar and Mangalore, show a lowering of the sea. Kandla, situated close to Indus Delta; Bombay, a highly industrialized and urbanized centre; and Cochin, another industrialized city adjacent to a lagoon, all may be undergoing subsidence. Nagar (1988) has presented evidence that the land is subsiding in the Kandla region. No precise information is available on the stability of bench marks situated close to other tide gauge stations. The pattern for Bhaunagar may indicate either a stable or a slightly rising trend for the coast. The data for Mangalore, however, reflects uplift of the coast, as discussed earlier. Nagar (1988) has expressed the opinion that on the average, there is an increase of 1 mm/year in the sea level along the Indian coast.

East Coast

The graphs for the east coast stations indicate a relative rise in sea level, except at Madras. The rise is more significant in the north, near the Bengal Delta than in the south. This apparent rise in part may be due to subsidence of the coast rather than a net increase in global sea level (Alam, this volume; Ahmad, 1972b).

Conclusions

A major part of the Indian peninsula is a shield area, which makes for fairly stable coastline. Precise data for sea-level fluctuations, although limited, indicate that the Quaternary sea-level changes have been mostly eustatic. Around 11,000 years

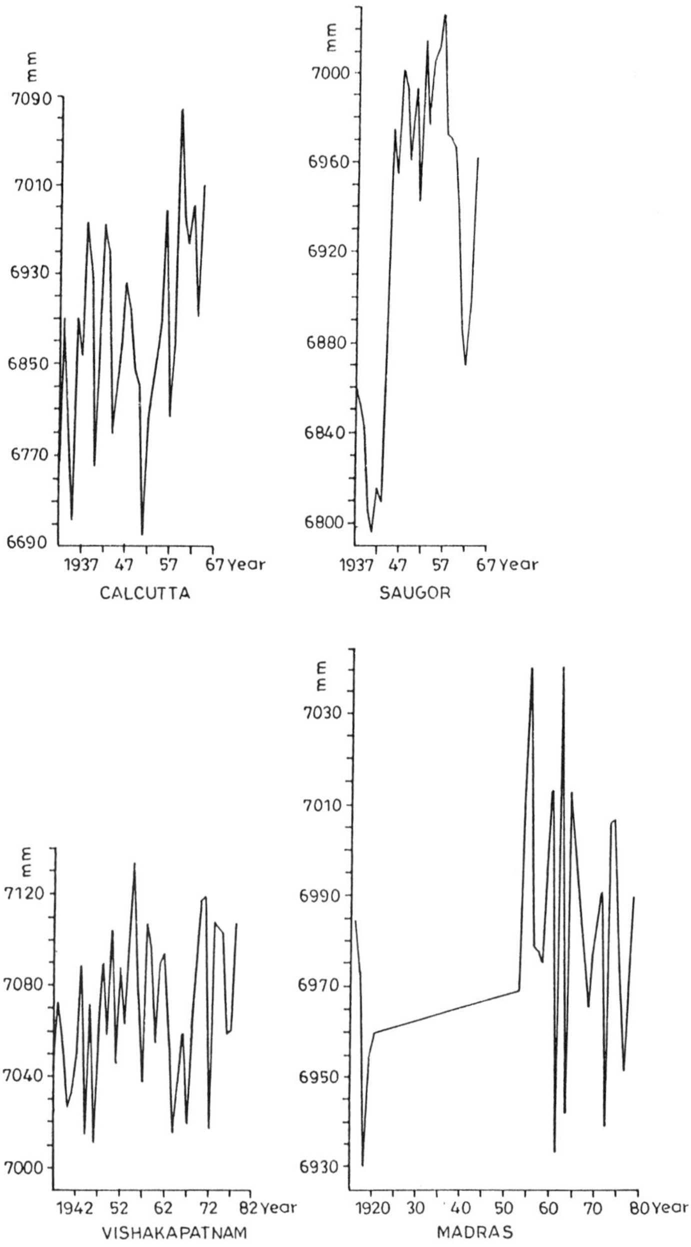

Fig. 4. Plot of the tide gauge records of the east coast of India.

BP sea level was at approximately −100 m, followed by a rapid rise up to 6,000 years BP, when it attained a level about 2 m above present. Between 6,000 years BP and the present, there was another oscillation of 5 to 6 m in the sea level. All these events have been noticed along the west and east coasts, though the available

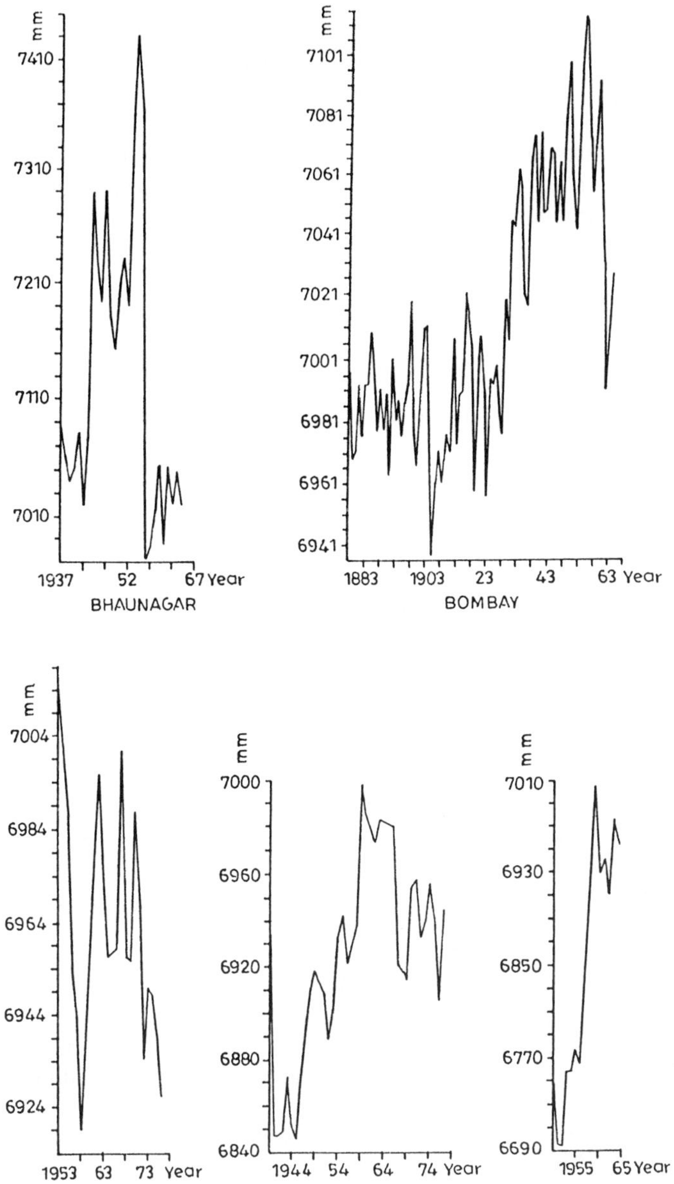

Fig. 5. Plot of the tide gauge records of the west coast of India.

data for the east coast are meager. Evidence in the Bengal Delta point out that the region is subsiding at a rate of 1.3 mm/year. The tide gauge data for the past few decades indicate that the land is subsiding along the Bengal and Saurashtra coasts. Minor subsidence is noticed around much of the rest of the Indian coastline, and this in part may be due to a global rise in sea level, which is estimated to be 1 to

1.5 mm/year. The only exception is in the region of 13° N latitude where there may be a relative fall in sea level due to an uplift of land at a rate of about 2 mm/year.

Preventive and remedial measures to counteract the problems resulting from subsidence and sea-level rise include controlling the overexploitation of ground water and hydrocarbons, reduction in the discharge of greenhouse gases, construction of coastal engineering structures based on cost-benefit ratio, and the redesigning of upstream reservoirs.

Acknowledgements

I am indebted to Dr. John Milliman for the invitations to the Bangkok workshop and to prepare this paper. SCOPE provided grants with allowed me to attend the workshop, and, as a result, my view on the topic of sea-level changes has been greatly enhanced. I benefited much from discussions with the participants of the workshop and from the help and encouragement my fellow scientists in India. My sincere thanks to all of them.

References

Ahmad, E., 1972a. *Coastal Geomorphology of India.* Orient Longman.

Ahmad, E., 1972b. Evidences of emergence and submergence on the Indian coast. In: K.R.Dikshit (ed.) *Contributions to Indian Geography, vol II - Geomorphology.* Heritage Publishers, 284-293

Alam, M., 1995. Subsidence of the Ganges-Brahmaputra Delta of Bangladesh, and Drainage, Sedimentation and Salinity Problems. (This volume).

Banerjee, M. and Sen, P.K., 1985. Occurrence of plants, mollusc shells and micro-foraminifera of mangrove ecology from Holocene sediments of Bengal basin. *Proc. Natl. Symp. Biol. Util. Cons. Mangroves*, 393-397.

Closs, H., Narain, H., and Garde, S.C., 1974. Continental margins of India. In: C.A. Burke and C.L. Drake (eds.), *The geology of Continental Margins*, Springer- Verlag, 629-639.

Curray, J.R. and Moore, D.G., 1974. Sedimentary and tectonic processes in the Bengal deep-sea fan and geosyncline. In: C.A. Burke and C.L.Drake (eds.) *The Geology of Continental Margins*, Springer-Verlag, 617-627.

Geological Survey of India, 1972. Map of coastal landforms of India, first edition. Geol. Surv. India, Calcutta. Geological Survey of India, 1972. Bottom sediment map of the sea bordering India. Geol.Surv. India, Calcutta.

Gupta, S.K., 1972. Chronology of raised beaches and inland coral reef of Saurashtra Coast. *J. Geol.*, 80, 357-361.

Gupta, S.K. and Amin, B.S., 1974. Io/U ages of corals from Saurashtra coast. *Mar. Geol.*, 16, 79-83.

Hashimi, N.H and Nair, R.R., 1988. Observed and predicted Holocene sea-level rise: A comparison (Abstract). In: Sea-level Rise due to Greenhouse Effect: Implications for India. Natl. Inst. Ocean. India.

Merh, S.S., 1987. Quaternary sea-level changes: The present status vis-a-vis records along coasts of India. *Indian J. Earth Sci.* , 14, 235-251

Nagar, V.K., 1988. Monitoring the sea-level variation along the Indian coast. Paper presented at the Workshop on Sea-level rise due to Greenhouse Effect: Implications for India. Natl. Inst. Ocean. India.

Naini, B.R. and Talwani, M., 1982. Structural frame work and the evolutionary history of the continental margin of Western India. In: J.S. Watkins and C.L. Drake (eds) Studies in Continental Margin Geology. Am. Assoc. Petrol. Geol. Mem. 34, 167-192.

Nair, R.R., 1975. Nature and origin of small scale topographic prominences on the western continental shelf of India. *Indian J. Mar. Sci.*, 4, 25-29.

Prudhvi Raju, K.N. and Vaidyanadhan, R., 1981. Geomorphology and evolution of Sarda river basin, Visakhapatnam dist., *A.P. Annals. Natl. Assoc. Geog.* , 1, 1-17.

Ramanathan, S., 1968. Stratigraphy of Cauvery basin with reference to its oil prospects. In: *Cretaceous-Tertiary Formations of South India.* Geol. Soc. India Mem. 2, 153-167.

Ramasamy, S.M., 1989. Morpho-tectonic Evolution of East and West Coasts of Indian Peninsula. Geol. Surv. India Spl. Pub. No.24, Arabian Sea Seminar, Mangalore. 333-339.

Rao, S.R., 1987. Marine archaeological explorations off Dwaraka, northwest coast of India. *Indian J. Mar. Sci.* , 16, 22-30.

Royer, Jean-Yves; Sclater, J.G.; and Sandwell, D.T., 1989. A preliminary tectonic fabric chart of the Indian Ocean. *Proc. Indian Acad. Sci.* (Earth and Planetary Sciences), 98, 7-24.

Somayajulu, B.L.K.; Broecker, W.S.; and Goddard, J., 1985. Dating Indian corals by U-decay series methods. *Quaternary Res.*, 24, 235-239.

Subba Rao, M., 1964. Some aspects of continental shelf sediments of the east coast of India. *Mar. Geol.*, 1, 59-87.

Subramanian, K.S., 1987. possible cause of easterly tilt of the southern part of India. *J. Geol. Soc. India*, 29, 362-363.

Subrahmanya, K.R., 1987. Origin of the Western Ghats, India – A simple model. in, Coastal Geomorphology of India, Spl. Issue, *J. Geol. Soc. India*, 29, 446-449.

Subrahmanya, K.R., 1994. Post-Gondwana tectonics of the Indian peninsula (in press).

Venkataraman, S. and Rangaraju, M.K., 1968. Oscillatory movements in Cretaceous-Tertiary basin of South India. In: Cretaceous-Tertiary Formations of South India. *Geol. Soc. India Mem.*, 2, 201-207.

Verma, K.K. and Mathur, U.B., 1979. Quaternary sea-level changes along west coast of India. *Geol. Surv. India, Misc. Publ.*, No. 45, 263–272.

CHAPTER 10

Effect of Sea-Level Rise and Human Activity on the Yangtze Delta, China

Ren Mei-e and John D. Milliman

ABSTRACT: The Yangtze Delta, with an area of about 40,000 km^2 and a population of about 40 million, is one of the largest and most heavily populated deltas in the world. The low-lying delta includes a considerable amount of reclaimed lands with elevations less than 2 m above mean sea level. Natural rates of subsidence throughout the delta appear to be less than 5 mm/year, but locally rates have exceeded 100 mm/year where groundwater pumping has been excessive.

Future sea-level rise will directly threaten the safety of the delta lowlands, particularly if subsidence accelerates in response to groundwater removal. The impact of rising sea level also may be exacerbated if water and sediment from the Yangtze River is prevented from reaching the East China Sea. A great part of the delta could be affected by salt-water intrusion, causing damage to agriculture, industry, water supply, and inland navigation.

The Yangtze River and its Delta

The Yangtze River is the ninth largest river in the world in terms of basin area (1.9×10^9 km^2), fifth largest in water discharge (920 km^3/year), and fourth largest with respect to sediment load (480×10^6 t/year). Much of the sediment load comes from the western part of the basin, while water input is more or less evenly distributed throughout the length of the river (Fig. 1).

With nearly 500 million people living in its drainage basin, the Yangtze may be the most human-impacted large river in the world (Fig. 2). Said another way, about one in every ten people on Earth live in the Yangtze drainage basin, and to some degree they affect and/or are affected by the river.

As recently as 2000 years ago, the subaerial portion of the Yangtze Delta was confined to a much smaller area west of the present-day shoreline, and much of the eastern part of the delta was subtidal. The northern part of the delta (Jiangsu Province) began accreting after a southern shift in the flow of the Yellow River in the late 13th century (Fig. 3A). Subsequently the northern banks of the Yangtze estuary prograded more than 100 km seaward. Thus, within historic times, much

J.D. Milliman and B.U. Haq (eds.), Sea-Level Rise and Coastal Subsidence, 205-214.

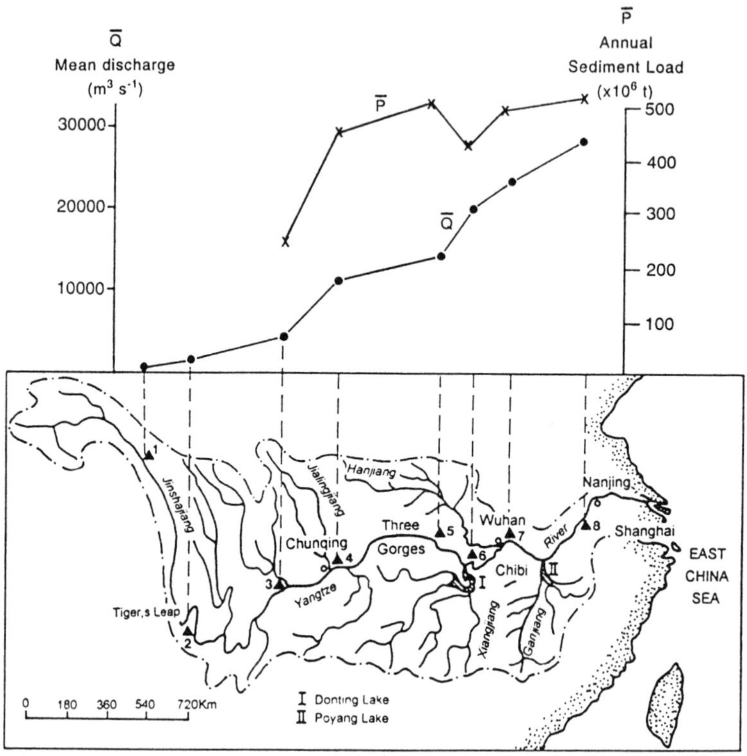

Fig. 1. Yangtze River drainage basin, showing the downstream change in sediment and water discharge. (After Milliman and Ren, 1995).

of the Yangtze Delta has been subtidal, and, in fact, the present sites of Shanghai and Nantong were both subtidal as recently as 2000 years ago. In the eastern part of the delta, a series of cheniers just west of Shanghai marks old shorelines of 6800–3200 BP (Liu et al., 1987). The low plain seaward of the chenier ridges began to be reclaimed in the 4th century (AD) by construction of three successive sea dikes (Fig. 4) (Tan, 1987).

West of the chenier ridges, the plain is low, swampy and dotted with innumerable small lakes and ponds. The plain is criss-crossed by a dense network of rivers and canals, averaging 3 km of waterway per km^2 of land. The lowest part of the plain is less than 2 m above sea level (Fig. 5) and poorly drained. Recent excessive reclamation of lacustrine flats along Tai Hu (Lake), west of the cheniers, has rendered the drainage of flood water more difficult, thus increasing the danger of delta flooding.

With an apex near Zhenjiang, the Yangtze Delta lies in a tectonically subsiding area where about 300–400 m of Quaternary sediment have accumulated on Pliocene basement. The evolution of the delta in the last 7500 years and the resultant sedimentary facies and sequences have been described in some detail by Li (1986). From the sedimentological point of view, the subaerial part of the delta occupies

Fig. 2. Map of the Yangtze Delta, showing major cities and recently reclaimed land.

about 22,800 km^2, but in terms of morphology as defined by the the 5-m contour, the delta has an area of about 40,000 km^2 and a total population of more than 40 million (as of 1990). To the north, the delta merges with the wide North Jiangsu coastal plain, formed primarily by sediment influx from the Yellow River since the 12th century.

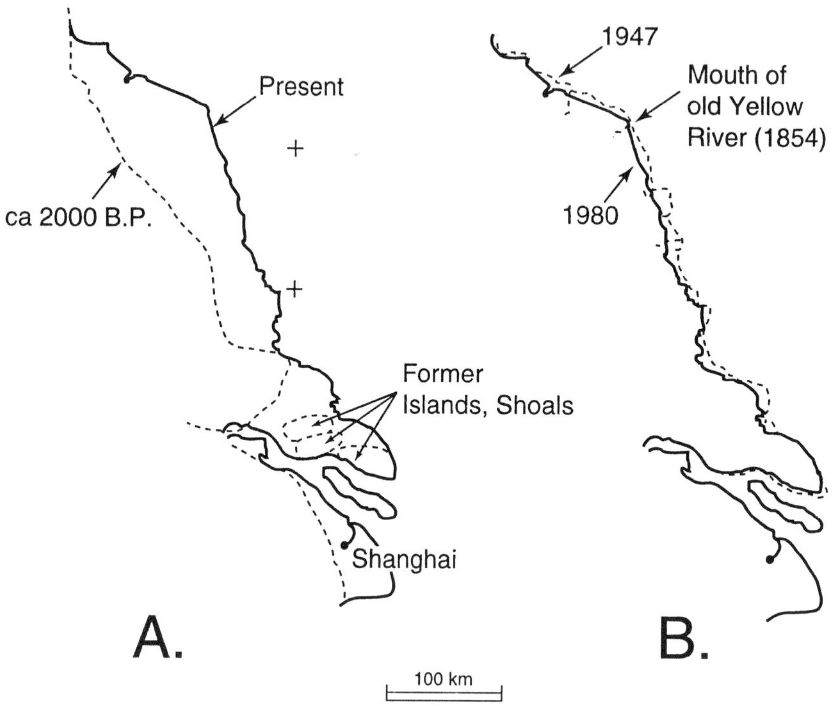

Fig. 3. (A) Position of the Yangtze Delta approximately 2000 years B.P. (B) Erosion of the Jiangsu coast between 1947 and 1980, the result of the northward shift of the Yellow River in the mid-19th century (after Milliman, 1992).

The delta is the most productive agricultural region in China, noted for its silk, rice, and aquaculture; because of the fertile flood plain soils along the river and its tributaries, the Yangtze is known in China as "Heaven on Earth". For several millennia the Yangtze River also has provided easy access to China's interior, and many of the country's largest cities lie along the river's path. As a result, the Yangtze Delta is the most industrialized region in China; no other region in China has so many large industrial centers clustered in so small an area (Fig. 2). In 1986, for instance, there were 20 cities in China with annual industrial outputs greater than 10 billion Yuan (RMB), six of which were located on the delta. Other than Shanghai, large industrial centers are located at Suzhou (population more than 1 million), Wuxi, Changzhou, Nantong, and Yangzhou (Fig. 2). The important ports of Nantong, Zhangjiagang, Zhenjiang, and Nanjing are all accessible by ocean-going cargo ships of 20,000 tons or less.

The eastern region is the most important part of the Yangtze Delta. Here, along the Huangpu River (which feeds into the Yangtze estuary), is the largest city in China, Shangahi, with an urban population of more than eight million, and an administrative population of more than 12 million; population density exceeds 2000/km^2. Shanghai is also the largest industrial center and largest port in China. The importance of Shanghai to the national economy can be seen from the fact

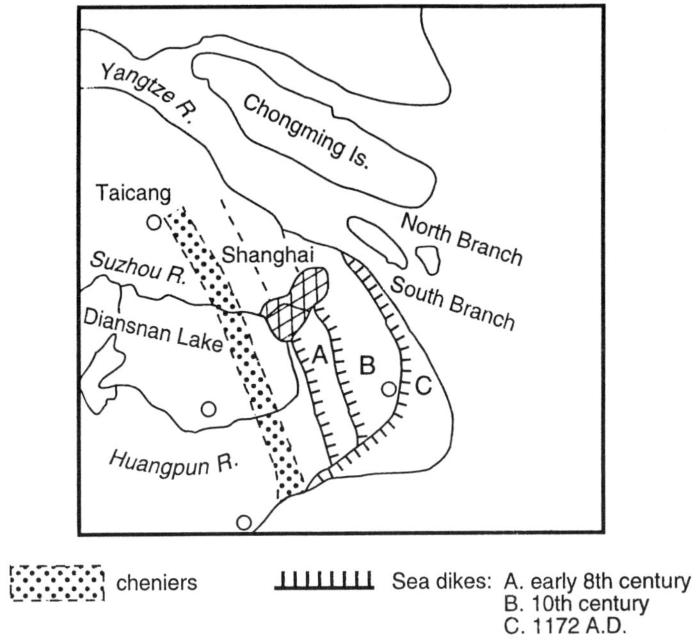

Fig. 4. Successive reclamation dikes in the Shanghai area. (After Tan, 1987).

that it accounts for one-third of China's total cargo, one-seventh of its revenue, one-eighth of its exports, and one-tenth of its industrial output.

Sea-Level Rise and Subsidence

Relative sea-level rise on the Yangtze Delta is poorly documented. Emery and Aubrey (1991), for example, cite the rise of sea level at Shanghai since 1956 to be 2.6 mm/year. Assuming a eustatic sea-level rise of 1.5 mm/year this would mean that subsidence is about 1 mm/year. However, subsidence has varied greatly with both location and time. Ren (1994) cites much greater rates of subsidence between 1957 and 1972, the result of overpumping of ground water; between 1957 and 1961, for instance, subsidence exceeded 100 mm/year (Fig. 6). Decreased pumping and recharging of the wells subsequently resulted in a 50 percent decrease in subsidence rates between 1962 and 1972, and a much lower rate since then (Fig. 6).

Even today, however, Ren (1994) cites four areas within Shanghai in which subsidence rates exceed 10 mm/year, one being along the Bund, Shanghai's major commercial and cultural thoroughfare. If these rates were to continue, relative sea level could rise locally by more than 1 m by 2100 irrespective of any rise in eustatic sea level.

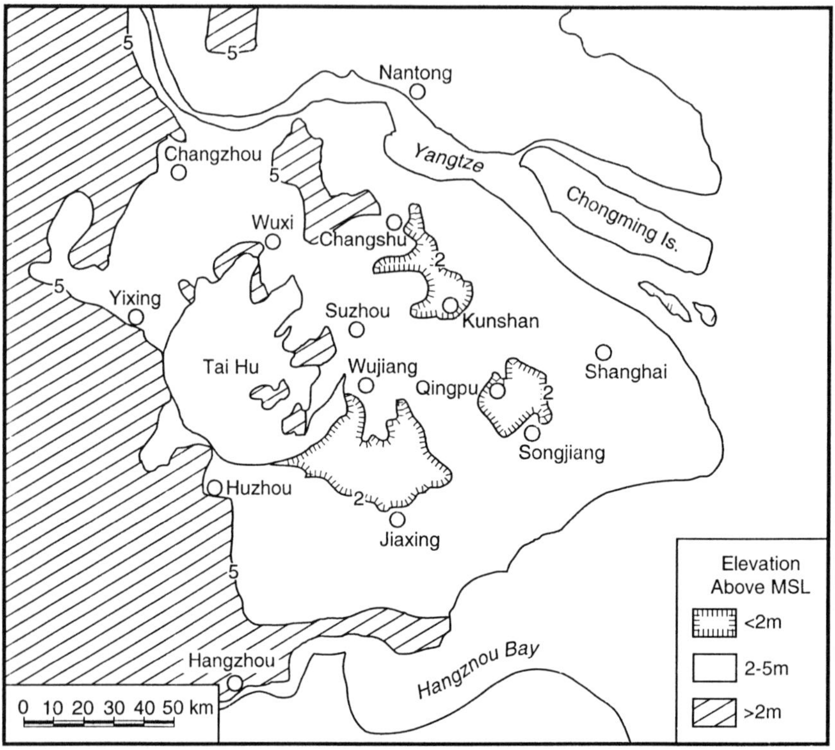

Fig. 5. General land elevations on the Yangtze Delta. Those areas less than 2 m above mean sea level will probably show the greatest impact to rising sea level.

Impact of a Future Sea-Level Rise and Changes in Landuse and the Yangtze River

Sea-Level Rise

A future sea-level rise of 0.5 to 1.0 m could have a serious impact on this rich and populous delta. Pending future studies, several major points can be discussed:

1) On the coast of the Yangtze Delta and adjacent Jiangsu and Zhejiang coasts, the total area of tidal mud flats reclaimed in the last 35 years amounts to 370,000 hectares (Table 1). If the North Branch of the Yangtze estuary were dammed, another 50,000 ha of new, low-lying land could be reclaimed. The newly reclaimed land in Jiangsu Province, however, is only 0–1.0 m above MSL; in Shanghai it is usually about 2 m above MSL; and in Zhejaing it often is 0.1–1.5 m above MSL.

These reclaimed tidal flats now are sites of rich cotton and wheat fields as well as dense areas of human habitation. The tops of sea dikes often are less than 6 m above sea level, which is barely sufficient to protect the land against violent storm surges. For example, at Xiaoyangkou in Jiangsu Province (Fig. 2), water level rose by 3.8 m above normal during a strong typhoon in 1981. If eustatic sea level were to rise 0.5–1 m, these dikes would need to be raised and strengthened.

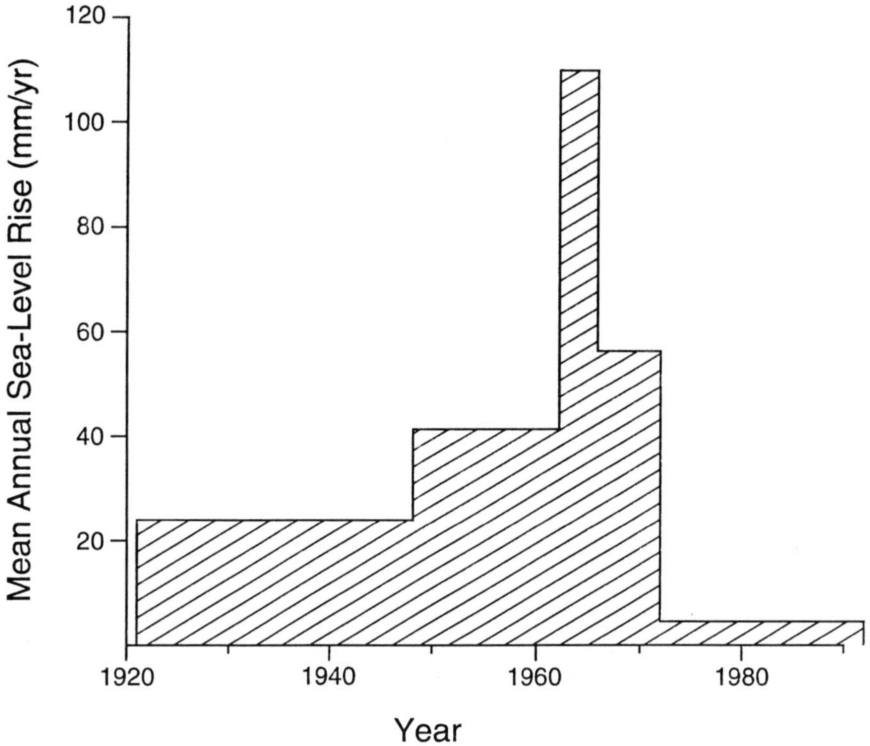

Fig. 6. Relative mean sea-level rise near Shanghai, 1927–1992.

TABLE 1
Area of tidal flats reclaimed (1955–1990) in and near the Yangtze Delta.

	hectares
Jiangsu	183,000
Shanghai	57,000
Zhejiang	130,000
Total	370,000

Also, measures must be taken to cope with increasing difficulty in drainage and in soil salinization in these newly reclaimed lands.

Considering the effects of maximum storm surge and salt-water intrusion, if all land less than 5 m in elevation is vulnerable to a 1-m sea-level rise, then a greater part of the entire Yangtze Delta (about 30,000 km^2) could be impacted in the next century — including Shanghai, Suzhou, Wuxi, Changzhou and Nantong.

The low-lying areas between Shanghai and Suzhou would be particularly affected (Fig. 5).

2) Mean tidal range in the Yangtze estuary is 2.7 m, although at spring tide during flood season it can reach 3 m at Wusong. However, during the 1981 typhoon, water levels at Wusong reached 5.7 m above sea level, coupled with storm waves as high as 3.2 m. In addition to the areas already mentioned, the islands of Chongming, Changxing and Hensha (combined area of 1185 km^2) are mostly reclaimed land, with dense populations and industrial centers. The mean elevation of these islands is only about or slightly above 2 m, meaning that the type of flooding seen at Wusong would have disastrous impacts.

These conditions make it imperative that solid stone embankments must be built to protect Shanghai and other cities against strong typhoons and spring tides. On the south bank of the Yangtze estuary, where the current is very strong, sea embankments are especially designed and built of large stone blocks. With a future sea-level rise of 0.5 to 1 m, large fiscal and human resources will be needed to raise and strengthen these embankments.

3) Shanghai port, handling 130 million tons of cargo annually, has long wharfs and warehouses that would have to be reconstructed or strengthened to cope with the rising sea level, particularly since some of these warehouses are located in areas with rapid local subsidence (Ren, 1994).

4) The municipal water supply of Shanghai largely comes from the Huangpu River, which presently is contaminated by salt-water intrusion. Future sea-level rise will allow salt water to intrude even further upstream, and a recently completed water supply plant would have to be moved further inland. A dam may have to be built across the Huangpu River to prevent the penetration of salt water, but care must be taken to prevent siltation below the dam (Li and Shi, 1986). Ironically, if the Huangpu's water becomes too saline, one solution would be to increase the pumping of ground water, which would accelerate subsidence.

5) Rising sea level will raise the salt-water head and increase salt-water seepage into the ground water, resulting in salinization of soils in the low-lying areas between Shanghai and Suzhou (Fig. 5), which presently are highly productive rice paddies.

6) A rising sea level also will reduce the surface gradient of the lower Yangtze and thus cause siltation. With the upstream movement of the salt wedge, there will be corresponding upstream shift of the turbidity maximum, where the greatest sedimentation will take place. This could result in the accretion of present-day sandy shoals in the lower reaches of the estuary, which could negatively affect the ability of the Yangtze to serve as an effective waterway to the interior of China.

Landuse and the Yangtze River

As mentioned previously, available data suggest that although natural subsidence in the Yangtze Delta is relatively low, as recently as 35 years ago subsidence rates in parts of Shanghai exceeded 100 mm/year, and even now there are local areas in Shanghai where annual subsidence exceeds 10 mm (Ren, 1994). Unless measures are taken to minimize the impacts from excessive groundwater pumping, local subsidence may have far greater impact upon the low-lying delta than the eustatic rise of sea level. Since groundwater is more likely to be pumped in areas of heavy habitation or industrial use, the economic implications of rapid subsidence in these areas would be accentuated.

Finally, it is interesting to speculate on the impact of the Three Gorges Dam on the future health of the Yangtze Delta. This dam, which will have a height of more than 180 m, may have a great effect on the freshwater discharge of the Yangtze River. Moreover, some or all of the coarse fraction of the sediment load (presently about 0.5×10^9 t/year) could be deposited in the reservoir behind the dam, although sluiceways are projected to minimize the amount sediment trapping. Almost invariably, however, river water leaving the dam will carry less sediment, which will mean increased erosion of the downstream bottom land and river bed.

If the Yangtze discharges appreciably less sediment to the estuary and East China Sea, erosion (at least locally) could have considerable impact, which could accentuate any sea-level rise. An example of the impact of decreased sediment flux to this area can be seen in the shift of the Yellow River in 1855, which has resulted in a 1400 km^2/year loss of coastal land around the former river mouth (Fig. 6). In fact, in the near future land bordering the northern edge of the Yangtze also may experience considerable erosion. Decreased sediment discharge from the Yangtze could result in erosion along the southern edge of the the delta and along the shoreline of Hangzhou Bay (e.g., Milliman et al., 1985), the site of considerable newly reclaimed land (Fig. 2).

Concluding Statement

One cannot ignore the potential effects of rising sea level and subsiding coastline in the Yangtze Delta. Depending on the discharge of the Yangtze River and the rates of local subsidence, however, environmental change may be relatively minor; nevertheless, the importance of this area to China means that even the small environmental change may have large economic impact. The extent of the impact in large part will depend on the ability of Chinese planners and authorities to manage both land use of the delta and discharge of the river. Measures must be taken to offset the potential adverse affects of sea-level rise and Yangtze River diversion, which will require huge investments and a long period of time for readjustment. It is hoped that Chinese policy-makers will take steps to cope with these potential environmental changes. In the face of such an important global, regional and local issue, the Yangtze Delta cannot afford to be unprepared.

References

Chen, J.Y., et al., 1981. Coastal Morphology. in Editorial Committee, *Physical Geography of China - Geomorphology* (in Chinese), Academia Sinica, Science Press, p. 319.

Emery, K.O. and Aubrey, D.G., 1991. *Sea Levels, Land Levels and Tide Gauges*. Springer-Verlag, New York, 237 p.

Li, C.X., 1986. Deltaic sedimentation. in M.-e Ren (ed.), *Modern Sedimentation in Coastal and Nearshore Zone of China*. Springer-Verlag, Berlin, 252-317.

Li, G.B. and Shih, Z.R., 1986. Study of the siltation and its control on the Yonjiang River (Ningbo), Donghai (East China Sea (in Chinese). *Marine Sciences*, 4(2), 1-7.

Liu, C.Z. et al., 1987. Sedimentary characteristics of cheniers in southern Changjiang Delta and their origin and age determination. *Acta Oceanologica Sinica*, 6, Suppl. 2, 405-412.

Milliman, J.D., 1992. Management of the coastal zone: Impact of onshore ctivities on the coastal environment. in K.J. Hsü and J. Theide (eds.), *Use and Misuse of the Seafloor*. John Wiley and Sons, Ltd., London, 213-227.

Milliman, J.D., Shen, H.T., Yang, Z.S. and Meade, R.H., 1985. Transport and deposition of river sediment in the Changjiang estuary and adjacent continental shelf. *Cont. Shelf Res.*, 4, 37-46.

Milliman, J.D. and Ren, M.-e, 1995. River flux to the sea: Impact of human intervention on river systems and adjacent coastal areas. In D. Eisma (ed.), *Climate Change: Impact on Coastal Habitation*. CRC Press, Boca Raton (Florida), U.S.A., 57-83.

Ren, M.-e, 1994. Relative sea level rise in China and its socioeconomic implications. *Marine Geodesy*, 17, 37-44.

Tan, Q.X., 1987. Tan Qi-Xiang's Selected Works (in Chinese). People's Press, Beijing, 2, 160-178.

CHAPTER 11

Rising Sea-Level and Subsidence of the Northern Nile Delta: A Case Study

Mahmoud Kh. El-Sayed

ABSTRACT: The northern Nile Delta is particularly vulnerable to a rise of sea level and subsidence. Due to its high population, key economic activities, agricultural and reclaimed lands, as well as lagoonal fish production, rising sea level will have major implications on Egypt's economic future. A 1-m rise in relative sea level, for example, may affect 15% of Egypt's current gross domestic product (GDP).

The northeastern region of the delta is more likely to be affected by a rise of sea level than other regions due to more rapid subsidence. Due to scanty or inaccessible information, however, the actual impact is difficult to quantify. Recommendations are outlined to overcome some of the information gaps.

Prevailing conditions and socio-economic realities in Egypt need to be taken into consideration in any future studies or development plans for the northern delta. Instead of imposing highly developed capital-intensive industrial and agricultural systems, for example, development should aim at controlling erosion, water conservation and engineering measures to protect this region from rising sea level. This approach should preferably last until a more meaningful prediction model is available for the region, one based on more concrete data.

Introduction

Many important economic and social decisions made today about long-term projections are based on the assumption that past climatic data can serve as a reliable guide to the future. However, estimates of future climatic conditions must be refined if we are to improve these decisions (UNEP, 1988). This is particularly true for rising sea level due to the greenhouse effect, where the public and scientific interest seems to be rising faster than the sea itself (J.D. Milliman, personal communication). Although rapid coastal inundation is predicted for many parts of the world, the rate of sea-level rise this century has only been 1 to 2 mm/year. Local changes in sea level in many subsiding deltaic areas over the next 50 to 60 years, however, may be far greater than the eustatic change.

J.D. Milliman and B.U. Haq (eds.), Sea-Level Rise and Coastal Subsidence, 215-233.

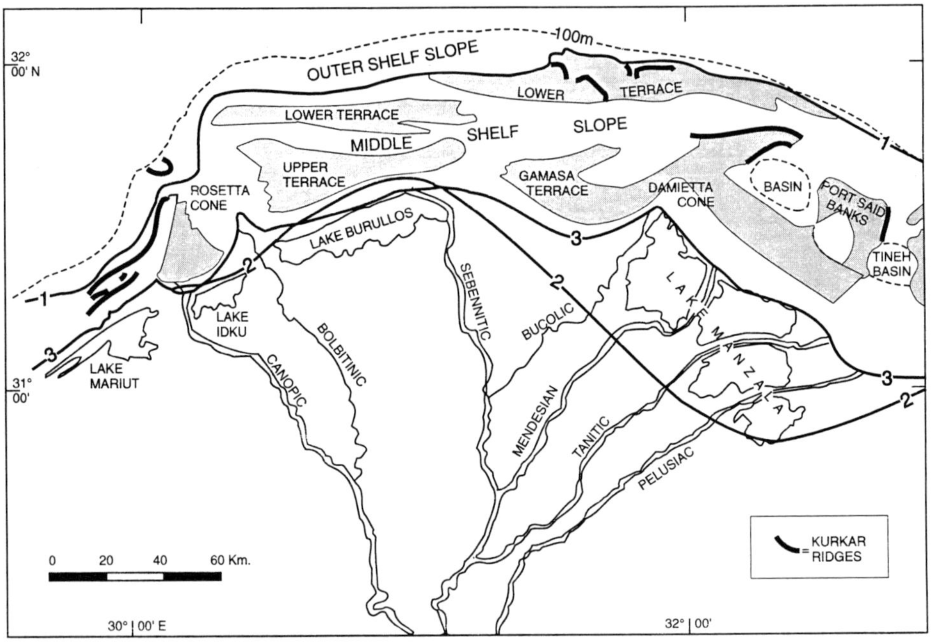

Fig. 1. Distributaries of the ancient river Nile and ancient shorelines of the delta. 1) beginning of the Holocene; 2) 6500 ka; 3) present day. Also shown are the historic branches of the Nile (after Said, 1981; Stanley and Warne, 1993).

The lower Nile Delta is one such area vulnerable to a rising sea level and land subsidence since it encompasses several economic and social foci. Ambitious plans for industrialization, land reclamation, communication, fishing and aquaculture activities are presently being implemented in this area. In contrast, scenarios presented by Broadus et al. (1986) speculate that 15% of Egypt's current gross domestic product (GDP) may be affected by the 1-m increase in relative sea level.

This article presents a synthesis of available data and information on the northern Nile Delta, with the aim to evaluate the possible impact of relative sea-level change on its morphology as well as the social, economic and demographic conditions.

The Nile Delta

General

The Nile Delta, covers an area of about 22,000 km^2, spreading into a plain studded with an intricate networks of canals and drains (Said, 1981). North of Cairo the river bifurcates into the western Rosetta branch (239 km long) and the eastern Damietta branch (245 km long). The former is the only branch presently open to the Mediterranean. Earlier in the Holocene, however, there were numerous distribu-

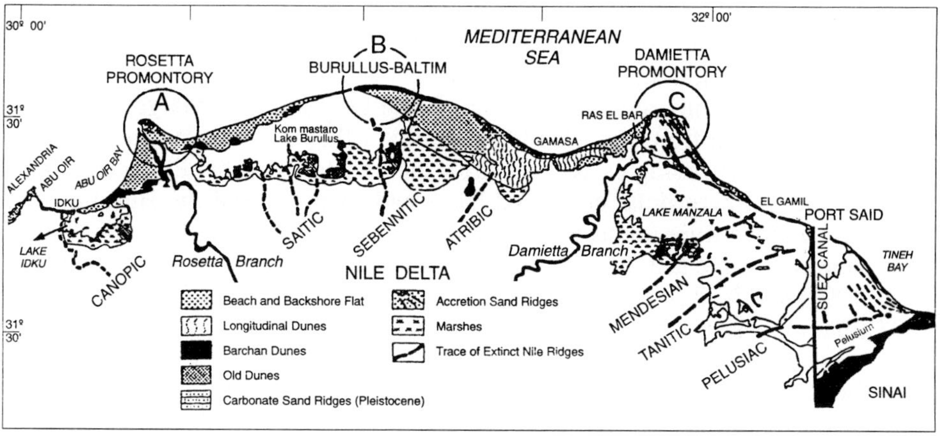

Fig. 2. Morphological features of the northern Nile Delta (after Frilly et al., 1988). Recent evolution of the circled Rosetta, Burullus-Baltîm and Damietta promontories are shown in Figure 9.

taries (Fig. 1), and 2000 years ago, the main flow was through the western (Canopic) and the central (Sebennytic) branches; the Rosetta and Damietta branches were then no more than canals. The disappearance of the older distributaries occurred mainly between the second and fifth centuries A.D, whereas the eastern branches (Tanitic and Pelusiac) and perhaps the Sebennytic branch persisted until the ninth century (Tousson, 1934).

The western terminus of the present delta is near Alexandria, while its eastern terminus is just east of the Suez Canal, at the Bay of Tineh where Pelusiac branch once entered the Mediterranean (Ross and Uchupi, 1977).

Despite its small area, which represents 2.3% of Egypt's total area, the delta comprises about 2/3 of Egypt's total cultivated land. It is the site of the major population, agricultural, fishing and aquaculture centres, as well as most of industrial communication links in the country.

Morphology

Much of the Nile Delta between Alexandria and Tineh lies within 2 m of mean sea level, with lagoons and agricultural lands (mostly recently reclaimed) bordered by 1–10 km wide sand belt of beaches, backshore plains and dunes (Sestini, 1976). The principal morphological units of the northern part of the Nile Delta recognized by Frihy et al. (1988) include (Fig. 2):

1) The nearshore zone;

2) The beach and coastal flat made up of fine and very fine sand, with limited patches of coarse and medium sand. This coastal zone contains Neonile sediments deposited from late Pleistocene to present (Said, 1981);

3) Coastal sand ridges that include old beaches fringing the deltaic plain in some localities and Pleistocene carbonate ridges west of Alexandria;

4) Coastal dunes, barchan and longitudinal belts of fine to very fine sand, usually located at the southern margin of the backshore parallel to the coast. Inactive sand and silt dunes lie further inland;

5) Coastal lagoons include Manzala, Burullus, Idku and Maryut Lagoons, generally separated from the Mediterranean by a 0.5–3-km wide sand belt and coastal dunes. Archaeological records show that in late Dynastic- Ptolemaic times much of the northern delta contained lakes and swamps, and the Burullus and Manzala Lagoons were not as extensive as today (Sestini, 1988). Lagoonal areas have increased in the last 2000 years due to diversion of the Nile discharge (Chen et al., 1992) and land subsidence behind a stable beach barrier (UNESCO/UNDP, 1978), but extensive reclamation also has decreased the area, particularly of the Idku Lagoon. Chen et al. (1992) estimate that the present Idku Lagoon occupies only about 30 percent of its former area. Abu Lagoon, just west of Idku Lagoon, disappeared completely in the late 19th century due to intensified land reclamation (Loizeau and Stanley, 1993).

GEOLOGY

The present Nile River valley was probably cut during the Miocene (Said, 1981). A Paleonile began to advance across the embayment in the late Pliocene and developed in the Pleistocene during the periods of major climatic changes. Said (1973) suggested that most (60%) of the Nile cone was deposited by the Paleonile system and that the Nile Delta was initiated at this time, taking form by the beginning of the Pleistocene.

The present river regime was established 9000 years ago. At about 5,000 years B.P., the time when sea-level rise had slowed, the Mediterranean had transgressed into the Nile Delta plain as much as 40–50 km south of the present coastline. Subsequently, the delta plain began to prograde seaward (Fig. 3; Coutellier and Stanley, 1987). By 25 A.D the shoreline off the Pelusiac branch stood about 10 km inland from the present shoreline (Sneh et al., 1975).

Although the Nile Delta occupies a key position in the context of the plate tectonic evolution of the eastern Mediterranean and the Red Sea, not all the tectonic features of the Egyptian margin may be related to plate tectonics (Ross and Uchupi, 1977). For example, the numerous small faults on the slope north of the Nile Delta may be related to downslope movement (Fig. 4). Stratigraphic and tectonic factors may cause rapid subsidence at Lake Manzala in northeastern region of the Nile Delta at Lake Manzala (Stanley, 1988; Fig. 5). This part of the delta contains one of the thickest (ca. 3,000 m) sections of lower Pliocene to Quaternary sediments in the delta (Said, 1981), which may account for differential loading. Geologic mapping of the delta has not identified any important faults in this region, but seaward of Lake Manzala geophysical surveys indicate that the northeastern African margin has been affected by listric faults and also by a dominant northeast fault system that extends from the Levant directly to the Nile Delta.

One of the major offshore faults, called the Pelusium Line, has been interpreted as an ancient, left-lateral, strike-slip fault that extends to the Gulf of Tineh and into the delta east of Lake Manzala. Another fault, the Damietta-Latakia fault, extends

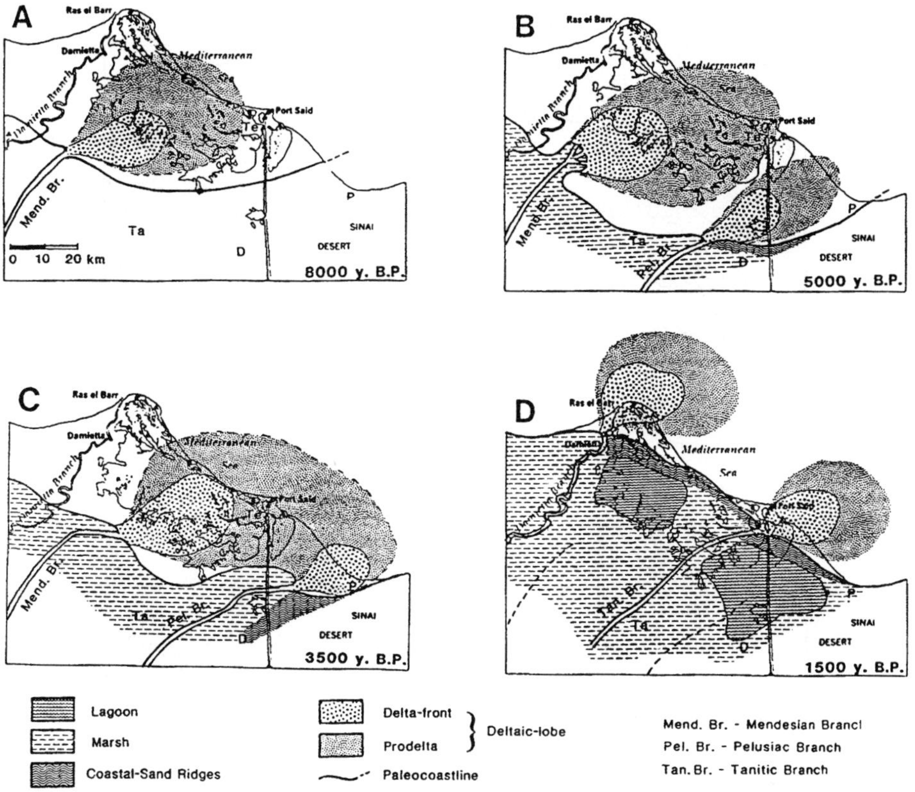

Fig. 3. Paleogeographic maps showing changes in coastline position with time (after Coutellier & Stanley, 1987).

toward the Damietta promontory and may continue inland (Fig. 5) These two faults, part of the Pelusium fault system, have been interpreted as large Mesozoic faults that played a major role in the formation of the southeastern Mediterranean basin. A northwest trending normal fault apparently related to reactivation of the above structures has been mapped along the outer shelf-upper slope boundary parallel to but northeast of the main axis of Lake Manzala. Continuous development of this complex structure during the late Quaternary could have caused the high subsidence rates.

OCEANOGRAPHY OF THE COASTAL AREA

The major agents affecting the coastal processes in the Nile Delta region are wind and waves, predominantly from the northwest. The delta coastline is shaped largely by waves (Coleman and Wright, 1974) that have a strongly seasonal character (Frilly, 1988), with storm waves in winter (periods 7–8 sec) and swells in summer (periods 8–9 sec). From wave refraction models and estimated transport values, the energy balance along the Nile Delta coast causes higher net transport at the

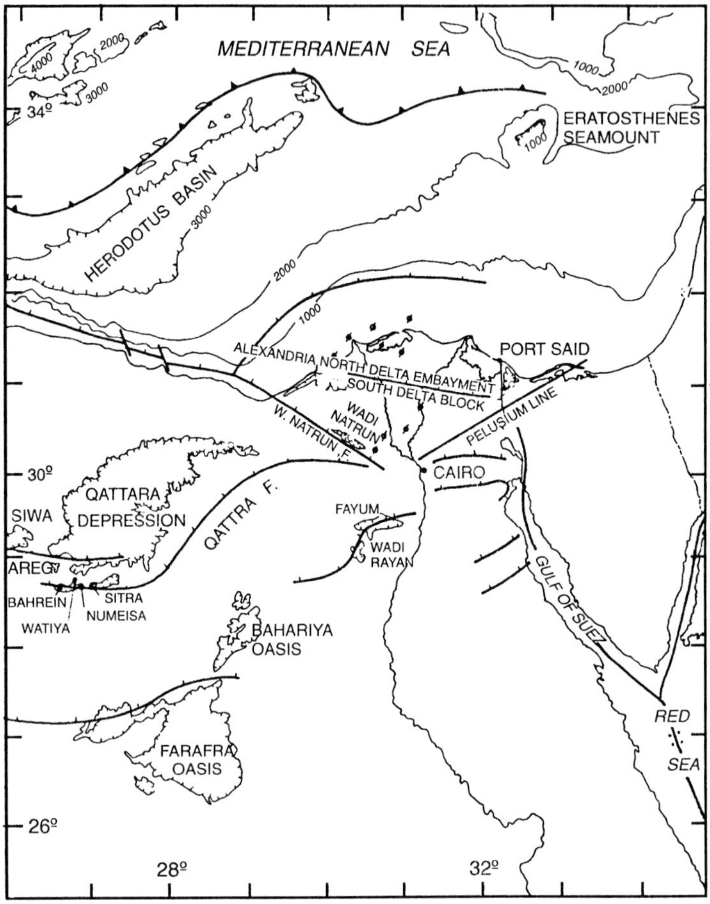

Fig. 4. Schematic map showing major tectonic elements of northern Egypt and the Nile Delta (after Said, 1981).

promontories and east-facing beaches. The dominant direction of the littoral current within the breaker zone near Burullus and Damietta is to the east, except during periodic current reversals. Along the delta coast, tides are semidiurnal and generally with a range of 25 to 30 cm.

THE NILE DELTA AQUIFER

The main delta aquifer is a confined aquifer, increasing in thickness from about 250 m in the south to about 900 m in the north near the coast (Fig. 6; Kashef, 1983). The base is a clay aquiclude with a slope of about 4 m/km. Clay layers and/or pockets encountered within the main aquifer presumably divide it into several partially connected smaller aquifers. These aquifers consist mainly of unconsolidated mixtures of sand and gravel.

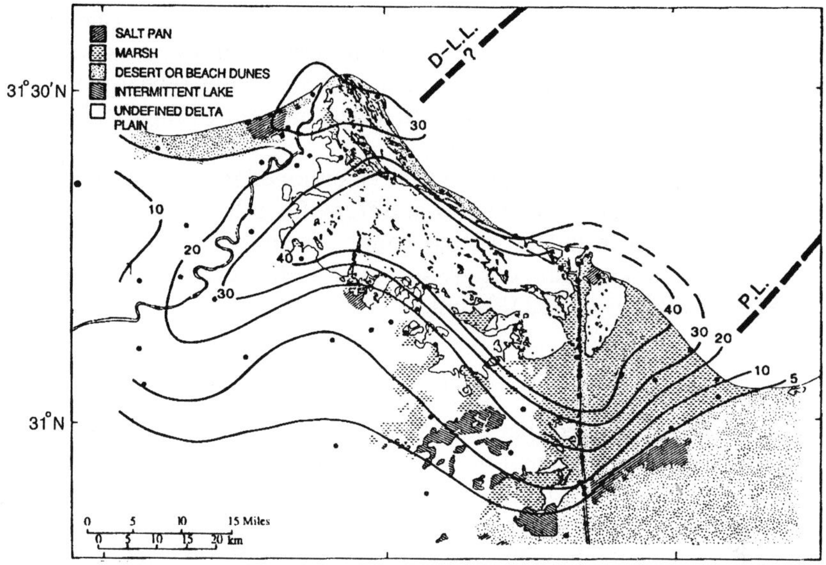

Fig. 5. Map showing two major strike-slip faults that extends to the delta, the Pelusium line (PL) and the Damietta-Latakia line (D-LL). Thickness of Holocene deposits in m (after Stanley, 1988).

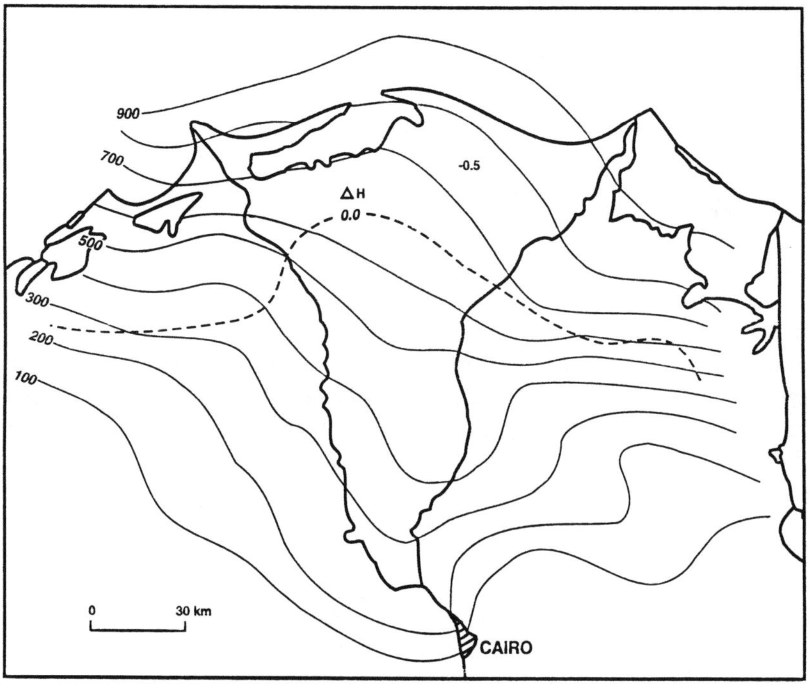

Fig. 6. Contour map of the delta aquifer; thickness in m; DH is the difference between the surface water and ground water levels (after Anonymous, 1980).

The delta presently receives about 30 km^3/year of surface water from the Nile, which recharges the aquifer through infiltration and excess irrigation as well as seepage from canals and drains. Rainfall also helps recharge the aquifer, particularly near the Mediterranean coast, where rainfall averages 180 mm/year (compared to 25 mm/year near Cairo).

Kashef (1983) and Zaghloul et al. (1988) mentioned that a wedge of salt water intrudes the huge artesian delta aquifer at different depths. Shata (1988) reported that within the delta basin and the western fringes, fresh ground water with salinities less than 100 ppm is dominant, but groundwater in the northwest near the Qattara depression has salinities of 30,000 ppm.

Relative Sea-Level Changes in the Nile Delta Region

It has been long recognized that both subsidence and eustatic sea-level change have influenced the development of the Nile Delta. None of the earlier studies of Holocene sea-level rise in Mediterranean (e.g Fairbridge, 1972; Flemming, 1972) provided any Holocene curve for the Nile Delta. More recent data from El-Gindy & Eid (1988), however, show that 2000 B.P. sea level in the Mediterranean was 50 cm lower than at present. In the last 100 years sea level has risen by 1.3 mm/year, resulting in a 10–15 m shoreline retreat of 10 to 15 m in Egypt (Inman and Jenkins, 1984).

Subsidence in the Nile Delta plain is the result of compaction and tectonics (Ross & Uchupi, 1977). In general, the eastern part of Lake Manzala and the northern part of the Suez Canal (including Port Said) has the highest rates of subsidence on the delta, accelerated by isostatic depression and faulting (Stanley and Warne, 1993); but even here the highest rates of Holocene subsidence are only about 4 mm/year (Stanley, 1990). In contrast, tide-gauge records at Alexandria show land uplift of +0.7 mm/year (Emery et al., 1988; Emery and Aubrey, 1991).

Present-Day Use of the Northern Nile Delta

The 0–2 m elevations of the northern delta occupies a 30–40 km-wide coastal stretch and covers an area 4500 km^2. But this relatively small area accounts for a disproportionately large part of Egypt's habitable land, and this is the area that will be most affected by sea-level rise (Milliman et al., 1989).

The present day usage of the northern Nile Delta most likely to be affected can summarized as follows (Fig. 7):

Population

The estimated population occupying the coastal stretch of the delta is about 6 million, about 10 percent of the present total population in Egypt. Major cities include Alexandria (about 4,000,000 inhabitants as of 1991), Port-Said (350,000), Rehid (Rosetta) and surroundings (200,000), Manzala and surroundings (300,000, Damietta (120,000), Ras El Bar and fishing villages (200,000), and Burullus and Baltîm (80,000). It is worth noting that presently Egypt's population increases by

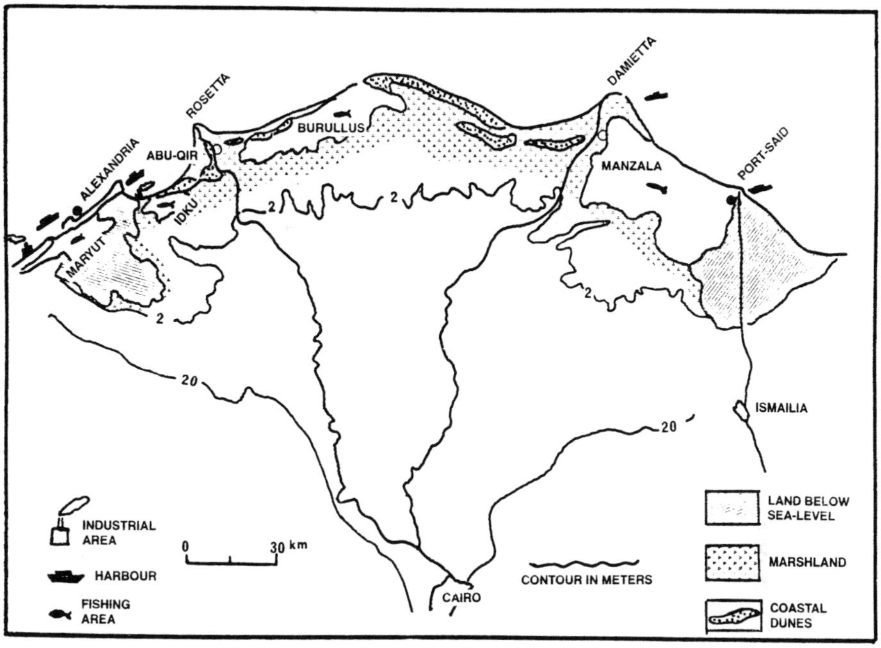

Fig. 7. Map of morphology and present-day usage of the northern Nile Delta.

more than 2 percent annually, 1 million every nine months (Stanley and Warne, 1993).

AGRICULTURE

The northern delta represents almost 20 percent of the total cultivated land in Egypt (Ministry of Agriculture; Bulletin of Economic Agriculture, 1987). The cultivated crops on the delta varies from summer to winter. In winter the wheat, ful (beans), barley, fennugree, flax, onions, lupine, chickpeas, berseem, garlic, vegetables are the major crops, while in summer major crops include maize, cotton, rice, sugarcane, peanuts, sesame, vegetables and fruit orchards (Kinawy, 1977).

LAND RECLAMATION

The current plans of the Egyptian government call for a massive expansion of irrigated land; about 0.85 to 1.25 × 10^4 km^2 by the year 2000. Most of this reclaimed area will be located immediately east and west of delta (some 43 areas out of the 59 areas to be reclaimed), and more than half of the total is located in the eastern delta and Sinai (Fig. 8).

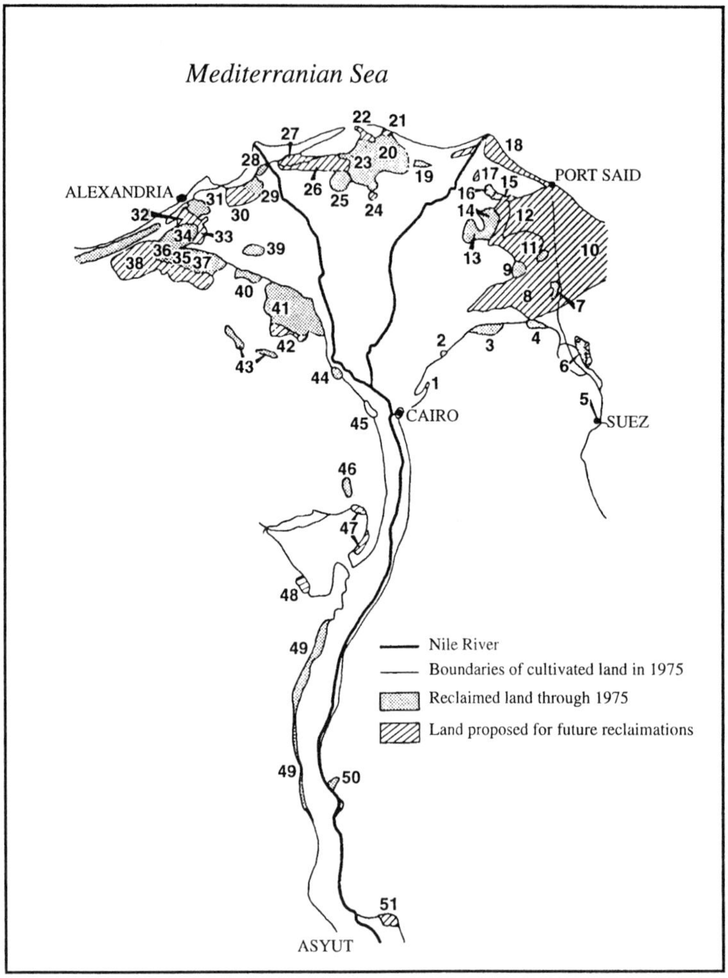

Fig. 8. Reclaimed lands in northern Egypt, with special reference to the northern delta (after USDA, 1976).

INDUSTRY

Forty percent of the nation's industry surrounds Alexandria. These industries are mostly petrochemicals, chemicals, refineries, textiles, steel and iron, pharmaceuticals and cement.

FISHERIES

The northern coastal lagoons of Egypt (Bardawil, Manzala, Burullus, Idku and Maryut) are considered the main fishing grounds in the country. According to older statistics, they produce nearly 50 percent of the fish catch in Egypt (Shaheen & Yousef, 1978). More recent statistics cite the five coastal lagoons as producing about 35 percent the total catch in Egypt, about 100,000 tons/year (Fisheries Statistics Department, General Authority for the Development of Fishery Resources).

COMMUNICATIONS AND HARBOURS

Egypt's main port along the Mediterranean is Alexandria, which handles more than 80 percent of the country's trade. Some shipments are also delivered to the newly constructed Damietta harbour and Port Said. A network of roads and railways connects the main cities, towns and villages along the northern Nile Delta.

Nile Damming and its Effect

Prior to the construction of the Aswan High Dam in 1964, the Nile River discharged about 85 km^3 of water annually to the Mediterranean and approximately 100 million tons of sediment (Milliman, 1985). Since the construction of the High Dam, the Nile transports virtually no sediment to the sea, however some discharge of water taes place through the Rosetta branch of the Nile as well as the lagoon outlets and drainage system. The average amount of water that flows out to the sea is 16 km^3/year.

Egypt has benefited greatly from the Aswan High Dam, as it has increased the arable land along the river by nearly 100 percent, increased dramatically the amount of available electric power, and has practically eliminated the disastrous effects of spring floods as well as prolonged droughts. However, damming the river also has had a number of severe environmental consequences, some of which are described in the following paragraphs.

GEOLOGICAL IMPACTS

The effect of decreased sediment flux has been difficult to quantify. One problem stems from the fact that the delta has been eroding since 1902 (following more than 100 years of continuous progradation) after construction of the first Aswan Dam. Still, the erosion of Rosetta branch between 1909 and 1945 was only a few hundred meters, while between 1945 and 1972 it was more than 1.5 km, presumably mostly after 1964 (Fig. 9a).

Analysis of a series of aerial photographs of the Nile Delta taken in 1955 and 1983 reveals a drastic change in the shoreline (Frihy, 1988). The outer margins of both promontories have been eroded (Figs. 9b and 9c), the greatest erosion rates being 114 m/year and 31 m/year at Rosetta and Damietta, respectively. Much of the eroded material from the tip of both promontories is transported and accreted eastward by littoral drift. The Burullus-Baltîm sector is also erosional (Fig. 9d), with a maximum erosion rate of 9 m/year.

Analysis of a series of landsat images of the Nile Delta by Klemas and Abdel-Kader (1982) and Smith and Abdel-Kader (1988) shows erosion along nearly the entire delta coast, with promontories being the most eroded. The Rosetta promontory had retreated 275 m/year between 1978 and 1984; in the 50 years between 1934 and 1984, total retreat was 3.7 km. In contrast, the Damietta does not show

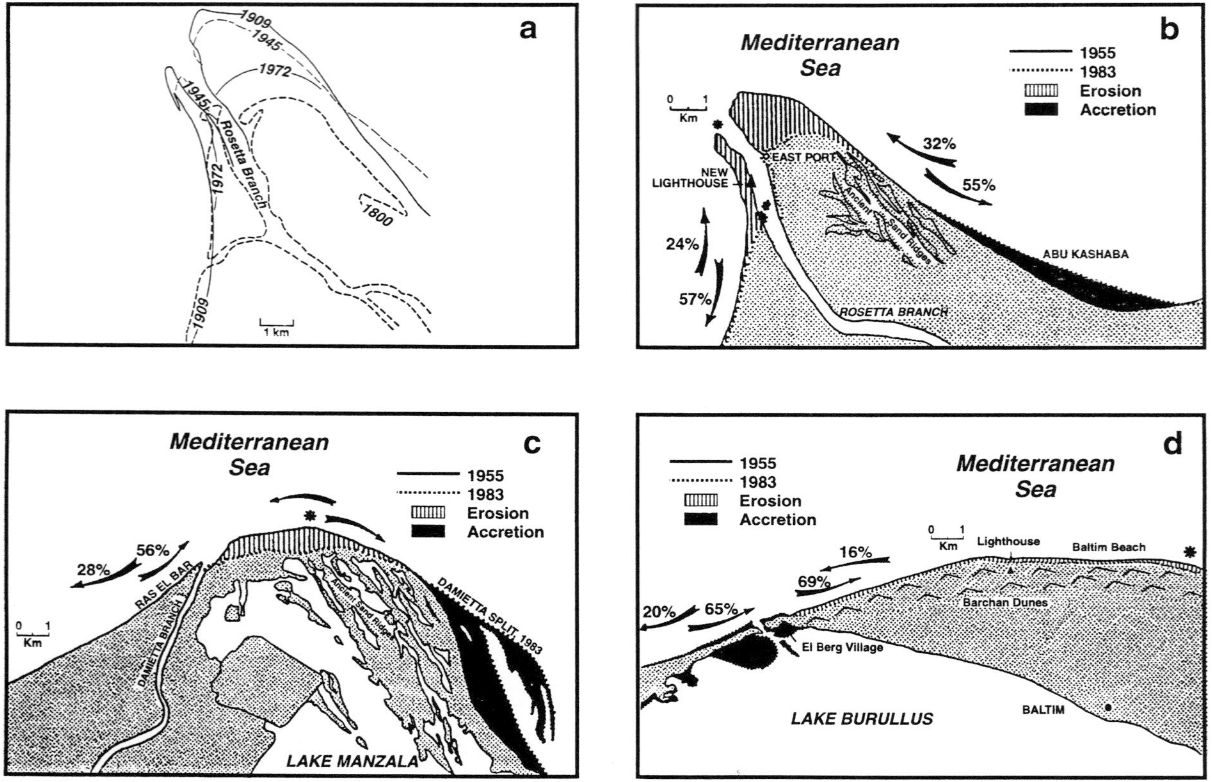

Fig. 9. (a) Changes in the Rosetta promontory since 1900 (after Sestini, 1976). (b) Shoreline changes at the Rosetta promontory during 1955–1983; * denotes areas of maximum erosion, arrows indicate the directions of dominant coastal currents, frequency is denoted by the percentages (after Frihy, 1988). (c) Shoreline changes at the Damietta promontory, 1955–1987; symbols as in (b) (after Frilly, 1988). (d) Shoreline changes along the Burullus-Baltîm sector, 1955-1983; symbols as in (b) (after, Frihy, 1988).

regular patterns of change; 35–50 m/year of erosion occurred between 1978 and 1984 just east of Damietta's mouth. The coastline between the two promontories was stable except for the Burullus headland, which eroded at 29 m/year between 1978 and 1984. The coastline between Damietta and Port-Said has experienced both erosion and accretion.

Despite the number of analyses and reports, the actual net erosion along the Nile Delta coast remains relatively imprecise.

BIOLOGICAL IMPACT

The biological productivity of the southeastern Mediterranean was greatly reduced after completion of the Aswan High Dam and the resultant cessation of the seasonal outflow of Nile flood waters (Dowidar, 1988). Immediate effects were reflected in the disappearance of the fall phytoplankton bloom and the sharp decline in sardine fisheries. The sardine industry reported a more than 90 percent decrease in catches within a few years after the completion of the high dam (Wahby and Bishara, 1981).

As a result of decreased influx of nutrients, the average standing crop of phytoplankton decreased to about 25 percent of its value in 1962. Total catch of Mediterranean fisheries by Egypt decreased from 37, 800 tons in 1962 to 12, 800 tons in 1983 (Dowidar, 1988). Available data indicate that the system has not yet reached biological equilibrium; long term effects at all trophic levels are still occurring, but the overall decline of the Mediterranean fisheries of Egypt is obvious (Dowidar, 1988).

Impact of Climate Changes and Sea-Level Rise on the Nile Delta

The first forecast of the impact of sea-level rise on the Nile Delta was presented by Broadus et al. (1986), in which the coastal area inundated in a 1-m sea-level rise represents approximately 12% to 15% of Egypt's arable land and contains about 16 percent of the nation's population. The area lost to the sea in the worst-case scenario (3-m rise in sea level) represents about 20 percent of the nation's arable land. In these scenarios Broadus et al. estimated that approximately 15 percent of Egypt's current gross domestic product (GDP) would likely to be affected by the 1-m rise in relative sea level, and approximately 20 percent of the nation's GDP would be affected by a 3-m rise in sea level. Large areas targeted for priority development in land reclamation projects, south of Lake Manzala and around Lake Maryut, would lie within the potentially flooded zone.

The second prognosis was presented by UNEP (1988) in which the main problem created by predicted climatic changes were discussed. These mostly represent a physical threat to the cities of Alexandria and Port Said, especially to harbour facilities, summer tourism, and residential areas. The shielding role of the sand belt might be effective where wide and high dunes (5–20 m) exist, but several narrow, low-lying stretches are a point of serious concern. Severe inundation could result from the undermining of a sea wall in the western part of Abu Qir Bay.

Burullus Lagoon is threatened in several parts. Strong erosion at Ras El Barr has been checked so far by groins and walls, but these probably could not withstand a higher sea level. Ground water, lagoonal fishing, agriculture, and soil characteristics also would be affected by rising of sea level.

In a third scenario, Stanley (1988) pointed out the continued land subsidence in the northeastern part of the Nile Delta, up to 40–50 cm/century. He projected eustatic rise in sea level conservatively estimated at 4 to 8 cm during the next 40 years and at least 50 cm by the year 2100, together with subsidence and reduced sediment supply, which collectively could cause a relative rise of sea level of 1 m or more by the end of the next century. This would lead to inundation of much of the delta region within 50 km of the coast. The area of Port Said and the northern Suez Canal-Lake Manzala region may be particularly susceptible to flooding because it lies in one of the more rapidly subsiding parts of the delta. Coupled land subsidence and sea-level rise almost certainly will affect some present and planned agricultural, irrigation, channelization, and land reclamation projects (Milliman et al., 1989).

Scenario Based on Present-Day Assumptions and Socio-Economic Conditions in the Northern Nile Delta

Previous scenarios were based on assumptions accepted at the International Conference in Villach, 1985, i.e. increased global temperature of 1.5 to 4.5°C and sea-level rise of 20–140 cm before the end of the 21st century. In these scenarios, the agricultural output in the delta is assumed to be distributed as arable area and all other sectors of economic activity are distributed in the same ratio as population. This scenario is based on the widely accepted assumptions of climatic impact and global sea-level rise (Raper et al., this volume) and on present-day use of the northern Nile Delta.

The most vulnerable region to rise of the sea level is presumably the one with the highest subsidence: the Manzala region, Port Said and the northern Suez Canal. Coastal erosion will be more significant along the narrow beaches of the summer resorts of Alexandria and Port Said. It has been predicted that some of the beaches of Alexandria will be entirely eroded within the next 20 years (El-Wakeel et al, 1980). The protective role of the sand barriers and dunes should continue to be effective in protecting most of the northern coastal delta plain from the 12 to 18 cm global rise of mean sea level estimated by Raper et al. (this volume), assuming no further acceleration in coastal erosion. The natural subsidence in the eastern region of the delta is unlikely to be accelerated by human activities, such as increased withdrawal of ground water, because in all low-lying areas of the delta plain below 5 m elevation, groundwater is already highly saline and therefore unsuitable for both irrigation and drinking.

Because of the lack of sediment influx from the dammed Nile River, coastal erosion almost certainly will increase in the coming years; sea-level rise and coastal subsidence will only exacerbate the problem (e.g., Stanley and Warne, 1993). One area of particular concern is the fate of the coastal dunes that presently help protect the northern confines of Egypt's coastal lagoons. If the dunes are eroded or breached, the lagoons could be threatened, and given the large yield of fish from these lagoons, loss of these habitats could be particularly disastrous.

The coastal lagoon ecosystem, in fact, is already under considerable stress, as they receive and increasing amount of land-derived pollutants. The northern lagoons also have shown considerable change over the last 30 years as revealed from landsat images (Klemas and Abdel Kader, 1982). Manzala (126,000 hectares) and Burullus (60, 830 hectares) are shrinking in size, a result of drying up, reclamation and siltation. Similarly, Idku (12, 500 hectares) and Maryut (2600 hectares) are shrinking. Algal productivity has increased with increased eutrophication, and the sand barriers separating the lakes from the Mediterranean are increasingly vulnerable to wave attack. On the other hand, according to Sestini (1992), the natural impacts on the lagoonal ecosystem could be minimal and possibly self-adjusting in the long term.

Recognizing that population, agriculture and other activities are not evenly distributed along the northern Nile Delta, the impact of rising sea level may be minimal on highly populated cities as Alexandria, which is situated above 3 to 5 m above sea level. The harbours of Port-Said, Damietta and the entrance of the Suez Canal , on the other hand, may be more affected because the these areas are already low-lying and are subsiding at rates greater than the rise of sea level.

Management of the Northern Nile Delta

WATER RESOURCES

As the population in and around Cairo continues to expand, less of the Nile's waters reach the northern delta. Moreover, water that does reach the northern delta is increasingly polluted (Stanley and Warne, 1993). One result of these trends almost certainly will be an increased reliance on groundwater in the northern delta.

Present-day use of groundwater in Egypt is limited. About 2.9 km^3/year of groundwater is presently used in the Nile valley and delta; this is within the safe yield because there is still 4×10 m^3/year of potential water surplus for future development. In the northern coast of delta only $0.07\times$ km^3/year is used, while the safe yield is $0.15\times$ km^3/year (Kashef, 1983). On the other hand, recharge of the aquifers will decrease as more Nile water is utilized upstream.

Groundwater investigations are presently being carried out throughout Egypt because of the urgent need to meet the demands of present and future land reclamation. The central African drought of 1978 provided an additional incentive for Egypt to seek alternative water sources to the Nile. There are problems in overpumping ground water, however: in the first stage of reclaiming 5000 acres, 0.5 km^3/year of water were needed from the delta aquifer (Anon., 1980), and delta subsidence was accelerated by increased water removal (Smith, 1986). Clearly the ground-water system must be monitored closely as pumping increases, particularly with respect to accelerated subsidence. Moreover, wells must be maintained and drainage systems installed that minimize salt-water intrusion into the ground-water system.

Coastal Erosion

Ambitious plans of coastal management, particularly protection against coastal erosion, have been proposed in cooperation with a number of international organisations, such as UNESCO/UNDP, since 1972. Originally the rate of erosion was monitored to provide a data base and to provide possible solutions to observed and predicted erosion. The coastal protection programme was started by providing simple and direct solutions to remedy eroded areas through the construction of groins, jetties, concrete blocks, breakwaters, etc. In 1987, a programme was adopted to artificially nourish eroded beaches in the summer resorts of Alexandria and Abu Qir. In 1988, engineering counter measures were taken to protect the Rosetta promontory against accelerated erosion. A gigantic sea wall was constructed along the western side of the promontory, with a length of 1.5 km, 12 m high of which 6 m were above the water line. A second 3.5 km-long sea wall was planned to be constructed in the 1990s along the eastern side of the promontory. It is still to soon to know if these engineering structures will prove effective in retarding long-term erosion of this area.

Experimental pilot projects to reinforce and stabilize the northern Nile Delta coast are under study by the Coastal Research Institute of Egypt, in which stabilisation of dunes is being considered, as well as the reinforcement of the low lying stretches of the coast. It is hoped that these activities will ultimately minimize the hazards caused by decreased sediment flux from the Nile and rising sea level.

Conclusions

The effort to integrate available data regarding the present utilization of the northern Nile Delta and the implications of rising sea level has identified the lack of information regarding some key environmental variables, such as local rates of present-day subsidence, verified geological, morphological, oceanographic data, etc. Scanty and/or unaccessible data necessarily result in superficial comprehension of environmental problems and their potential solutions. New inventories are clearly necessary to furnish a broad data base for hazard assessment and potential loss of agricultural and reclaimed lands, lagoonal fisheries, ground water and industrial and social resources.

Some of the conclusions and recommendations can be summarized as follows:

1) Although fluctuations of Holocene sea level have now been established for much of coastal Egypt, we still do not have a sufficiently good data base to define present-day rates of uplift (Alexandria?) or subsidence (Port Said).

2) There is still a considerable gap of information regarding the paleoenvironmental and paleoceanographic conditions of the Nile Delta. Regional models of climatic changes, past present and future, for the eastern Mediterranean are clearly needed.

3) Careful comparison of maps, aerial photographs and satellite images can help in accessing the overall retreat of the coastal deltaic area.

4) Rational water management must be predicated on a thorough understanding of water availability and movement. Rational management of the northern delta

also should consider the possible gradual landward migration of barrier beach-lagoon systems with future coastal retreat.

5) Coincident with any discussion of future oceanographic conditions and beach dynamics must be the realization that the negative impact of the Aswan High Dam will continue to be felt for many years to come. Increased erosion coupled with the relative rise of sea level, for example, could destroy or severely alter the coastal lagoons that presently account for much of Egypt's fish catch.

6) Prevailing environmental conditions and demographic/social realities should be considered in any future development of the northern delta. Instead of imposing highly developed capital investment industrial and agricultural systems, development should focus on smaller scale projects that take into account erosion control and water conservation. This stage should preferably last until a meaningful predictive model of the likely impact of sea-level rise becomes available.

7) Finally, the above mentioned recommendations underline the need for further action on the part of scientists, planners and land users to develop sound land-use strategies. Without this critical step, the other recommendations will be difficult to implement and probably will have minimal positive impact.

Acknowledgements

I am grateful to Dr. John Milliman for providing me the opportunity to participate in the SCOPE Workshop in Bangkok. He and Dr. Bilal Haq offered many constructive remarks and helpful suggestions regarding this manuscript.

References

Anonymous, 1980. Investigation project of sustained yield of aquifers. Part 1. Ground water (in Arabic). Res. Inst./ASRT, Cairo, 342 p.

Broadus, J.M., Milliman, J.D., Edwards, S., Aubrey, D.G., and Gable, F., 1986. Rising sea level and damming of rivers: possible effects in Egypt and Bangladesh. in J.G. Titus (ed.), Effects of Changes in Stratospheric *Ozone and Global Climate*. UNEP/EPA, v. 4, 165-189.

Chen, Z., Warne, A.G. and Stanley, D.J., 1992. Late Quaternary evolution of the northwestern Nile Delta between the Rosetta promontory and Alexandria, *Egypt. J. Coast. Res.*, 8, 527-561.

Coleman, J.M. and Wright, L.D., 1974. Modern river deltas: variability of processes and sand bodies. in L.M. Broussard (ed.), *Deltas*. Houston Geol.Sc.: 99 - 150.

Coutellier, V. and Stanley, D.J., 1987. Late Quaternary stratigraphy and paleogeography of the eastern Nile Delta, Egypt. *Mar. Geol.*, 77, 257-275.

Dowidar, N.M., 1988. Effect of Aswan High Dam on the biological productivity of southeastern Mediterranean. in M.I. El-Sabh and T.S. Murty (eds.), *Natural and Man-made Hazards*. D. Reidel, Dordrecht, 477-498.

El-Gindy, A.A. & Eid. F., 1988. Sea level variation in the Mediterranean. *Rapp. Comm. Int. Mer. Medit.*, 31, 196.

El-Wakeel, S.K, El-Sayed, M.Kh., and Mahmoud, B., 1980. The evolution of Alexandria beaches: A preliminary study. *Thalassia Jugosl.*, 16, 1-8.

Emery, K.O., Aubrey, D.G., and Goldsmith, V., 1988. Coastal neotectonics of the Mediterranean from tide-gauge records. *Marine Geol.*, 81, 41-52.

Emery, K.O. and Aubrey, D.G., 1991. *Sea Levels, Land Levels and Tide Gauges*. Springer-Verlag, New York, 237 p.

Fairbridge, R.W., 1972. Quaternary sedimentation in the Mediterranean region controlled by tectonics, paleoclimates and sea level. In D.J. Stanley (ed.), *The Mediterranean Sea: A Natural Sedimentation Laboratory*. Hutchinson and Ross, Stroudsburg (Pa), 99 - 113.

Flemming, N.C., 1972. Eustatic and tectonic factors in the relative vertical displacement of the Aegean coast. In D.J. Stanley (ed.), The *Mediterranean Sea: A Natural sedimentation Laboratory*. Hutchinson and Ross, Stroudsburg (Pa), 189-201.

Frihy, O.E., 1988. Nile Delta shoreline changes: Aerial photographic study. *J. Coast. Res.*, 4 : 597-606.

Frihy, O.E., El-Fishawi, N. M., and El-Askary, M.A., 1988. Geomorphical features of the Nile Delta coastal plain. A review. *Acta Adriatica.*

Inman, D. and Jenkins, S., 1984. The Nile littoral cell and man's impact on the coastal zone of the southeastern Mediterranean. Proc. 19th Coast. Eng. Conf., Houston, Texas, 1600 - 1617.

Kashef, A A. 1983. Salt-water intrusion in the Nile Delta. *Groundwater*, 21, 160-167.

Kinawy, I., 1977. The efficiency of water use in irrigation in Egypt. Proc. Conf. Arid Land Irrig. in Developing Countries. UNESC/UNDP/ASRT, Cairo.

Klemas, V. and Abdel-Kader, A.M.F., 1982. Remote sensing of coastal processes with emphasis on the Nile Delta. Int. Symp. Remote Sensing of Environment, Cairo.

Loizeau, J.-L. and Stanley, D.J., 1993. Petrological-statistical approach to interpret Recent and sub-recent lagoon subfacies, Idku, Nile Delta of Egypt. *Mar. Geol.*, 111, 55-81.

Milliman, J.D., 1985. Tropical river discharge to the sea: Present and future impacts from man's activities. IOC-UNESCO Workshop on regional co-operation in Marine Science in the central Indian Ocean and adjacent seas and gulfs, Colombo, Sri Lanka.

Milliman, J.D., Broadus, J.M. and Gable, F., 1989. Environmental and economic implications of rising sea level and subsiding deltas - the Nile and Bengal examples. *Ambio*, 18, 340-345.

Raper, S.C.B. Wigley, T.M.L. and Warrick, R.A., 1995. Global sea level rise: past and future. (this volume)

Ross, D.A. and Uchupi, E., 1977. Structure and sedimentary history of southeastern Mediterranean Sea-Nile Cone area. *Am. Assoc. Petrol. Geol. Bull.*, 61, 872-902.

Said, R., 1973. The geological evolution of the river Nile. Int.Conf.Northeast African and Levant Pleistocene Prehistory, Unpubl. Rep., 118 pp.

Said, R., 1981. The Geological Evolution of the River Nile. Springer-Verlag, New York, 151 pp.

Sestini, G., 1976. Geomorphology of the Nile Delta. Proc. Seminar Sedimentology Nile Delta, UNESCO/UNDP/ASRT, Alexandria, 12 - 24.

Sestini, G., 1992. Implications of climatic changes for the Nile Delta. in L. Jeftic, J.D. Milliman and G. Sestini (eds.), *Climate Change and the Mediterranean*. Edward Arnold Publ., London, 535-601.

Shaheen, A.H. and Yousef, S.F., 1978. The effect of the cessation of Nile flood on the hydrographic features of lake Manzala, *Egypt. Hydrobiol.*, 84, 339-367.

Shatal A.A., 1988. The Quaternary aquifer in the Nile valley and delta (abs.). 7th Symp. Quaternary Develop. Egypt, Mansoura Univ./ASRT, 9.

Smith, S., 1986. Effect of Ethiopian drought on water resource management in *Egypt. J. Soil Water Conserv.*, 4, 297-300.

Smith, S.E. and Abdel Kader, A., 1988. Coastal erosion along the Nile Delta. *J. Coast. Res.*, 4, 245-255.

Sneh, Y., Weissbrod, T., and Perath, I., 1975. Evidence from an ancient frontier canal. *Amer. Scientist*, 63 542-548.

Stanley, D.J., 1988. Subsidence in the northeastern Nile Delta: rapid rates, possible causes and consequences. *Science*, 240, 497-500.

Stanley, D.J., 1990.Recent subsidence and northeast tilting in the Nile Delta, *Egypt. Mar. Geol.*, 94, 147-154.

Stanley, D.J. and Warne, A.G., 1993. Nile Delta: recent geological evolution and human impact. Science, 260, 628-634.

Tousson, O., 1934. Les ruines sous marines de la baie d'Abukir. *Bull. Soc. Archaeol. di Alexandrie*, 29.

UNEP, 1988. Implications of climatic changes in the Mediterranean. UNEP/MAP, Athens, 27 pp.

USDA, 1976. Major constraint to increasing agricultural productivity. USDA / AID / Ministry of Agriculture, Egypt. Foreign Agricultural Economic Report 120.

Wahby, S.K. and Bishara, N.F., 1981. The effect of the river Nile on Mediterranean water, before and after the construction of the high dam at Aswan. in River Input to the Ocean Systems, UNESCO Publ., 311-318.

Zaghloul, Z.M., El-Fayoumi, I.F., and Serag El Din, H., 1988. On the problem of sea water invasion in the northeastern part of the delta (abs.). 7th Symp. Quaternary Develop. Egypt. Mansoura Univ./ASRT, 10.

CHAPTER 12

Land Subsidence and Sea-Level Rise: The Case of the Po Delta Region, Italy

G. Sestini

ABSTRACT: The Italian lowland in the NW Adriatic Sea includes the Po River Delta and the Lagoon of Venice, both of which experienced rapid subsidence in the 1950's and 60' in response to the excessive withdrawal of groundwater. Even though overpumping has mostly ceased, the large area of coastal lowland lying below sea level suggests that any rise in local sea level could have severe environmental and economic consequences. Exacerbating the problem is the erosion of local beaches, largely the result of decreased fluvial sediment input. Unless defense structures are built or "accommodation" is practiced, one can expect considerable loss of coastal land and the corresponding impact on local and regional socio-economic well being.

Introduction

The 300 km long coast from Monfalcone (near Trieste) to Rimini (Italy) borders a lowland 20–50 km wide that is generally under 2 m in elevation. This belt is contains lagoons, salt and fresh-water marshes, and reclaimed land (the remnants of former lagoons and distributary bays), separated by channel systems of rivers that flow from the Alps and the Apennines (Figs. 1 and 2).

Future sea-level rise may have considerable impact in this area, because much of the Venetian and Romagna coastal plain lies below sea level (Fig. 2). This situation stems not only from natural subsidence of the Po Basin, but by the drying of the former lagoons and marshes resulting from land reclamation over the last century. Substantial subsidence occurred in the Venice and Ravenna areas, by as much as 1–2 m in the Po Delta, following excessive water extraction between 1950 and 1970. The lowering of ground level in Venice has contributed to more frequent and higher tidal floods that have afflicted the city in recent years, although other causative factors include the artificial reduction of the lagoonal surface through land reclamation, the building of fixed dykes around aquaculture basins, and the armouring and deepening of the lagoon outlets (Sestini, 1992).

J.D. Milliman and B.U. Haq (eds.), Sea-Level Rise and Coastal Subsidence, 235-248.

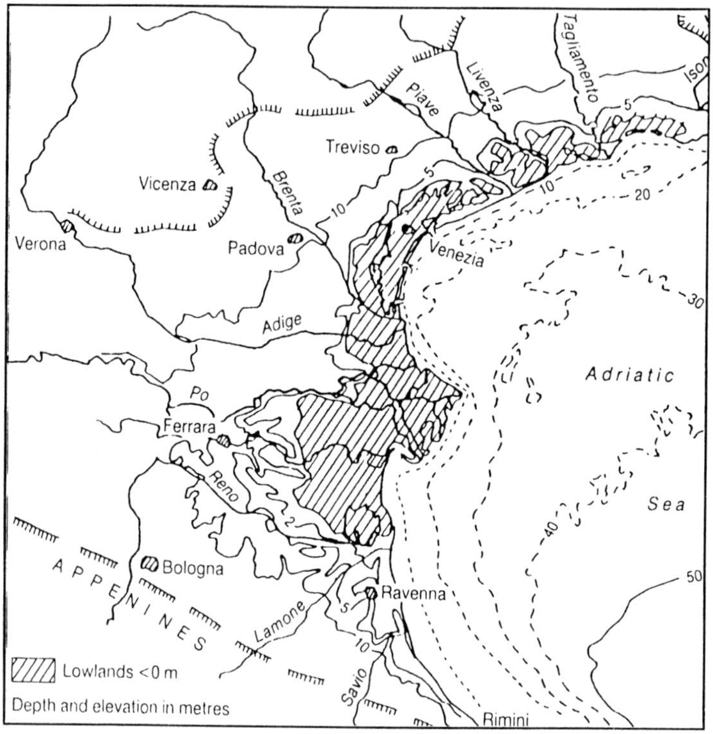

Fig. 1. Northwestern Adriatic Sea, showing major rivers, Po Delta and coastal lowlands (after Sestini, 1992).

Another element of coastal weakness has been the reduction of present-day fluvial influx of sand due to its mining from river beds. As a result, many beaches are retreating, but the natural defences formerly afforded by wooded dunes have been minimized by economic development on the coast. Artificial coastal defence structures also have contributed to beach instability by their interference with sand movement.

Historical Evolution of the NW Adriatic Coast

Deltas are generally unstable geomorphic features, dependent on a balance between terrestrial processes within the river basin (that contribute sediment) and marine processes (that redistribute the sediment). The present morphology of the Po Delta, as well as that of the lagoons and of the rest of the Veneto-Romagna coast, largely results from human intervention that has acted on an ever increasing scale since the 14th century. The present-day Po Delta is a recent geomorphic development, the product of constricted river advance during the last four centuries. In fact, one could consider it almost an artificial body resulting from human activities in the basin; deforestation, expansion of agriculture, confining the river banks, and managing river discharge are the most obvious culprits.

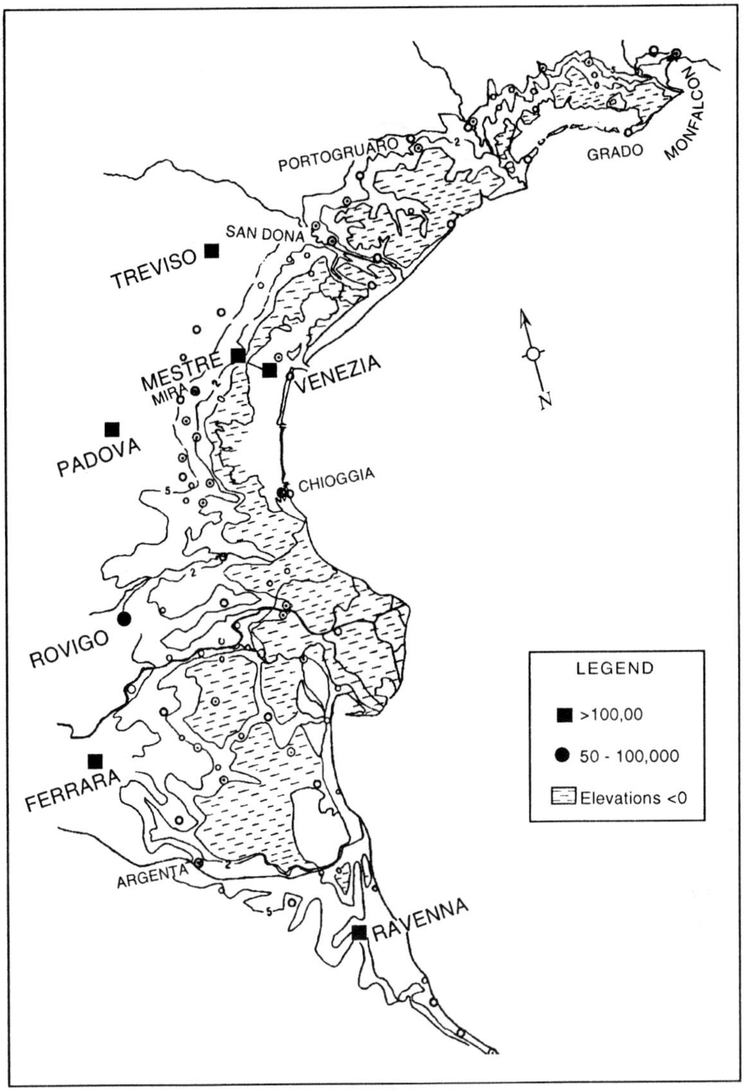

Fig. 2. Venice Lagoon, Po Delta and environs. Note the large amount of coastal land that presently lies below sea level (after Sestini, 1992).

At the peak of the Holocene transgression (6000–7000 BP) the sea stood 5–20 km inland of the present coast (Fig. 3). As sea-level rise moderated, the coast prograded, with the outbuilding of cuspate deltas and gradual enclosure of lagoons by beach ridges and the development of barrier islands (Fig. 4). The Venice Lagoon extended from the Piave to the Brenta Rivers; the Grado Lagoon was formed after the gradual eastward shift of the Isonzo River in the 10th Century (Brambati, 1987). Between Venice and Ravenna in Etruscan and Roman times (2000 to 3000 BP), several branches of the Po, Adige and Brenta Rivers formed modest triangular

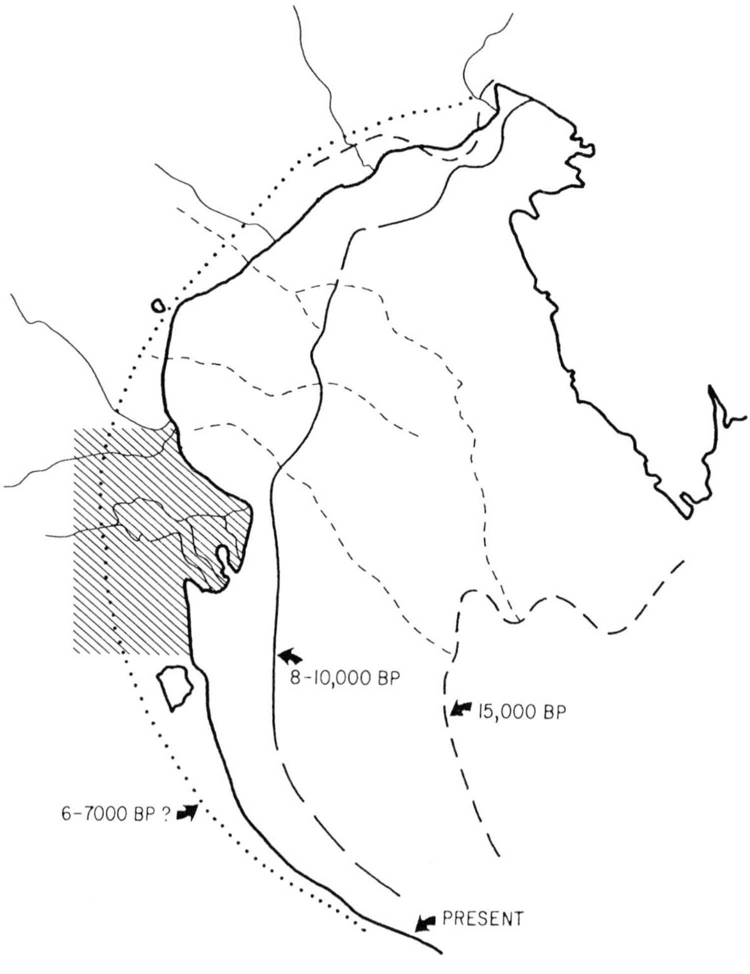

Fig. 3. Approximate Holocene shorelines of the northwestern Adriatic (after Sestini, 1992).

deltas that prograded at fairly modest rates (4 m/year; Veggiani, 1974; Ciabatti, 1979; Fabbri, 1985). Until the early 12th Century the Po discharged south of the present delta through what is now the River Reno and the Volano branch (Fig. 4). A flood west of Ferrara in the mid 12th century diverted the river to the east, and then, in the 14–17th centuries, there was a gradual increase in discharge and sediment load in response to increased upland rainfall (the result of the "little ice age"), deforestation, and increasing flood control (Zunica, 1978). River flow was more confined because of channel deviations and dykes erected by the Republic of Venice to prevent the Po and other rivers from silting in the Venice Lagoon. Consequently, the present lobate delta of the Po began to form and grow at a much faster rate (70–130 m/year), first to the south and then to the east, with the gradual enclosure of elongated marine bays (Fig. 4).

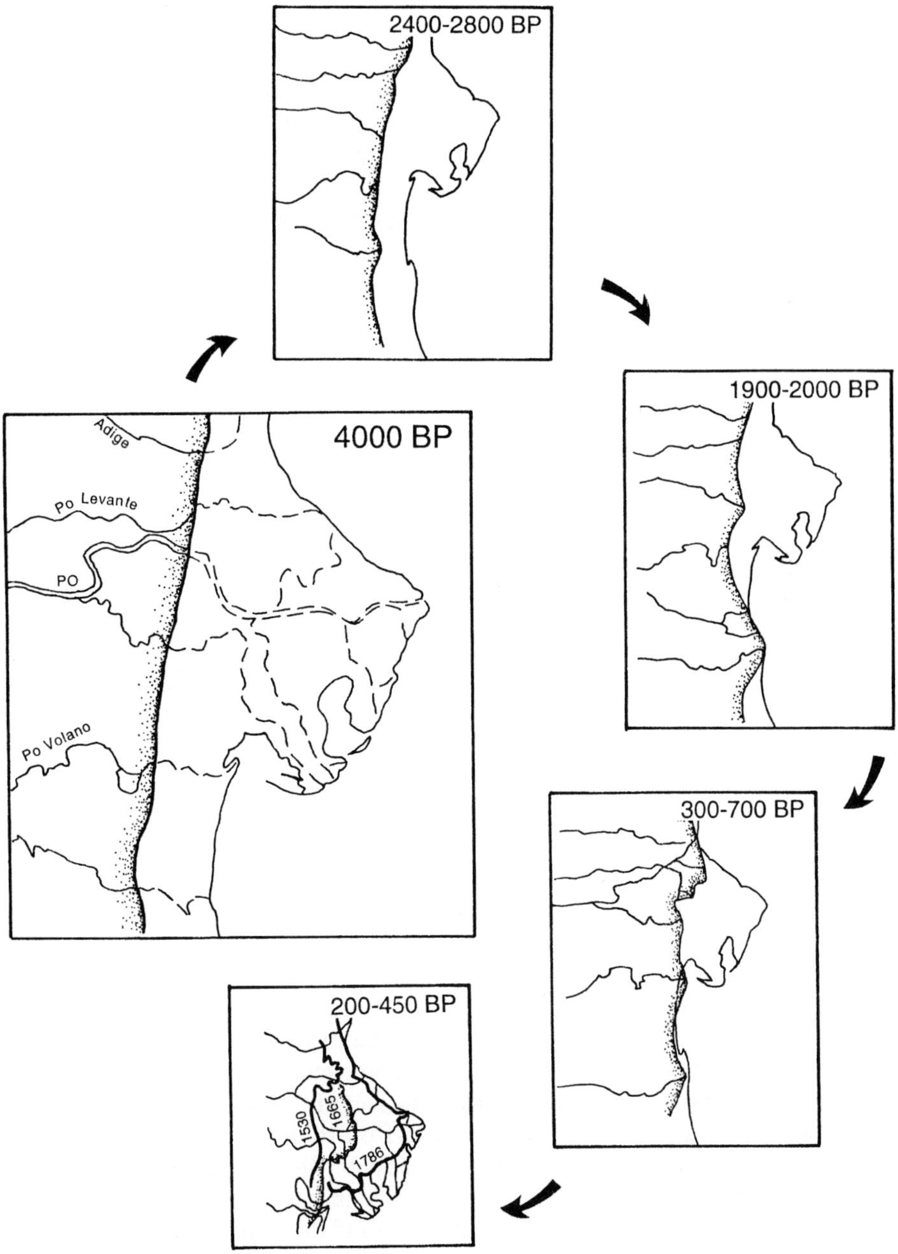

Fig. 4. Evolution of the Po Delta and environs from 2000 BC to present. The boxes on the right show morphological evolution of the delta itself over the past 2800 years (derived from Sestini, 1992).

North of Venice and south of the Po, drainage modifications over the past three centuries have included the diversion and straightening of the lower reaches of Sile, Piave and Reno Rivers, mainly for flood control. These rivers, as well as the

Po and those of Romagna, have always tended to overflow easily due to elevated channels. As a consequence of concentrated discharge, several rivers have built cuspate or digitate deltas. Altogether, riverine and coastal advance during the last 5000 years has resulted in deposition of a coast-parallel sedimentary prism, 10 to 25 m thick, that extends to the 20-m isobath in the Adriatic Sea (Colantani et al., 1979).

The anthropogenic manipulation of the Adriatic coast accelerated drastically in the early 1950's with the development of the harbours and industrial complexes of Porto Marghera (Mestre) and Ravenna, including the destruction of tidal flats, infilling of marshes, construction of deep access-canals, and the seaward extension of long jetties. A further deterioration followed the explosion in tourist facilities. Many summer resorts are essentially narrow urban strips populated with high-rise buildings. Except for the Po Delta, which has remained relatively undeveloped, more than 50 percent of the Isonzo-Rimini shore line is now developed.

The coastline also has a large number of fixed structures, such as seawalls, jetties, groins and breakwaters, mostly build since the end of WW II, to protect new agricultural lands and beaches from erosion and to regulate the entrances of several canalized harbours. Longshore transport of sand has been particularly hampered by the groins and jetties, which have trapped sand upstream and caused erosion downstream. Offshore breakwaters along the Romagna coast initially retarded coastal erosion, but water tended to stagnate behind them, resulting in the formation of tombolos, gradual infilling, and the eventual resumption of beach erosion (Cencini et al., 1979).

Subsidence and its Impact

Natural Subsidence

At the beginning of the Quaternary, the sea formed a wide gulf extending far into the Po valley. Active subsidence resulted in the accumulation of 1500–2000 m of fluvial-lacustrine-littoral sediments (Sestini, 1992). In addition to sediment compaction, the basin was influenced by deeper structural processes, with deformation progressing towards the north and northeast. The average computed Pleistocene subsidence is 0–5 mm/year for the present compacted section, but as much 1 mm/year for uncompacted sediments (Bondesan et al., 1986).

Radiometric dates of material recovered from borings indicate that for the last 6000 years subsidence rates have been about 1.0–1.5 mm/year (Fontes and Bertolami, 1972). The effects of geological subsidence are obvious in the presence of Roman pavements at water depths of 3 to 4 m off Ravenna, Aquilei, and Spina, and by the low elevation of ancient dune ridges south of the Po River.

Although geological subsidence may have been responsible for the development of the coastal lagoons, this and sea-level rise of 1.2 mm/year did not prevent the extension of deltas and a general shoreline advance until the first part of this century. In the Venice Lagoon, natural subsidence is thought to have caused a ground lowering of 3 cm since the early 1900's (Gatto, 1984); 20–30 cm in 50

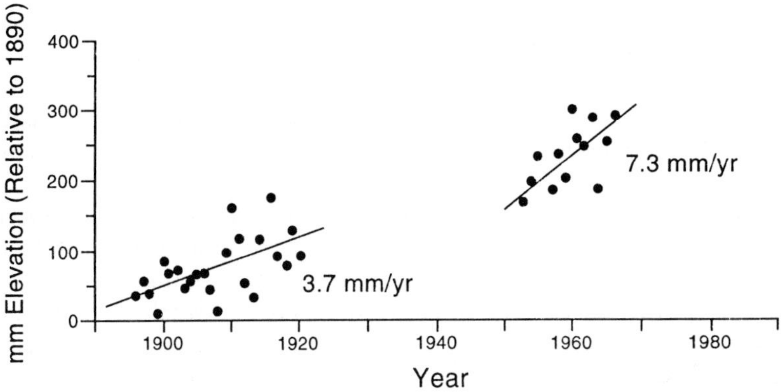

Fig. 5. Sea-level rise in Venice as measured by tide-gauge records. Note the abrupt increase in the rate of rise after WW II, from an average of 3.7 to 7.3 mm/year; locally , however, subsidence was much greater (see Fig. 6) (derived from Emery et al., 1988).

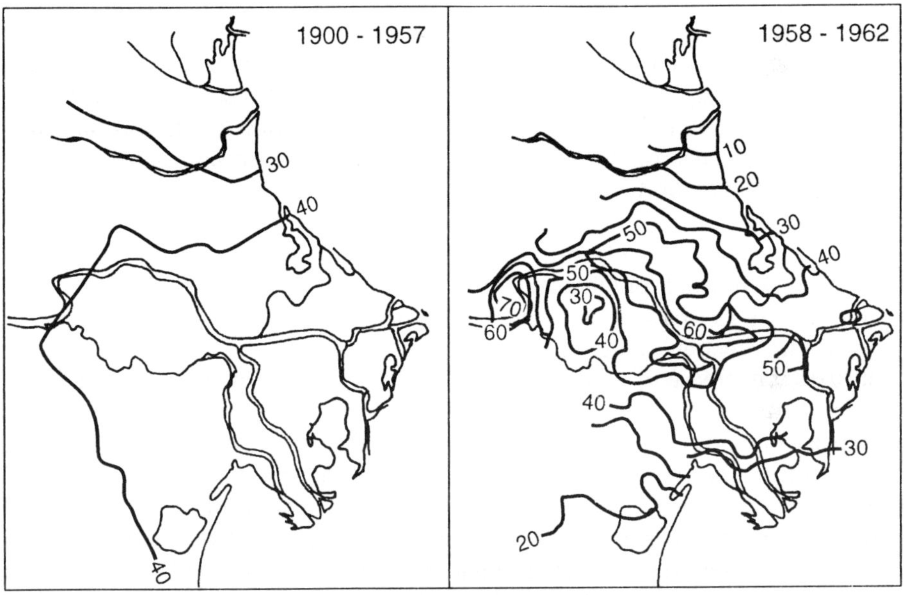

Fig. 6. Subsidence on the Po Delta between 1900 and 1962 (in cm); note the dramatic increase in subsidenc between 1958 and 1962 (after Caputo et al., 1970).

years in the Po Delta (Fig. 6); and more than 40 cm since 1902 in the Ravenna area (Carbognin et al., 1984).

ANTHROPICALLY-CAUSED SUBSIDENCE

Subsidence due to Land Reclamation – Reclamation of lagoons and marshes was generally followed by a rapid sinking of the newly drained terrain, as much as

1–2 m. A number of phenomena are thought to have been responsible: compaction of clays and silts, oxidation of peats, volume reductions in clays due to chemical reactions, and lowering of the water table (Bondesan et al., 1986). This type of subsidence, however, decreased and mostly ceased within one or two decades after reclamation. Its is probably no longer significant in areas drained before the early 1960's.

Subsidence due to Groundwater Extraction – Starting in 1969, and through the 1970's, intensive hydrological research was carried out by the C.N.R. Institute for the Study of the Dynamics of Large Masses, in Venice, with the purpose of investigating the relation between accelerated subsidence and groundwater extraction in Venice and Ravenna (see Fig. 5). This included detailed inventory and analysis of stratigraphic, hydrological, geodetic and soil mechanics data. Statistical analyses unequivocally showed that 70% of the 1952–1969 land subsidence was due to excessive withdrawal of groundwater (Carbognin et al., 1981).

In the Venice region the extraction of artesian water began modestly in the 1930's with the first industrial developments at Marghera. With increased pumping in the early 1950's, the hydraulic heads (piezometric levels), which had been above ground level (except in Venice during World War II), declined below the surface. Water withdrawal from 50 wells, with a maximum rate of 460 liters/sec, occurred at four out of six aquifers in the upper 350 m of the confined Pleistocene aquifer. In Venice, water extraction was about 50 times less, but subsidence was almost the same magnitude because the geological section has more compressible soils.

Between 1952–1968 geodetic surveys showed an average subsidence of 6.5 mm/year in the Mestre area, 5 mm/year in Venice, with maxima of 17 and 14 mm/year in 1968–69, respectively. Overall, between 1952 and 1969 average subsidence was more than 110 mm at Marghera and 90 mm in Venice (Gatto and Carbognin, 1981).

In the early 1970's decreased industrial activity, restricted drilling for irrigation water in the Venetian plain, and the prevention of well use in Venice resulted in a substantial decrease of ground-water exploitation. With diversification of water supplies, there was a rapid recovery of piezometric levels and a drastic reduction in subsidence rates. There also was a surface rebound in Venice of 2 cm, which slowed the relative rise of sea level. By 1980 artesian heads were back to their pre-1952 values, and since then subsidence has stabilized to its natural level. Numerical models, run on the basis of a statistically valid correlation between subsidence and piezometric depression, showed a remarkable ability to simulate past behaviour, thus allowing for a valid prediction. A stabilization of subsidence is thus predicted at the rate of 3mm/year until 2000 (Gatto and Carbognin, 1981).

In the Po Delta natural subsidence was balanced by sedimentation until 1950. The exploitation of subsurface methane-bearing waters had started in the 1940's, but increased dramatically in the 1950's and early 1960's. Total withdrawal from 1400 wells increased from 60×10^6 m^3 in 1950 to 300×10^6 m^3 in 1959, which drilled progressively deeper aquifers (5 beds between 100 and 600 m). Whereas subsidence until 1950 was estimated to have been <1 cm/year, afterwards it was 25–30 cm/year in certain locations; between 1958 and 1962 subsidence exceeded 50 cm (Fig. 6) and locally was several meters. When a trial cessation of pumping in 1960 resulted in a return to natural subsidence rates, pumping for methane-bearing waters was stopped over the entire delta in 1965. However, local subsidence in 1982 was still as high as 2 cm/year, the result of the considerable extraction of groundwater by industry and agriculture in the eastern parts of Rovigo and Ferrara provinces.

Land subsidence had long been noted round the city of Ravenna, but it became particularly apparent in the early 1960's after development of industrial sites. By 1972, 877 wells had penetrated 9 confined aquifers at depths between 90 and 430 m. Maximum drawdown of 40 m was observed in 1977. An asymmetric zone of depression developed around the industrial area, with subsidence exceeding 3 cm/year between 1972 and 1977, about 30 times greater than in the 1950's. Exploitation of natural gas reservoirs, quite intensive in the Ravenna region until early 1970's, contributed somewhat to land subsidence, although not nearly as much as related to water extraction. Accelerated subsidence also was noted in the adjacent reclaimed marshlands, 90 percent of which lie more than 1 m below sea level. To maintain buildings and historical monuments, continuous pumping of these low-lying areas had to be maintained. Maximum annual subsidence was 8 cm in the Ravenna historical district, and total subsidence between 1949 and 1977 was 1.20 m (Carbognin et al., 1984).

The impact of subsidence on the shoreline has been considerable. Since the early 1970's, coastal subsidence between the Reno and Rimini Rivers locally has exceeded 25 cm, resulting in several tens of meters of shoreline retreat. Subsidence, however, is not the only cause of shoreline retreat; at Lido Adriano, for instance, the beach retreated 126 m between 1957 and 1977. Another consequence of aquifer depression was an upwelling of salt water from the deeper aquifers, causing irreversible pollution to the system.

In 1980 a large area of lowered piezometric levels extended over the western and southern parts of the Ravenna region due to increased groundwater withdrawal for irrigation, seaside resorts, and new industrial complexes established in the outskirts of Ravenna. Measures to control the pumping have been introduced. Maximum subsidence in 1982, for instance, was "only" 2 cm/year, with a further attenuation of subsidence by 1986.

Impact of Changes to River Sediment Discharge

A number of detailed studies have been made on the state of the environment in NW Adriatic coast in recent years. Included in these efforts have been studies dealing with shoreline retreat northeast of Venice (Brambati, 1984, 1987; Brambati et al., 1978), the Po Delta (Bondesan and Simoni, 1983), and Romagna (Cencini et al., 1979, Bondesan et al., 1978). At present all the Veneto and Romagna beaches are unstable, and most began retreating in the 1950's (Fig. 7). Retreat has been significant, especially since the stormy years of 1960–70. In Romagna, shoreline advance was common until about 1915, and since 1935 retreat has prevailed, especially at the Reno and Fiumi Uniti Deltas, which have experienced a straightening of the shoreline and a steepening of the foreshore (Bondesan et al., 1978).

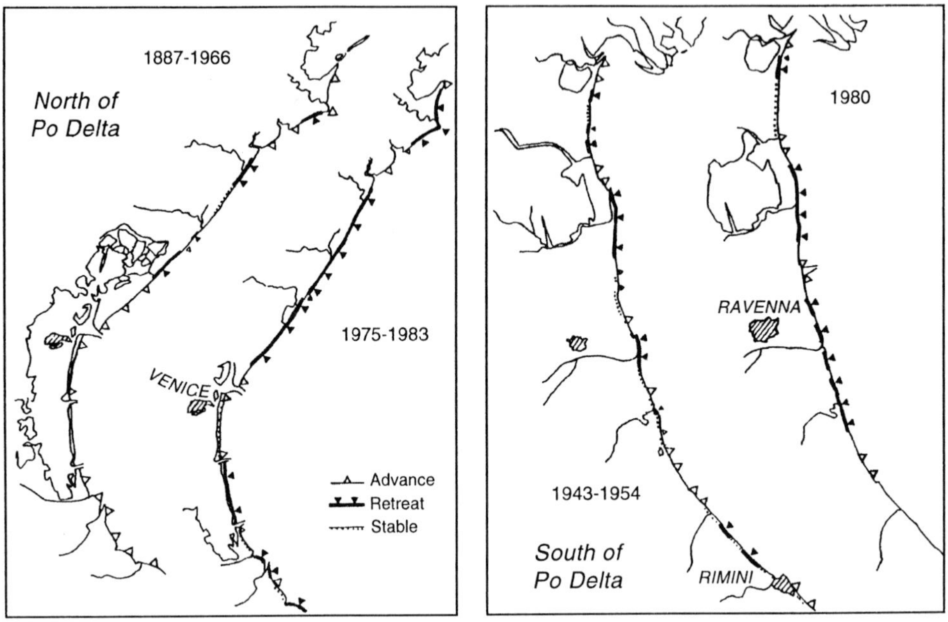

Fig. 7. Stability of the Veneto and Romagna coasts; coastal retreat is far more common that coastal advance (after Sestini, 1992).

The instability and retreat of these shorelines are partly the result of natural phenomena, but they also reflect the consequences of human activities on the coast and hinterland. Natural erosion occurs through strong littoral drift generated by the high (>1.5 m) oblique SE-NE waves and storm surges (often associated with high tides) and/or (aeolian) beach deflation during the winter season. Unfortunately sand removal has not been offset by fluvial input of new sand. In part the decreased fluvial sediment flux stems from decreased up-basin erosion, the result of decreased farming and increased reforestation of the catchment basins. Much sediment also is retained in reservoirs behind hydroelectric dams, and until the late 1970s industrial dredging of sand removed large quantities of sand from river beds, especially in the flood plains. Between 1965 and 1973, for example, the Po's sediment load decreased from 16.9 to 10.5×10^6 tons/year. The decreased bed load, about 25 percent of the total suspended load, was barely sufficient to satisfy the estimated 3.1×10^6 tons/year necessary to maintain the delta margin in equilibrium (Bondesan and Dal Cin, 1975; Dal Cin, 1983). Similarly, the solid discharge of the Isonzo, Tagliamento and Romagna Rivers has almost halved in the last few decades compared to the years before 1950.

Sand dredging from the Po River bed, officially reported as 100×10^6 tons between 1958 and 1961, was as much as six times greater (Dal Cin, 1983). In Romagna, the official 8×10^6 tons of sand extraction between 1957 and 1971 was also short of reality (Cencini et al., 1979). The sand still delivered by the Po River now partly comes from the erosion of the river bed itself.

Sand input must increase substantially if this situation is to be remedied, as in places the river bed is nearing or falling below sea level. Even if all dredging operations were to cease immediately, it is estimated that the sand deficit would last another 30 to 100 years (Dal Cin, 1983).

Discussion

The rate of natural subsidence is generally small but not insignificant when considered over periods of decades. Assuming that the pre-WW II sea-level trend for Venice (3.7 mm/year; Fig. 5) attained as groundwater withdrawal is reduced, sea level in Venice over the next 40 years should rise by about 15 cm, of which 9 cm will be due to subsidence (assuming that eustatic rise of sea level remains at 1.5 mm/year). Sea-level rise will be correspondingly greater on the Po Delta and at Revenna, where subsidence rates are higher. An accelerated rise in eustatic sea level (Raper et al., this volume), of course, would only increase the local rise of sea level. Therefore it is not unlikely that sea level in the Venice may exceed 17 cm by the year 2035, and in Ravenna it may be as much as 35 cm.

Because reclamation is no longer allowed, it no longer can be considered a cause of subsidence; however, large areas of former lagoons and marshlands now lie 1 to 2 m below MSL. The experience of the Po Delta and the Venice-Mestre areas clearly indicates that land subsidence due to the over-pumping of groundwater can be controlled with better management of water resources. The consequences of previous subsidence, however, can only be partly reversed; thus large parts of the Veneto-Romagna lowlands will remain below sea-level. Marine flooding of these low-lying areas can be prevented only by the artificial protection afforded by dykes and embankments and by mechanical drainage facilities.

In the Po Delta and in Ravenna the negative effects of subsidence have been considerable. In the Po Delta, much inconvenience was caused by the inversion of land gradients towards the interior; the sinking of river and canal levees, bridges and pumping stations; and by the marine flooding at the edges of the delta, especially in the bays (Carbognin et al., 1978; Bondesan, 1988). In Venice the problem is compounded by other phenomena that have led to more frequent and higher tidal flooding over the last 20 years. A gradually rising sea level would aggravate the impact of storms (e.g., Emanuel, 1987), and storm surges in fact have damaged parts of the NW Adriatic coast several times in the past.

Taken alone, a 15–20 cm rise in local sea level might not have a major effect on the beach resorts, but given the expected negative budget of river and coastal sands for years to come, the current trends of coastal erosion should continue. To mitigate this erosion, some protective sea walls (e.g. Grado, Caorle, Pellestrina) will have to be raised. Nevertheless, lagoons and inland area may be flooded more frequently. In addition, higher tidal levels could mean greater erosion of tidal channels and flats as well as greater salinity, and eventually gradual alteration of ecosytems that have been vital for biological resources (aquaculture, wildlife) in this area. Increased infiltration of saline waters into the adjacent low lands, in fact, may be more serious than the sea-level rise itself.

The course of future societial and governmental responses to sea-level rise and adaptation of effective counter-measures, of course, is still speculative. Evaluation of economic impacts on the coastal lowlands will depend largely on land use and economic activities at that time, as well as projected demographic and economic trends.

Much of the present economy of the Veneto-Romagna coastal belt lying below sea level is based on the exploitation of ecologically vulnerable lowlands: the shoreline (e.g. tourism), lands lying below sea level (agriculture) and lagoons (fishing and aquaculture). These activities lie at the root of much of the present deterioration, yet they also are most at risk. Whereas population in Italy has reached a level of zero-growth (and, in fact, may decrease in the future), the economy has reached a stage of high complexity that involves immovable commercial and industrial investments that are interrelated with external trade as well as social and political factors such as tourism and EC policies.

The question, therefore, may be whether total protection by walls, dykes, etc. is rational or economically feasible? Coastal defenses in some cases are unavoidable, e.g. Venice and Ravenna. At other places, however, retreat or redeployment of some landuses might be a better option, and in some cases certain industries (such as power plants, factories, coastal facilities) may have to move if or when they no longer can be sustained economically.

New approaches and adaptions, however, might be considered. The rate shoreline retreat, for example, can be minimized no matter the rate of sea-level rise if the sand supply is maintained, which, among other things, could facilitate artificial nourishment of some beaches. However, adaption will be easiest for undeveloped beaches, which may necessitate that future tourist accomodations be located further inland.

With regard to agriculture in the reclaimed lands lying below sea level, it might be more economical if some areas are allowed to return to their original lagoonal state; fishing in such lagoons, in fact, is a more efficient and remunerative activity than cereal farming. In the Po Delta and elsewhere, these lagoons and marshes could act as buffer zones between the open sea and higher land, and in parts could become nature reserves. The Camargue in the Rhone Delta (e.g. Corre, 1992) is an excellent example of how eco-tourism can create a market value greater than agricultural output.

Conclusions

1. Natural subsidence in the coastal belt of the NE Adriatic (Italy) amounts to 1.5–5 mm/year. This could add several cm to a possible 10–20-cm sea-level rise by the year 2030. As a result, relative sea-level rise would be 17–23 cm in Venice, 45–50 cm in the Po Delta, and 30–40 cm in the Ravenna region.

2. Reclamation and excessive ground-water pumping have locally increased subsidence by 1–3 m. Although stopped and no longer a problem, this type of subsidence has produced large areas are now below sea level and thus more vulnerable to river and storm surge flooding. At present, these low-lying have to be drained mechanically.

3. The shoreline in many places is retreating due to decreased sediment input by rivers and the disruption of coastal sediment movement by fixed structures. Even if dredging is stopped now, sand deficits are expected to last several decades. Locally, beach erosion has been aggravated by land subsidence.

4. The impact of a relative rise of sea level on the Romagna and Venetian coasts could be very serious for the maintenance of recreational beaches, for the safety of towns located on the coast and within the lagoons (especially Venice), for the hydrology and ecology of the lagoons, and for water and soil quality (i.e., increased salinity).

5. Extensive protection measures must be undertaken, if economically justified. Alternative responses, however, could be sought, at least in some areas, such as the return of low-lying reclaimed lands to lagoonal conditions, forming buffer zones dedicated to fishing, aquaculture and wildlife protection.

References

Bondesan, M. and Simeoni, U., 1983. Dinamica e analisi morfologica statistica dei litorali del delta del Po e alle foci dell-Adige e del Brenta. *Mem. Scienze Geol.*, 36, 1-48.

Bondesan, M., Calderoni, G., Dal Cin, R., 1978. Il litorale delle province di Ferrara e di Ravenna (Alto Adriatico); evoluzione morfologica e distribuzione dei sedimenti. *Boll. Soc. Geol. Ital.*, 97, 247-287.

Bondesan, M. and Dal Cin, R., 1975. Rapporto fra erosione lungo i littorali emiliano romagnoli e del delta de Po, e attivita estrattiva negli alvei fluviali. *Cave e Assetto del Territorio*. Itlaia Nostra e Regione Emilia Romagna, 126-137.

Bondesan, M., Minarelli, A. and Rosso, D. 1896. Studio dei movimenti verticali del suolo nella provincia di Ferrara. *Studi Idrogeo. sulla Pianura Padana*, 2, CLUP, Milan.

Brambati, A., Erosione e difesa delle spiagge adriatiche. Boll. *Ocean. Teoricae Applicata*, 2, 91-184.

Brambati, A., 1987. Studio sedimentologico e marittimo-costiero dei litorali del Friuli-Venezia Giulia. Regione Autonoma Friuli Venezia Giulia. Direz. Regionale Lavori Pubblici, Trieste, 67 pp.

Brambati, A., Marocco, R.G., Catani, V., Carobene, L., and Lenardon, G., 1978. Stato delle conoscenze dei litorali dell'Alto Adriatico e criteri di intervento per la loro difesa. Mem. Soc. Geol. Ital., 19, 389-398.

Caputo, M., Pieri, L., and Unguendoli, M., 1970. Geometric investigations of the subsidence in the Po delta. CNR, 1st Studio Dinamica Grandi Masse, Venezia, TR 2, 37 pp.

Carbognin, L., Gatto, P. and Mozzi, G. 1981. La riduzione altimetrica del territorio veneziano. 1st Veneto Sci. Lettere, Arti, Rapp. Studi, VIII, 55–83.

Carbognin, L., Gatto, P., and Mozzi, G., 1984. Case history 9.15. Ravenna, Italy. in J.F. Polland (ed.), *Guidebook to studies of land subsidence due to groundwater withdrawal.* UNESCO, 291-305.

Catani, G., Marocco, R., Brambati, A., Carobene, L., and Lenardon, G., 1978. Indagini sulle causes dell-erosione nel tratto orientale di Valle Vecchia (Caorle, Adriatico Sett.). *Mem. Soc. Geol. Ital.*, 19, 399-405.

Cencini, C., Cuccoli, L., Fabbri, P., Montanari, F., Semeloni, F., Torresani, S., and Varani, L., 1979. Le spiagge di Romagna: uno spazio da proteggere. *Consiglio Nazionale delle Ricerche*, I., Bologna.

Ciabatti, M., 1979. Ricerche sull'evoluzione del delta Padano. *Giornale di Geologia*, Sr. 2, 34, 26 pp.
Colantoni, P., Gallignani, P. and Lenaz, R., 1979. Late Pleistocene and Holocene evolution of the North Adriatic continental shelf. *Mar. Geol.*, 33, 41–50.
Corre, J.-J., 1992. Implications des changements climatiques etude de cas: le Golfe du Lion (France). In: L. Jeftic, J.D. Milliman and G. Sestini (eds.), *Climatic Change and the Mediterranean*. Edward Arnold, London, 328-427.
Dal Cin, R., 1983. I litorali del delta del Po e alle foci dell'Adige e del Brenta caratteri tessiturali e dispersione dei sedimenti, cause dell'arrentramento e previsioni sull'evoluzione futura. *Boll. Soc. Geol. Ital.*, 102, 9–56.
Emery, K.O., Aubrey, D.G., and Goldsmith, V., 1988. Coastal no-tectonics of the Mediterranean from tide-gauge records. *Mar. Geol.*, 81, 41-52.
Emanuel, K.A., 1987. The dependence of hurricane intensity on climate. *Nature*, 326, 483-485.
Fabbri, P.C., 1985. Coastline variations in the Po delta since 2500 BP. *Zeitschr. Geomorph. Suppl. Bd.*, 57, 155-167.
Fontes, J.C.H. and Bertolami, G., 1973. Subsidence of the Venice area during the past 40,000 years. *Nature*, 244, 339-341.
Gatto, P., 1984. Il cordone litorale della laguna de Venezia e le cause de suo degrado. *1st Veneto Sci. Lett. Arti. Rapp. Studi.*, 9, 163-193.
Gatto, P. and Carbognin, L., 1981. The Lagoon of Venice: natural environmental trend and man-induced modification. *Hydrol. Sci. Bull.*, 26, 179–391.
Raper, S.C.B., Wigley, T.M.L., and Warrick, R.A., 1995. Global Sea-level rise: past and future. (This volume).
Sestini, G., 1992. Implications of climatic changes for the Po delta and Venice Lagoon. In: L. Jeftic, J.D. Milliman and G. Sestini (eds.), *Climatic Change and the Mediterranean*. Edward Arnold, London, 428-494.
Veggiani, A., 1974. Le variazioni idrografiche del basso corso de Fiume Po negli ultimi 3,000 anni. Padusa, Rovigo, 1-2, 1-22.
Zunica, M., 1978. Il Delta de Po, elementi per un approccio ecologico. *Boll. Mus. Civico Storia Naturale, Venice*, 24, 19-30.

CHAPTER 13

The Niger Delta and Sea-Level Rise

A. Chidi Ibe

ABSTRACT: The Niger Delta is arcuate in plan form and low lying, with a maximum elevation about 3m above mean sea level on the sandy barrier islands that border the sea. The barrier islands, approximately 2,000 km^2 in area, are cut by tidal channels through which oceanic waters and tides gain access to an extensive mangrove swamp (7,000 km^2). Presently, freshwater swamps are flooded for 3 to 5 months of the year while the barrier islands are in retreat, with rates of erosion reaching several tens of meters annually, the result of both natural and man-induced factors. Coastal settlements are continually displaced, agricultural and recreational grounds destroyed, harbour and navigational structures disrupted, and oil-producing and export-handling facilities dislodged.

With the continuing subsidence of the delta and the decreased sediment input into the coastal area from fluvial sources, the projected acceleration of eustatic rise in sea level will accentuate wave erosion and acute flooding, leading to the devastation of vast areas of coastal land. Additional effects may include loss of wetlands, salinisation of coastal aquifers and rivers, and modification of the fragile ecosystems that make up the delta. When these effects are viewed against the background of the Niger Delta as home of an oil industry, which contributes over 90% of Nigeria's export and foreign exchange income, as well as the nation's new industrial "heartland" and established 'bread basket' of the south, Nigeria's response to rising sea in this low-lying and subsiding terrain must be addressed.

Introduction

The Niger Delta in Nigeria extends along the Gulf of Guinea coast, from the mouth of the Benin River for a distance of about 450 km, with its eastern flank ending at the Imo River in the east (Fig. 1). Inland, the delta begins where the River Niger, rising from the Futa 'Jalon Highlands in Guinea and flowing through its confluence with the Benue, bifurcates into the Nun and Forcados Rivers and forms a system of natural waterways through which the Niger River winds its way to the sea. There are 20 river mouths on the delta coast (Fig. 1).

This definition treats the Niger Delta as a geomorphic unit. However, due to the general association of the Niger Delta with petroleum occurrence and the fact that

J.D. Milliman and B.U. Haq (eds.), Sea-Level Rise and Coastal Subsidence, 249-267.

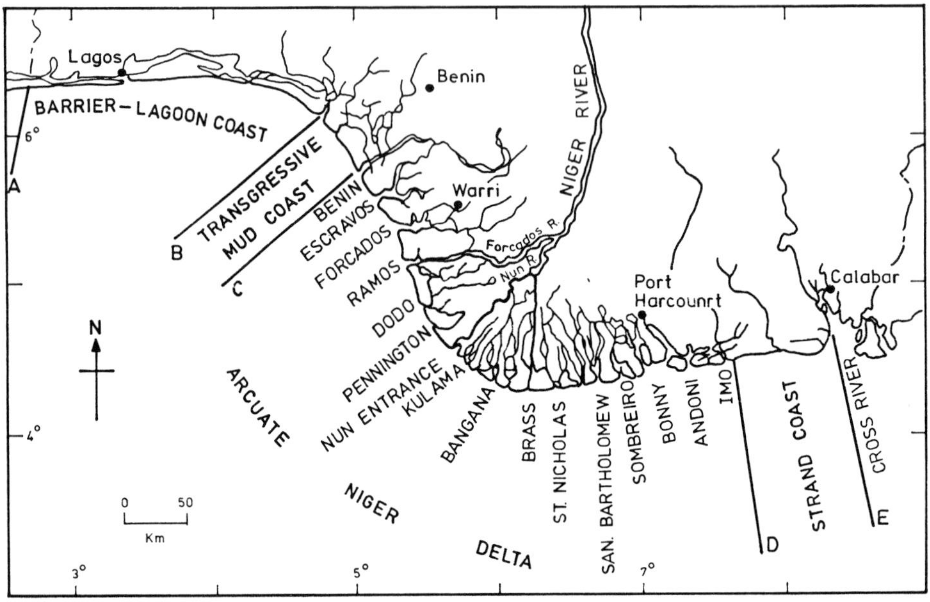

Fig. 1. Location and regional map of coastal Nigeria (from Ibe, 1988).

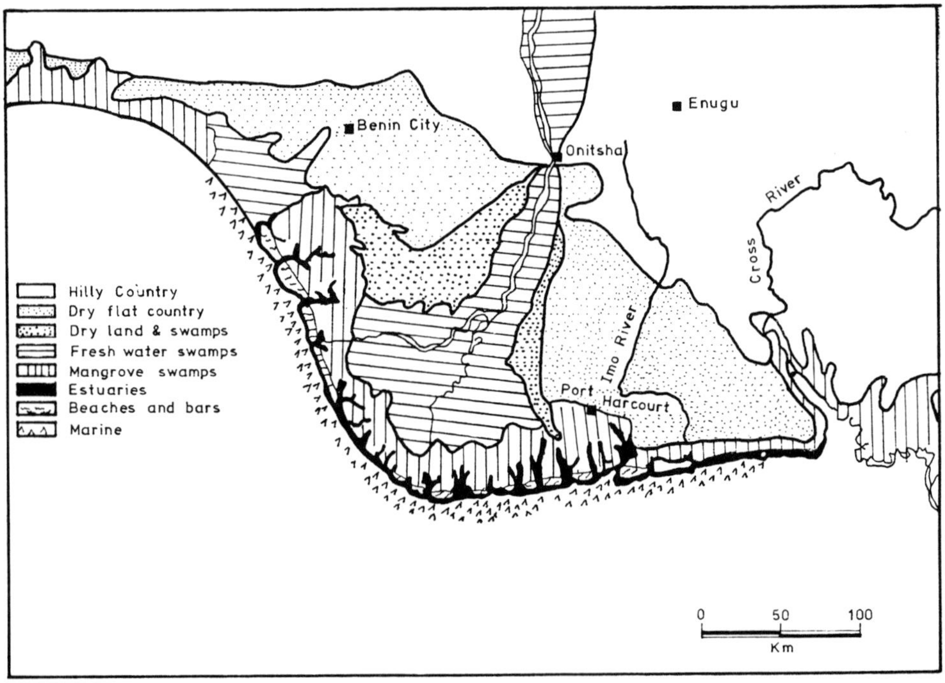

Fig. 2. Sketch map of the main physiographic features of the Niger Delta (after Short and Stauble, 1967).

early studies (e.g. Allen, 1965a, 1965b; Short and Stauble, 1967; Frank and Cordry, 1967; Weber, 1971) were related to hydrocarbon exploration and production, there has been a tendency to regard all hydrocarbon areas in the coastal zone as being within the Niger Delta. Thus defined, the Niger Delta also would include the transgressive mud coast and the strand coast (Fig. 1). Inland, it would include many of the fresh water swamps and parts of the upper flood plain up to Onitsha (Fig. 2), which is loosely described as the northern Niger Delta. While there may be merits (particularly relating to the tectonic evolution and paleostratigraphy of the delta) in such a broad classification, the more strictly geomorphic classification is adhered to in this paper.

The Niger Delta is in the humid equatorial forest zone with a large annual rainfall of more than 350 cm and temperatures that sometimes exceed 38°C. Humidity is consistently greater than 80%.

The delta is a complex, dynamic environment, particularly because of the interface of a large river and an ocean-dominated coast (Dangana, 1981). The Niger River and its distributaries flood, erode, transport and shape their channels and the surrounding land, while the sea scours and floods. Erosion dominates, although deposition occurs locally. The lithological characteristics of the delta are essentially alluvial and friable, which render them vulnerable and compound the problems of erosion and flooding.

The main relief of this area is provided by the river levees as the highest land sloping gradually away from the banks to the backswamp (Egborge, 1981). Along the coast, beach ridges, rising about 3 m, constitute the highest ground on barrier islands. At present, the vast expanse of low lying backswamps is inundated by river bank overflow at least thee months annually (sometimes five months depending on the yearly rainfall pattern) and periodically inundated during spring high tides. The barrier islands that rim the delta are continually under attack by wave erosion and tidal flooding, with rates of shoreline retreat locally reaching several tens of meters.

It is against this background that this paper examines the spectre of the projected increase in eustatic rise in sea level, which in the Niger Delta will be accentuated by both natural and man-induced subsidence as well as other effects such as decreased input of fluvial sediment into the coastal area. The socioeconomic implications of rising sea level are discussed as a prelude to suggesting possible mediation for the predicted phenomena.

Tectonic and Sedimentary Evolution

The tectonic evolution of the Tertiary Niger Delta is closely related to events dating back to the Cretaceous Period, which culminated in the opening of the South Atlantic (Stonely, 1966; Burke et al., 1971; Wright, 1970; Grant, 1971; Manchens, 1973; see Fig. 3). Three Cretaceous rift systems, described as a ridge-ridge-ridge triple junction (Burke et al., 1970) and also as a ridge-ridge-fault junction (Grant, 1971), occupied the position of the present delta. Two arms of the triple junction have continued to spread, bringing the two continents of Africa and South America

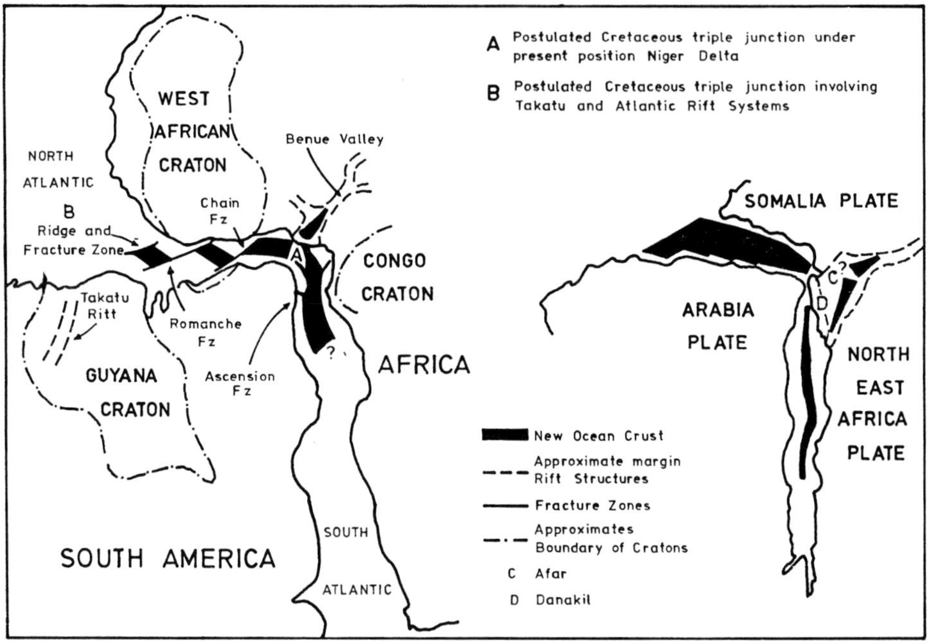

Fig. 3. Cartoon showing the Gulf of Guinea in the early Cretaceous compared with the present-day Red Sea. The Atlantic opens about triple RRR junctions situated under the Niger Delta and north of the Takatu Rift. Craton boundaries based on Hurley and Rand (1969); fracture zones at the African coast are near their margins (after Burke et al., 1971).

to their present positions. The third arm (the Benue Trough) is believed to have spread for 30 million years and closed at the end of the Cretaceous.

The delta is framed by a set of older stable megatectonic elements (Merki, 1970). Hospers (1965, 1971) described the various basement blocks and pointed out the predominance of northeast-southwest and northwest-southeast trends in the megatectonic framework.

The post-drift development of the Niger Delta was marked by instability and subsidence along the rift margins. A series of transgressions and regressions dating from the Paleocene to recent are related to the Santonian folding in the Benue Trough, fluvial input from coastal rivers, and clastic material derived from the trough and environs. Short and Stauble (1967) discussed the advance of the delta through paleogeographic shorelines (Fig. 4). Another account of this growth was presented by Evamy et al. (1978).

General Geomorphology

Each of the barrier islands that rims the delta is composed of alternating sandy ridges and muddy hollows, indicative of earlier or successive strand lines or barrier growth. These lie behind an active beach that is typically narrow (about 50 m) because of the relatively small tidal range (microtidal coast). The beach sands

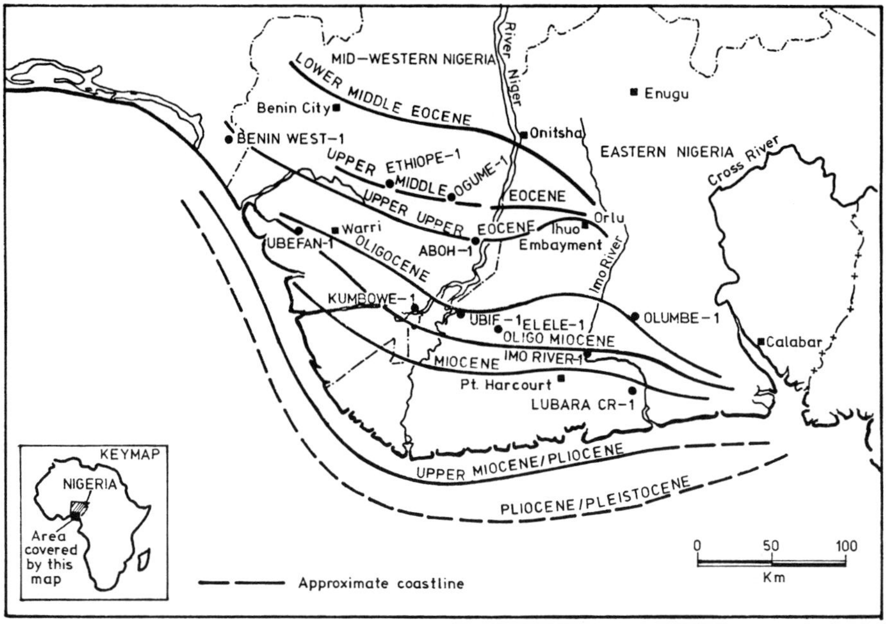

Fig. 4. Paleogeography of the Tertiary Niger Delta, showing stages of delta growth (after Short and Stauble, 1967).

consist almost invariably of fine-grained quartz sand (Ibe et al., 1984). Because of the sandy ridges, the barrier islands stand a maximum of 2 to 4 m above mean sea level. The lengths of the islands range from 5 to 36 km, while widths range from a few hundred meters to about 11 km.

The delta is characterised by trains of long spits that grow in the direction of longshore movement of sand. Allen (1965b) observed that the advance of most spits along the shore partly is a response to erosion. Mouth bars in front of almost all the river mouths of the Niger Delta often are associated with the development of micro-deltas. Ebb tide flows diverging from the river entrances, flood currents converging into them, and wave- generated longshore drifts all interact to determine the shapes of the mouth bars (Allen 1965a).

The barrier islands of the Niger Delta are backed by an extensive region of mangrove swamps. This zone has evolved primarily through the bifurcation of the Niger and subsequent splitting of the bifurcations. The swamp extends over an area of 9000 km^2 between the Benin River in the west and the Calabar-Rio del Rey estuary beyond the eastern limit of the delta. A maximum width of 30 to 40 km is attainted on the flanks of the delta where the swamps fill the area between the boundary of the river floodplain in the north and the terrace of late Tertiary sediments.

The mangrove swamp is essentially a vegetated tidal flat, lying between mean low and high tides, and thus regularly inundated. Morphological features in the swamp have been shaped largely by the above forces as well as the rivers that criss-cross it from the hinterland.

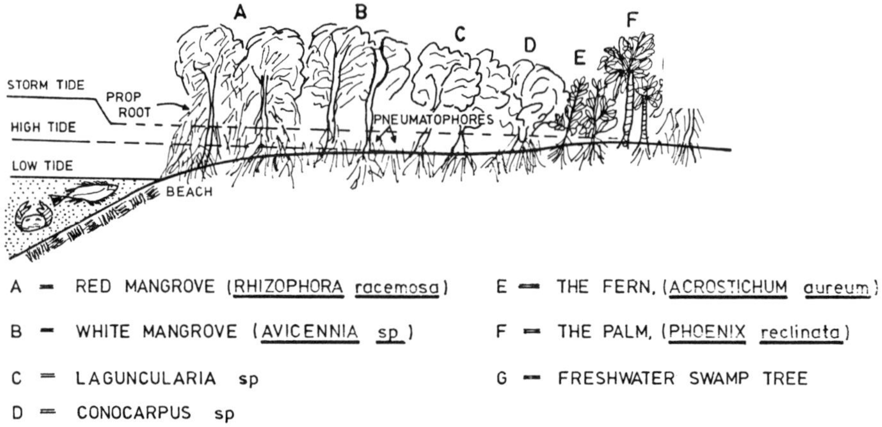

Fig. 5. Natural zonation of a tropical mangrove swamp in the Niger Delta (after Ibe et al., 1985). A = red mangrove (*Rhisophora racemosa*); B = White mangrove (*Avicennia* sp.); C = *Laguncularia* sp.; D = *Conocarpus* sp.; E = the fern *Acrostichum aureum*; F = the palm *Phoenix reclinata*; G = freshwater swamp trees.

The swamp is composed of three elements (Allen, 1965b): the main feeding channels, which are mainly extensions of fresh water streams passing through to the sea enroute to the barrier islands; smaller channels that connect the feeding channels; and elevated flats exposed at low tide. Drainage on each flat is centripetal due to the rims being higher than the centers. The main feeders in the western flank of the delta receive substantial freshwater from the major rivers (Nedeco, 1954), while those in the east do not obtain much fresh water, sediments being acquired chiefly from tidal inflow. Inland of the mangrove swamp, a freshwater zone consists of progressively narrower, shallower, and more numerous rivers, lakes and swamps.

The delta has provided the best conditions for luxurient vegetation on the Nigerian coast, and thus characteristically hosts the widest variety of trees and plants. Figure 5 shows the natural zonation of a tropical mangrove swamp in the Niger Delta.

Coastal Processes

Analysis of wind measurements from coastal stations in the Niger Delta shows that onshore southwesterly winds are the most frequent and strongest, even during the dry season (November–March) when northeast winds reach the coastal areas as a consequence of the downward migration of the Inter Tropical Zone (ITZ). Wind velocities are usually between 2.5 and 5 m/sec. except during the frequent thunderstorms and line squalls in the rainy season (April–October) when winds can exceed 12 m/sec.

Waves in the Niger Delta include less regular local wind waves and seas (period 3–8 sec), which are more regular (Ibe et al., 1984). The swells are thought to originate in storm centres in the "roaring forties" in the Atlantic (Ibe and Antia, 1983). The waves, which approach typically at angles of 3°–12° to the shoreline, are generally 45–210 cm high and break in water depths of 52–270 cm (Ibe, 1985a). Everywhere on the low gradient beaches, spilling breakers are more common (85%) than plunging waves except during storms and near storm conditions, particularly in the meteorologically harsh rainy season, when plunging waves tend to dominate. Collapsing or surging breakers are rare. The breakers generate longshore currents, generally less than l m/sec., that move large quantities of sand. Longshore drift is dominantly eastwards in the eastern delta and north-northwestwards. Calculated longshore transport rates are of the order of 0.35 to 0.74 × $10^6 m^3/y$, which agree with the 0.5 × $10^6 m^3/y$ given by Nedeco (1961) and Burke (1972). Another effect of the breaking waves is to set up localised rip currents, estimated at 0.5 m/sec. (Ibe, 1987) that carry sediments away from shore at steep angles for distances up to 1.6 km.

The tides in the Niger Delta, like elsewhere on the Nigerian coastline, are semi-diurnal. Mean tidal range varies from 1.3 m at the Benin River to 2.0 m at the Imo (Opobo) River. The tides approach the coast from a southwesterly direction and create strong semi-permanent currents that occur at high angles to the coast, carrying sediments away from seaward in foam-bounded plumes of turbid water. In the inlet tide-generated currents are strong enough to move gravel and sand-sized materials, but maximum velocities reaching 2.8 m/sec. are attained at mouth bars during ebb tides. Flood currents are usually of shorter duration, set along different line, and attain velocities, between 0.3 and 1.5 m/sec. (Allen, 1965a).

Fluvial currents are active in the mangrove swamp, with mean velocities of 0.37 m/sec., and 0.8 m/sec. under non-flood conditions (Allen, 1965b). The Guinea current moves across the delta coast in a dominantly west-east direction (with some reversals) at speeds up to 0.3 m/sec., but it has little impact on the littoral dynamics. The north-northwest currents of the underlying Central South Atlantic Water occur at depth (700 m) and are important only in bringing fine-grained materials from the outer shelf into coastal waters.

Erosion Scenario

An analysis of historical hydrographic charts and aerial photographs as well as data from an ongoing study (1981-present) of the coast of Nigeria by the author, point to widespread erosion along the 800 km national coastline (Ibe and Antia, 1983a, 1983b; Ibe, 1984-198; Ibe and Awani, 1986; Ibe and Awosika, 1986; Ibe and Murday, 1989; Oguara and Ibe, 1989; Stein, et al., 1986).

In the Niger Delta, the rates of erosion can reach several few tens of meters per year. Typical measured annual rates on the delta include 18–24 m at Ogboiodo/Escravos, 20–22 m at Forcados, 16–19 m at Brass, 15-20 m at Kulama, 20-24 m at Bonny, and 10-14 m at Imo (Opobo) river entrances. Despite several remedial measures (mostly by sand replenishment) at various coastal sites by state and federal agencies as well as oil companies, erosion has continued to devastate large areas of the delta. In 1987 alone, the subcommittee on erosion and flooding disbursed over 100 million Naira (about 20 million 1987 U.S. Dollars) for various erosion control measures within the delta. Both administrators and scientists maintain that this level of expenditure, though considerable, is much too small, and that billions of Naira would be needed annually to mitigate the problems caused by coastal erosion and flooding.

Retreat of the coastline, with the concomitant flooding causes hazards through uprooting settlements, destroying agricultural and recreational grounds, disrupting harbour and navigational structures, and dislodging oil producing and oil export handling facilities, which are mostly located near or in coastal towns such as Escravos, Forcados, Brass and Bonny.

Natural factors identified as causing the rapid erosion include (1) an intense wave climate that not only unleashes great physical energy on the beaches but also generates destructive littoral currents; (2) the micro-tidal coastline (1–2 m tidal range), which ensures that the waves impinge in the coast at consistent levels; (3) the low lying coastal plain, which is prone to constant flooding or concomitant erosion; (4) the fine grained compact nature of the sandy beaches, which makes them susceptible to erosion from backwash; (5) the tides that set up semi-permanent reversible currents; (6) the narrow continental shelf, 50–75 km in width, that allows waves to arrive at the coast at greater heights and steeper angles than on a wider shelf because waves do not shoal or refract as much; (7) the presence of offshore canyons and gullies through which sediments transported seaward are lost irretrievably to deeper waters; (8) the continued subsidence of the Nigerian coastal areas and, in particular, the Niger Delta due to natural dewatering and compaction of its sediments, but also to other effects; (9) the global eustatic rise in sea level, which on a low lying coastline is manifested as a landward translation of the coastline several orders of magnitude when compared with the vertical rise; etc.

In many places, man's intervention in the natural environment by way of construction in the littoral zone, erecting of harbour-protecting breakwaters, dredging of port channels (e.g., Escravos and Bonny beaches), the mining of sand either from the coastal rivers or from the beaches themselves, withdrawal of enormous quantities of petroleum and water from coastal and nearshore reservoirs and aquifers, the structural modification of coastal land through canalisation for transportation and hydrocarbon extraction, devegetation of primeval forests (in particular mangroves), and the uncontrolled construction of dams across rivers that normally would supply sediment to the coastal zone, have served to exacerbate the impact of natural forces.

A review of the natural and man-made factors outlined above show that in the next several decades, the eustatic sea-level rise will be accentuated in subsiding areas through both natural and anthropogenic forces. These may well accelerate erosion and flooding, and adversely affect the social and economic life of not only the delta province but, indeed, the whole country. These factors are briefly discussed below.

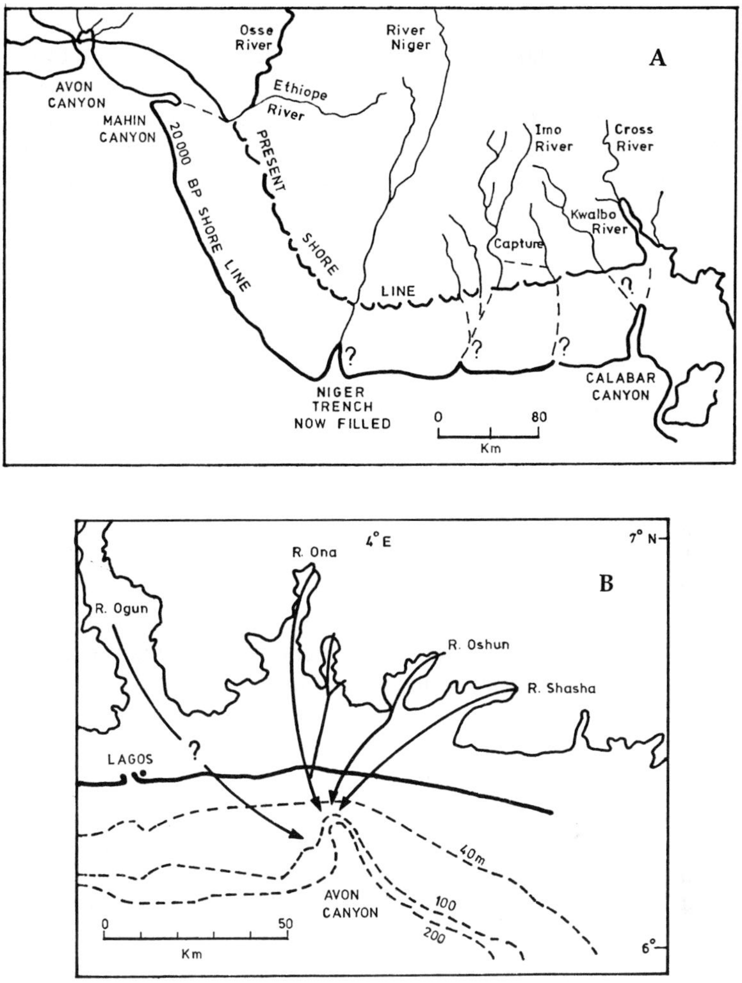

Fig. 6. (A) Possible river courses across the Nigerian continental shelf during low sea level, 20,000 years B.P. Only canyons kept open by scouring have not yet been filled (after Burke, 1972). (B) River courses heading toward Avon Canyon at low sea level. The Ogun River, however, may have fed its own canyon (after Burke, 1972).

SEA-LEVEL CHANGE

Changing sea level constitutes an important single process that shapes coastal areas world wide. The past several million years have been characterized by glacial and interglacial epochs, with eustatic sea level fluctuating over a 100-m range, as rapid and extreme a shift as observed in geological history (Brunn, 1962). During the Wisconsin lowstand of sea level, the Niger Delta shoreline extended to the outer shelf, some 80 km seaward of present (Fig. 6A). Allen and Wells (1962) used a sequence of dead coralline banks in shallow water off the Nigerian coast to indicate rise in sea level and/or shelf subsidence over the last four thousand years.

More specifically, Burke (1973) attributed all the observed subsidence (approximately 80 m in about 15,000 years) in southwestern Niger Delta to the eustatic sea-level rise and the isostatic adjustment to water load. The old river valleys are thought to have been drowned, resulting in an estuarine coastline (Fig. 6B). Ibe et al (1984) concluded that sea-level rise must contribute to recent erratic retreat of the Nigerian shoreline. At present there are no reliable estimates of recent rise in sea level, but tide record data from Dakar and Takoradi on the west African coast suggest an average rise of about 2 mm/year over the past 30 to 40 years; it probably is much greater in the subsiding Niger Delta (Ibe and Quellennec, 1988).

According to various projections, the rise of eustatic sea level will accelerate due to global warming and subsequent deglaciation. The seriousness of the projected rise in sea level with respect to increased erosion and flooding of the low lying and marshy Niger Delta can be deduced from the data of Brunn (1962), which coastal retreat in marshes and other low lying areas of hundreds to thousands of times the vertical distance.

SUBSIDENCE

The Nigerian coastal areas continue to subside. A sequence of dead coralline banks in shallow waters off the Nigerian coast, for instance, indicates subsidence and/or sea-level rise during the past four thousand years (Allen and Wells, 1962). Sinking is not only the result of tectonic subsidence, but also continued dewatering and compaction of rapidly deposited sediments. This setting is not atypical of most large deltas.

The present rates of subsidence are still inconclusive, and the only reliable figures to date reveal subsidence rates of more than 2.5 cm/year at a tank farm along the delta coast, after correcting for loading effects by the oil in the tanks (Pender Awani, personal communication, 1987). There is a suspicion that the extraction of petroleum since the late 1950's (which at peak production in 1979 averaged 2.4 mbd) from the porous reservoir sands of the underlying Agbada Formation is accelerating the subsidence of the delta (Ibe, 1985, 1986; Ebisemiju, 1985). An additional factor derives from the extraction of potable water from wells, which has led to the haphazard drilling of boreholes and unregulated drawdown of water from coastal aquifers in the Niger Delta. The main effect of fluid extraction, as described by Cooke and Doornkamp (1974, pp. 170–171), is the reduction of fluid pressure in the reservoir or aquifer, thus leading to an increase of effective stress (or grain to grain stress) in the system. Compaction results and the basin subsides.

Continued natural and man-induced subsidence of the Niger Delta coastal zone would exacerbate the erosion and flooding scenario consequent upon the global sea-level rise in this area.

Change in Fluvial Sediment/Water Input to the Delta

The Niger Delta coast is fed by rivers that have their source in the hinterland and debouch large quantities of sediments to coastline where they are redistributed

TABLE 1
Freshwater discharged by distributaries in the eastern Niger Delta (NEDECO, 1961).

River Entrance	Average Discharge Tidal and Freshwater (m^3/sec)	Average Discharge Fresh water (m^3/sec)
Nun	10300	1400
Brass	16670	460
St. Nicholas	4700	60
Santa Barbara	5320	–
San Eartholomeu	5070	200
Sombreiro	10530	196
New Calabar	8800	–
Bonny	18750	–
Andoni	10560	–
Opobo	6990	230

by waves and tidally generated currents. Fluvial sediment supply has been identified as the most important source of beach sediments on the Nigerian coastline (Ibe and Awosika, 1986) where they provide natural sediment replenishment for the beaches. Other sources are longshore-drifted sand and sand transported from offshore.

The Niger-Benue drainage system is clearly the greatest fluvial contributor. According to Nedeco (1954, 1959, 1961), the Niger River, which drains a catchment dominated by Paleozoic and Cretaceous metamorphics and igneous rocks as well as Pleistocene sediments, annually carries a bed load of mostly medium to coarse sand amounting to about 0.3×10^6 m^3 and an additional 4.6×10^6 m^3 of suspended sediment at its confluence with the Benue River. The Benue river, in contrast, drains an area dominated by basement rocks, Cretaceous sediments, and basic volcanics; it transports mainly a coarse sand bed load, amounting annually to about 0.6×10^6 m^3 while suspended load is about 11×10^6 m^3, exceeding that supplied by the Niger. Many other independent rivers in the western and eastern delta also contribute sediments to the delta coast. These include the Benin and Escravos Rivers in the western delta, and the New Calabar-Bonny and Imo Rivers in the eastern delta. Table 1 shows the discharge of some rivers in the eastern delta. The Benin River, recently investigated by Ibe et al. (1988), has sediment discharge of 1×10^6 m^6/year.

From its early history, the Niger Delta has received more than enough sediment to offset the effects of crustal subsidence, sediment dewatering and compaction, and eustatic rises in sea level; thus the delta has been able to prograde and attain its present configuration. However, certain recent developments threaten to upset the status quo and some adverse effects, in fact. are already apparent. One impact results from the construction of several dams, particularly in the 1970's, on major rivers and their tributaries for hydroelectric, reservoir and/or irrigation. This has led to a drastic reduction in fluvial discharge and thereby sediment supplied to the coast. The River Niger alone has two very large concrete dams at Kainji and Jebba, and a third is planned for Lokoja. The River Benue has two very large dams, and in addition there are numerous earth dams either directly on these rivers or their tributaries.

Scientists at the Nigerian Institute for Oceanography and Marine Research are presently studying the current discharges of the rivers that reach the delta coast, with a view towards quantifying their effects on coastal sediment supply. But even now impacts can be deduced from the studies of Olofin (1985) for two rivers in northern Nigeria. That author found that the mean discharge of River Kano, completely controlled by the Tiga Dam since 1974/75, decreased during the humid months from a predam 100 m^3/sec. to a post-dam discharge of 14 m^3/sec. An adjacent river decreased its flow from 383 m^3/sec to a post-dam 56 m^3/sec, subsequently increasing to 80 m^3/sec. The drastic reduction in natural sediment nourishment reaching the coast after dam construction has accelerated coastline erosion.

A second impact is that the development of new, and expansion of old, coastal towns and construction of oil-producing and export facilities on the coast following the discovery of oil in late 1950's has led to large civil engineering projects, including land reclamation schemes. Much of the sand and gravel used in such construction have been mined from coastal rivers, beaches or the shallow nearshore ocean.

Sand and gravel extraction from coastal rivers, particularly in estuaries, has an acceleration of coastal erosion because it have decreased the sediment supply that normally would renourish the beaches. Mining of sand directly from the beaches depletes the available sediment, directly resulting in erosion, whereas upstream mining enhances the erosion of downstream sections of the river. Sand and gravel dredged from the coastal ocean also have been a major source of constructional materials, particularly for land reclamation and beach replenishment, although such dredging almost invariably affects the dynamic equilibrium of the nearshore, thereby enhancing shoreline retreat.

The third impact has been increased canalisation associated with oil activity. Canal construction and river diversion for transportation and hydrocarbon extraction have had the effect of trapping sediments that otherwise would be discharged from upland rivers.

A further problem is that drought and desertification have affected the upper reaches of river drainage basins. According to Langbein and Schumm (1968), as climate becomes more arid, sediment supply to rivers and the coastal system will decrease substantially. The drought and desertification that have ravaged much of Sahelian and sub-Sahelian Africa have meant a drastic reduction in the sediment load of the Niger River (and Benue), which, by implication, has helped decrease the quantity of sediment reaching the coastal zone.

Socio-Economic Setting

The Niger Delta, encompassing the entire Rivers State and a large part of Bendel State is a petroliferous province that accounts for over 90% of Nigeria's export and foreign exchange earnings (Ibe et al., 1988). Apart from oil drilling rigs, production platforms, flow stations and tank farms that dot the delta, it is the home of two of the country's 3 petroleum refineries (Port Harcourt and Warri) and is criss-crossed by pipelines that transport both crude oil and refined petroleum products. The value of petroleum-related facilities on the delta many billions of Naira. Projected investments in the planned Liquefied Natural Gas (LNG) project will further increase the value of petroleum facilities in the area.

Arising from the presence of the oil industry has been the development of large coastal settlements at Warri, Port Harcourt, Sapele, Forcados, Escravos, Brass, Akassa, Bonny, etc. The first two, Warri and Port Harcourt, western and eastern headquarters of petroleum exploration and exploitation operations, each have populations of over 1 million people. The population of the whole delta is more than 6 million (projected from 1963 census and recent demographic trends). A broad range of manufacturing companies has taken advantage of the burgeoning populations, increased prosperity, sales opportunities and the relatively easy access to international shipping to create new industries on the delta. Since the 1970's, with the emphasis on decongesting Lagos, most industries have been preferably located in the Niger Delta, including one of the two major iron and steel complexes in the country and two of the country's three petrochemical plants. With tyre, glass, fertiliser, shoes, plastics, boat-building, brewing, food (particularly seafood) processing, the Niger Delta is fast growing into the industrial 'heartland' of Nigeria.

Before the advent of the oil industry in the late 1950's, the Niger Delta was as a base for much of the country's artisan fisheries. A myriad of fishing settlements established on protected, well drained land or on stilts in river estuaries and beaches were (and still are) the dominant feature of the delta. Artisan fishermen and women, who harvest fish, shrimps and molluscs in the fresh, brackish and the coastal waters with set nets, traps and other passive gear, use crafts ranging from dugouts to motorised large canoes. Ajayi and Talabi (1984), based on an earlier survey by Tobor et al. (1976), estimated the yield of the coastal and brackish water artisan fisheries to be between 128,000 and 170,000 tonnes/year.

Bonga (*Ethmalosa fimbriata*), Sardines (*Sardinella madarensis = Sardinella eba = Sardinella cameronensis*) and shad (*Ilisha africana*) are the principal components of the pelagic and semi-pelagic coastal artisanal fishery. The demersal components of this fishery are croaker (*Pseudotolithus elongatus*, *P. typus* and *P. senegalensis*), catfish (*Aurius* spp.), sole (*Cynoglossus* spp.), shineynose (*Polydactylus quadrifilis*), grunt (*Pomadasys* spp.), snappers(*Lutjanus* spp.) and grouper (*Epinephelus* spp.). Large size tarpons (*Megalops atlantica*), bill fish, shark and rays are also caught. Shellfish harvested by artisanal fishermen include white shrimp (*Nematopalaemon hastatus = Palaernon hastatus*), brackish prawn (*Macrobrachium machrobrachion*), river prawn (*Macrabrachium vollenhovenii*) and juvenile pink shrimp (*Penaeus notialis = Penaeus duorarum*). The mangrove oyster, *Crassostrea gasar*, and other molluscs (e.g., *Pachyacllion*)are delicacies in high demand.

The artisan fishery revolving around the above resources not only contributes to the nation's march towards self-sufficiency in its protein needs, but also provides, particularly in the case of shrimp and oysters (annual production approximately 48,000 tonnes), exportable resources with high foreign exchange potential. For example, one tonne of prawns in 1990 fetched US $ 12,000. The mangrove swamps in the delta serve as the spawning and nursery grounds for most of the finfish and shell fish resources caught by the artisan fishery. In addition, the swamps provide premium grounds for brackish water fish aquaculture whose productive potential is conservatively estimated at 175,150 to 525,000 tonnes annually (Ajayi and Talabi, 1974).

Apart from fisheries, an extensive rain forest and the vast, fertile river flood plains have made the delta the food basket of southern Nigeria. Ongoing programmes to cultivate the swamps with swamp rice could strengthen its role as a food basket. The rain and mangrove forests also serve as the basis of a flourishing lumbering industry, and Sapele in the western delta is the home of the African timber and plywood industry, which is one of the largest of its kind south of the Sahara. At least a score of such companies exist throughout the delta, supplying logs for both local use and export. Game animals, which provide supplementary protein for the rural populations, are an important resource of these forests and are 'exported' to the hinterland where they attract handsome prices.

The Niger Delta has five ocean ports at Forcados, Warri, Koko, Sapele and Burutu, collectively known as the delta Ports, and a sixth port in Port Harcourt. Only two other ports at Lagos (Apapa/Tin Can complex) and Calabar exist outside of the delta region. Of the delta ports, Forcados is presently in disuse, but the remaining four and Port Harcourt handle about 60% of the country's shipping (both export and import).

The Niger Delta has long beaches of scenic beauty that are part of about 20 barrier islands that rim the delta. These beaches, which are under attack by marine erosion, are the focus of present government efforts to turn the delta front into a tourist area. The potential earnings, particularly in foreign currency, from this tourist industry warrants particular attention in the preservation of these beaches.

Owing to economic activities, an extensive network of roads (about 4000 km in Bendel State; 2500 km in Rivers State) has been developed on the mainland, while the extensive creeks, channels, rivers and estuaries provide an excellent water communication network. One of three functional international airports in the country is sited in Port Harcourt. Smaller airports are at Warri, Escravos, Forcados and Brass. In addition, heliports dot the delta landscape. There are also numerous social amenities in the delta, including 12 tertiary institutions, hundreds of secondary schools, thousands of primary schools, banks, commercial houses, hospitals, luxury hotels, etc. in which billions of Naira have been invested.

The above socioeconomic setting is predicated on the Niger Delta as a geomorphic zone. When the delta is viewed in more general terms, the human and economic resources of this zone more than quadruple. For example, the population of this larger delta exceeds 30 million of Nigeria's estimated population of 100 million.

If the barrier islands that absorb the waves were to disappear due to erosion and accelerated subsidence as well as increased rise in sea level, the swamps behind them also would be lost. Acute flooding and perhaps total destruction of settlements would result and relocation of large populations will be necessary, large investments by the oil industry would be dislodged, shipping activities would be threatened, businesses would have to be relocated, the export oriented shrimp and oyster fishery would probably collapse, and the vast socio-economic structure in the area would be rendered unusable. Money, jobs, lives and the prodigious development opportunities that the Niger Delta offers would be lost also.

Concluding Statements and Suggestions

Erosion and flooding are ever-present phenomena in the Niger Delta, with rates of coastline retreat reaching several tens of meters locally. Today, these twin problems are of paramount concern given the continual drive to optimize the economic, social and cultural potential of this coastal zone.

A review of the natural and man-induced factors presently responsible for shoreline retreat and flooding shows that some of the factors may grow in intensity. Particularly disturbing is the projected increase in eustatic, sea-level rise by 0.56 to 3.45 m in the next century, given the low-lying morphology of the deltaic coast, with maximum elevations about 3 m above MSL on the barrier islands and barely above MSL in the mangrove and freshwater swamps.

Locally, the projected rise in sea level will be accentuated by the continued subsidence of the delta which presently is being exacerbated by human impact through the uncontrolled withdrawal of fluids (petroleum and water) from the porous reservoirs underlying the surface delta. Although subsidence has occurred since the early days of its formation, the Niger Delta was able to grow and prograde because the Niger River and smaller rivers brought enormous volumes of sediment to the area. Today, however, dam construction Niger and smaller rivers, increased canalisation of the coastal zone and other diversions of coastal rivers, continued mining of sand from coastal rivers, and, of course, drought have meant a drastic reduction in the fluvial sediment supply to the coast. The loss of this natural source of sediment replenishment will threaten the delta's ability to replenish itself and keep pace with eustatic rise in sea level and local subsidence. Consequently, vast areas of coastal land and (in particular) mangrove swamp will be devastated.

The socio-economic effects of such devastation would include the disruption of oil-related activities, which presently account for over 90 per cent of Nigeria's export and foreign exchange earnings, the displacement of 6 to 10 million people, which would be particularly difficult for those to whom the delta is their ancestral homeland. The destruction of the traditional food basket of southern Nigeria and the dislodgement of strategic industries within the new industrial 'heartland' of Nigeria would add to the problems. Land, resources, jobs and lives would be lost. When the Niger Delta is viewed in terms of its liberal definition the physical and human losses highlighted above could be multiplied four or five fold.

The seriousness of these perceived effects make it imperative that future plans for economic and social development of the Niger Delta include consideration of landuse, sea-level rise, and coastal subsidence. The predictions as to the mechanism and effects of the projected rise are far from exact, but the near dearth of information in Nigeria makes predictions problematical, thereby complicating the planning process. For example, there are no long term data on the behaviour of mean sea level along the delta coastline because of the lack of installed tide gauges in the area; a tide gauge has been in place in Bonny since 1958, and short term sea-level measurements have been made for specific tasks (such as harbour development). The lack of precise information on rates of subsidence or the extent of diminished fluvial sediment supply to the coast complicate attempts to compute the overall impact of the expected increase in eustatic rise in sea level.

Though it may seem late to begin to initiate data acquisition for a process already perceived to have begun, it is better done late than never. A usable data base must include adequate hydrological, meteorological, geological, economic and demographic information. In the meantime, certain policy options should be implemented quickly. One is the adoption of the concept of set back lines along the national coastline. Ibe (1987, 1988) drawing from published knowledge, suggested a set back line at least 20 times the estimated annual rate of erosion. Another is a determined effort to exclude or minimise human interventions that accelerate natural subsidence and decrease the sediment from fluvial sources that normally would replenish beaches and marsh land. A third is that any future coastal structures or development must be planned in harmony with the future perspective of the coastal zone highlighted in this paper.

Environmental problems can be solved either 'surgically' or 'therapeutically' (Triscart, 1976). The present author believes that plans necessary for the Niger Delta to cope with continued (or accelerated) subsidence, decreased sediment supply, and the projected accelerated rise in sea level must be 'therapeutic', taking into account all the various environmental elements, both present and projected. This calls for an integrated development of the Niger Delta under an authority or commission that has legal status and adequate funding. As presently constituted, the Niger Delta Basin Development Authority is ill conceived, ill funded, ineffective, and its actions are too *ad hoc* and restricted to have any meaningful impact on the future developmental needs of this area in the face of perceived increased global and local environmental stress. A redress of this situation is imperative.

References

Allen, J.R.L., 1965a. Late Quaternary Niger Delta and adjacent areas; sedimentary environments and lithofacies. *Bull. Amer. Assoc. Petrol. Geol.*, 49, 547 - 600.

Allen, J. R. L., 1965b. Coastal geomorphology of Eastern Nigeria: beach ridge barrier islands and vegetated flats. *Geol. Mijnbouw*, 44,1-21.

Burke, K.,1972. Longshore drift, submarine canyons and submarine fans in the development of Niger Delta. *Bull. Amer. Assoc. Petrol. Geol.*, 233, 1975-1983.

Burke, K., Dessauvagie, T. F. J. and Whiteman, A. J., 1971. Opening of the Gulf of Guinea and geological history of the Benue depression and Niger Delta. *Nature*, Phys. Sc. 233, 51-55.

Brunn, P., 1977. Practical solution to a beach erosion problem. *Coastal Engineering*, 1. 3-316.
Cooke, R. U. and Doornkamp, J. C., 1974. *Geomorphology in Environmental Management: An Introduction*. Clarendon Press, Oxford.
Dangana, L. B., 1981. Ecological Dynamics and flood control in the Niger Delta. in Proc. Seminar on Flood and Erosion Control in the Niger Delta. Port Harcourt, 25 - 26 March 1981.
Egborge, C. E., 1981. Hydrological Consequences of proposed Flood Control Works in the Niger Delta. in Proc. Seminar on Flood and Erosion Control in the Niger Delta. Port Harcourt, 25-26 March 1981.
Ebisemiju, F. S., 1985. Human impact on marine processes in western Niger Delta. in Proc. Policy Seminar on Environmental Issues and Management in Nigerian Development, University of Benin, Benin, Nov. 25 - 27 1985.
Evamy, B. D., Haremboure, J., Kermerling, P., Knaap, W. A., Morley, F. A. and Rowlands, P. H., 1978. Hydrocarbon habitat of Tertiary Niger Delta. *Bull. Amer Assoc. Petrol. Geol.*, 62,1-39.
Grabt, N. K., 1971. South Atlantic Benue trough and Gulf of Guinea Cretaceous triple junction. *Geol. Soc. Amer. Bull.*, 82, 2295-2298.
Hosper, J., 1965. Gravity field and structure of the Niger Delta, Nigeria. *Geol. Soc. Amer. Bull.*, 76, 40.7-422.
Hosper, J., 1971. The Geology of the Niger Delta area. In: F. M. Delany (ed.), *The Geology of the East Atlantic Continental Margin*, Part 4. Africa. Inst. Geol. Sc. Report 70/16, 121 - 142.
Ibe, A. C., 1984a. Defending Victoria Island against sea incursion. A special report for the Federal Ministry of Education, Science and Technology, 7pp.
Ibe, A. C., 1984b. Protecting Victoria Island against sea incursion. A position paper submitted to the Ad Hoc Committee on the Victoria Island Erosion Problem, August 1984. 5pp.
Ibe, A. C., 1984c. A rational strategy for defending Victoria Island against oceanic surges. in Proc. Conference on the Bar Beach Surge. Federation of Building and Civil Engineering Contractors of Nigeria Ltd.
Ibe, A. C., 1985a. Nearshore Dynamics and Coastal Erosion in Nigeria. Paper presented at the UNESCO Expert Workshop on WACAF/3 Project; Dakar, Senegal; March 11-19, 1985.
Ibe, A. C., 1985b. Harbour development related erosion at Victoria Island, Lagos. First International Conference on Geomorphology, University of Manchester, England; September 15 - 21, 1985. In: V. Gardiner (ed.), *International Geomorphology* 1986, 165-181.
Ibe, A. C., 1986. Port Development related erosion at Escravos, Bendel State, Nigeria. In Proc. Man's Impact on the Coastal Environment. Barcelona, Spain, 6-13 September, 1986. in Thallasa, Revista De Ciencias Del Mar 4, 91-96.
Ibe, A. C., 1987a. Marine erosion on a transgressive mud beach in Western Nigeria. In: Proc. Intern. Symposium on Geomorphology and Environmental Management, Alahabad, India, Jan. 17-20, eds. S. Sighn and R. G. Tiwari.
Ibe, A. C., 1987b. Collective Response to Erosion Hazards along the Nigerian Coastline. In: Proc. Coastal Zone '87, America Soc. Civil Engrs., Seattle, Washington, 741-754.
Ibe, A. C., 1987c. Human Impact on the coastal erosion problem in the West aand Central Africa (WACAF) region. In: Proc. International Sea Conference, University of Mauritius, Reduit, Mauritius, 7-12 September 1987.
Ibe, A. C., 1988. Nigeria. In: H. J. Walker (ed.), *Artificial Structures and Shorelines*. Kluwer Academic Publ., 287-294.

Ibe, A. C. and E. E. Antia, 1983a. Preliminary study of the impact of erosion along the Nigerian coastline. In: Proc. First International Conference on Flooding, Desertification and Erosion in Africa. Rivers State University of Science and Technology, Port Harcourt, May 2-6, 1983.

Ibe, A. C. and E. E. Antia, 1983b. Preliminary assessment of the impact of erosion along the Nigerian shoreline. NIOMR Tech. Paper No.17.

Ibe, A. C. and P. E. Awani, 1986. Erosion management strategies for the Mahin mud beach, Ondo State. In: Proc. National Seminar on Flooding and Erosion along the Nigerian Coastline. University of Lagos, May 1986. Contrib. No. 2.

Ibe, A. C. and L. F. Awosika, 1986. Sedimentology of the barrier bar complexes in Nigeria. NIOMR Tech. Paper No.26.

Ibe, A. C.; Awosika, L. F. and E. E. Antia. 1984. Progress Report No.2, Coastal Erosion Research Projects. NIOMR Special Publ., 106 p.

Ibe, A.C., Awosika, L.F., Ihenyen, A.E., Ibe, C.E., and Tiamiyu, A.I., 1985a. Coastal erosion at Awoye and Molume, Ondo State, Nigeria. Report for Gulf Oil Co. (Nigeria) Ltd., 123 p.

Ibe, A.C., Awosika, L.F., Ihenyen, A.E., Ibe, C.E., Tiamiyu, A.I., Okonya, E.C., and Orekoya, T., 1985b. A study of currents and scouring effects at the proposed Davy Banks 'A' location in OML 14. Report for Shell Petrol. Development Co. of Nigeria Ltd., 56 p.

Ibe, A.C., Awosika, L.F., Ibe, C.E., Tiamiyu, A.I., Egerongbe, F.O., and Orupabo, S., 1986. The erosion problem at Victoria Island (1900 to present) and its solutions. in Proc. National Seminar on Flooding and Erosion Along the Nigeria and Similar Coastlines. Univ. Lagos, May 1986.

Ibe A.C., Ihenyen, A.E. and Ibe, C.E., 1985c. A hydrographic survey of the proposed Benin estuary location OML 43. Report for Shell Petrol. Development Co. of Nigeria Ltd., 25 p.

Ibe, A.C. and Murday, M ., 1986. Aerial Photostudy of recent migration of the Lagos shoreline. NIOMR Tech. Paper.

Ibe, A.C., Ajao, E.A. and Oyewo, E.O., 1988. The oil industry and pollution in the Nigerian environment. In: Proc. Seminar on the Oil Industry and Nigeria - Current Issues and Perspectives. Nigerian Inst. International Affairs, Lagos, 27 Oct 1988.

Ibe, A.C. and Quellennec, R.J., 1988. Methodologie d'Inventaire et de Controle de l'Erosion Cotiere dans la Region de l'Afrique de l'Ouest et du Centre. 154 p.

Langbein, W.B. and Schumm, S.A., 1968. Yield of sediment in relation to mean annual precipitation. *Amer. Geophys. Union Trans.*, 30, 1076-1084.

Merki, J.P., 1972. Structural geology of the Cenozoic Niger Delta. in T.F.J. Dessauvagir and A.J. Whiteman (eds.), *African Geology*. Univ. Ibadan Press, 635-646.

Murat, R.C., 1970. Stratigraphy and paleo-geography of the Cretaceous and lower Tertiary in southern Nigeria. African Geology, Ibadan.

Manchense, E., 1973. The geological history of the marginal basins along the north shore of the Gulf of guinea. in A.E. Nairn and F.G. Stehli (eds.), *Ocean Basins and Margins*, v. 1, South Atlantic. Plenum Press, New York, 351-390.

NEDECO, 1959. River studies and recommendations in improvement of Niger and Benue. Amsterdam, North-Holland Publ. Co., 1000 p.

NEDECO, 1961. The waters of the Niger Delta. The Hague, 317 p.

Oguara, T.M. and A.C. Ibe, 1989. Decision analysis for the selection of erosion control measures. NIOMR Tech. Paper.

Stein, D., G. Echwebber and A. C. Ibe, 1986. Investigation on the Bar Beach erosion and a proposal on the maintenance of the coast. National Seminar on Flooding and Erosion along the Nigerian and related coastlines. May 1986, University of Lagos.

Stoneley, R., 1966. The Niger Delta region in the light of the theory of continental drift. *Geol. Mag.*, 103, 385-396.

Tricart, J., 1976. Ecodynamique et l'Amenagement" RGD XXV. No.1, "Sedimentological aspects of Oil Fields in Niger Delta". *Geol. Mjnbouw*, 50, 559-576.

Weber, K. J., 1971. Sedimentological aspects of oil fields on the Niger Delta: *Geol. Mijnbouw*, 50, 559-576.

Wright, J. B., 1970. The Benue Valley - a rift in the African continent. Nigerian Field. V. XXXV, 78 - 85.

CHAPTER 14

Sea-Level Rise and the Stability of Barrier Islands, with special reference to the Wadden Sea

H. Postma

Introduction

Chains of sand barrier islands are common morphologic features along many shores, occupying about 8% of the world's coastlines. The Wadden Sea, which stretches along the eastern side of the North Sea, has one of the best studied barrier island systems in the world (Fig. 1).

Primary conditions for barrier island existence are a gently sloping sandy shelf, a modest vertical tide, strong incoming waves and, of course, supply of sand. In a simplified sense, where the tide is only a few decimeters in range, barriers are long with only a few tidal inlets, as for example in the Gulf of Mexico; obviously the tidal prism in the lagoon behind the barrier would be too small to maintain more than the occasional tidal inlet. In contrast, if the vertical tide is more than a few meters, no barrier is formed, as exemplified in the English Channel, the Gulf of Maine and the East China Sea; the energy of the tidal flow is so much greater than the wave energy that no sand bank can form perpendicular to the flow field. Another type of environment with no barrier islands is a coast with a large mud supply. Although beach ridges can be present, any small lagoon in between them is rapidly filled with fine grained sediment. Such coasts mainly occur in tropical areas near large rivers such as in Southeast Asia and along northeastern South America.

Post-Glacial Evolution

During the last post-glacial rise of sea level, systems of barrier islands and tidal channels formed on the present continental shelf. Locally drowned barriers and

J.D. Milliman and B.U. Haq (eds.), Sea-Level Rise and Coastal Subsidence, 269-280.

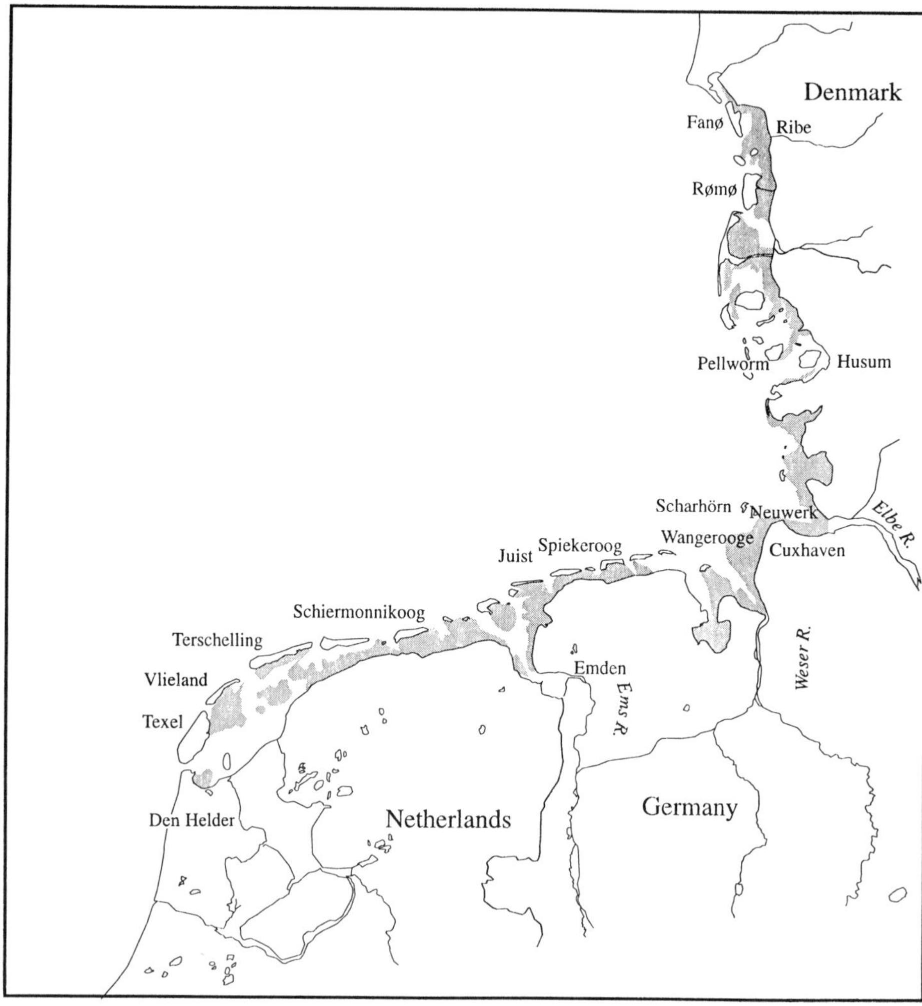

Fig. 1. Map of the Wadden Sea with the names of the islands, tidal inlets and intertidal flats. Mean tidal ranges are 135 cm at Texel; 230 cm at Borkum; 290 cm at Wangerooge; 300 cm at Scharhorn; 280 cm at Pellworm; 180 cm at Sylt; and 150 at Fanø. Islands are absent where tidal ranges exceed 250 cm.

channels and relict sediments with tidal flat characteristics can still be discerned, but these have been flattened and no longer survive as structural units. They have been superseded by new structures such as large sand waves and submarine tidal ridges. A rate of sea level rise of a few meters per century, which prevailed during much of the Holocene transgression (Fig. 2), did not allow sufficient time for the build-up of more prominent barriers.

Judging from the history of the Wadden Sea system, stable barriers began to develop around 6000 BP when sea level was about 20 m below its present

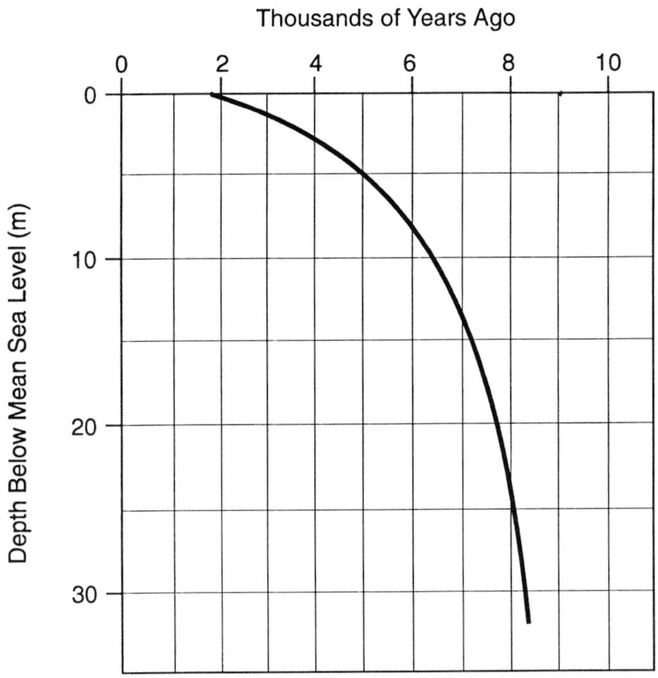

Fig. 2. Relative rise of Holocene sea level off The Netherlands. After Jelgersma (1979).

level and the transgression had slowed to less than 1 cm/year; the sea was about 10 km seaward of its present location. At that time islands were firmly anchored on Pleistocene strata. A further sea-level rise of 10–15 m over the next millennium displaced the barrier system to approximately its present position (Fig. 3) and loosened most of the islands from their bases. As a result, the Wadden Sea has only a few islands (Texel and Sylt) that have outcrops of boulder clays.

As sea level slowly continued to rise over the following centuries, shallow lagoons were formed between barrier islands and the coast, which, depending on the climate, were subsequently filled by marsh and moors. Often these features disappeared, and barren sand and mud flats took their place. Sand was carried into the lagoons through inlets and by wash-over of the barriers. With sufficient sand supply, new ridges were formed against the seaward side of the islands, steepening their seaward profiles. As a result, the originally low dunes increased considerably in height to several tens of meters (Fig. 4). The total amount of sand in the Dutch barrier system is on the order of 10^{10} m^3.

Present-Day Conditions

Depending on the amount of sand available, the strength of longshore drift, and the tidal prism (difference between high and low tide volumes) in the lagoon, inlets could be closed and opened. When inlets grew above a certain size, however,

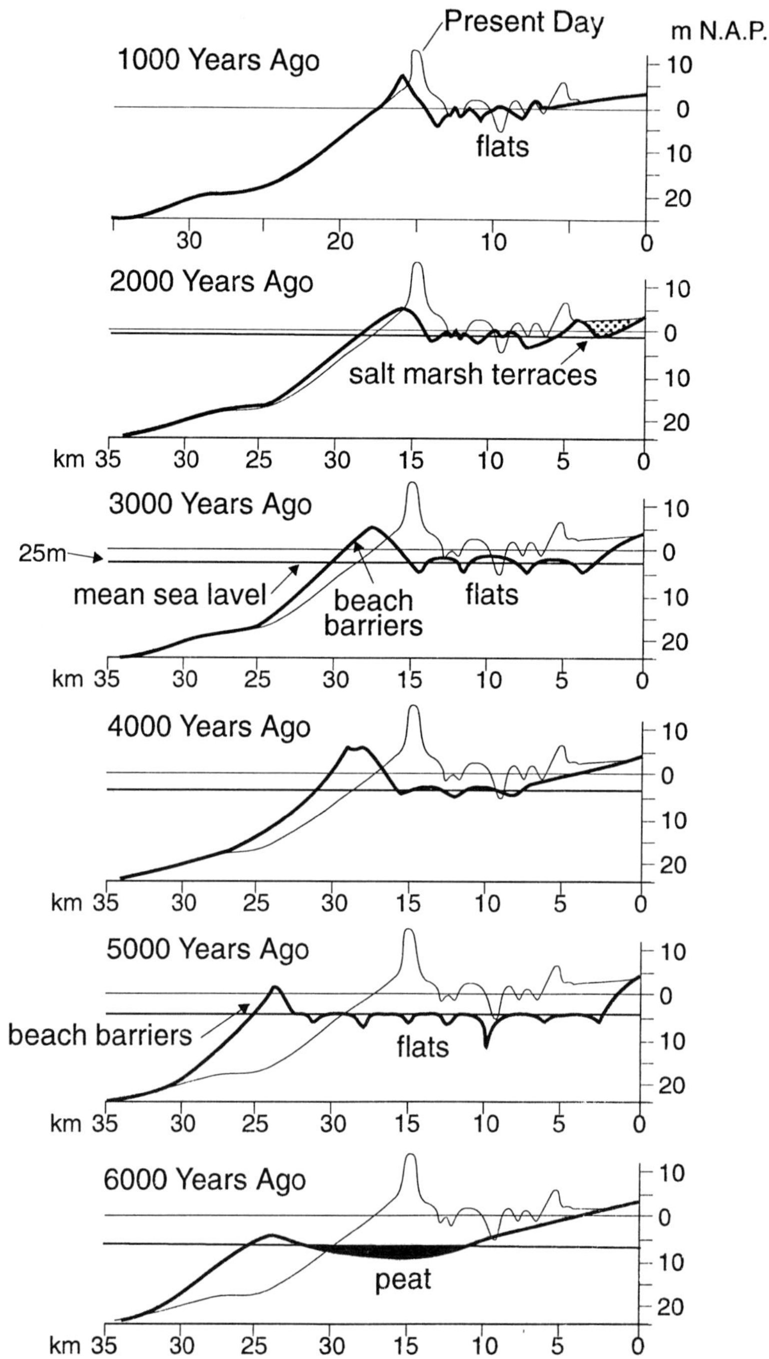

Fig. 3. Evolution of the Wadden Sea land forms, 6000–1000 B.P. NAP = Dutch Ordinance Level. After Veenstra (1980).

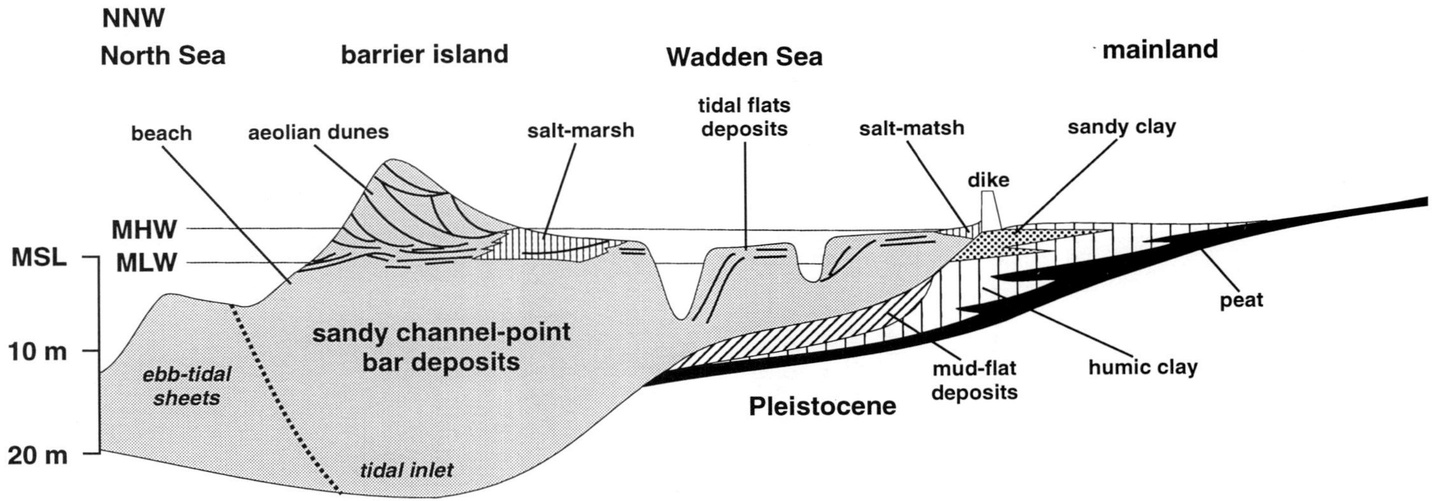

Fig. 4. Schematic cross section from the Dutch Wadden Sea to the Frisian mainland, showing the sequence of late Quaternary deposits. After van der Spek (1994).

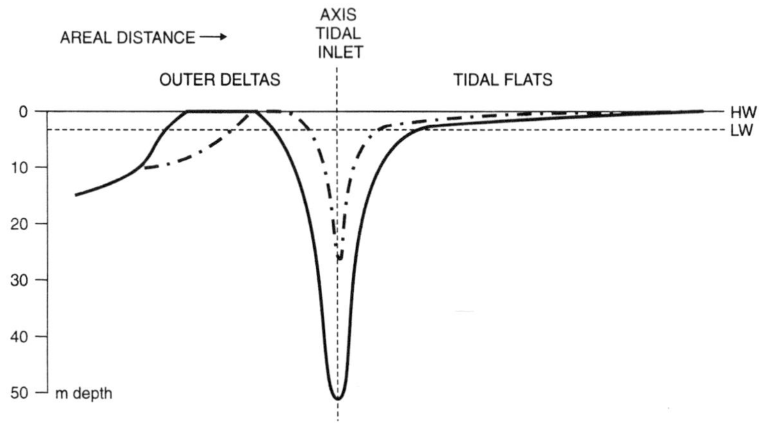

Fig. 5. Relation between depth and areal distance of two Wadden Sea systems with tidal inlet depths of 25 and 50 m. After Postma (1982).

they became a permanent feature and became difficult to close or move. With few exceptions, inlets in the Wadden Sea beyond the island of Texel have been in existence and stable since at least the Roman times. As the islands became more stable, the inlets narrowed and deepened, cutting into the underlying Pleistocene substrata. The size of an inlet is closely related to the tidal prism of the basin it is feeding (Fig. 5).

For a barrier to be stable, the loss of sand should be equal to or smaller than its supply. It helps when a beach section has a concave curvature since an incoming wave front directs sand transport to the middle of this section rather than to both ends. Where this is not the case, there usually is a historical reason, such as conditions during a lower stand of the sea. This may hold for the Dutch Wadden Sea island chain, since the underlying Pleistocene profile has a convex curvature. However, the underlying geology, as mentioned above, no longer contributes to the stability of the chain at present.

It is assumed that island stability is nevertheless maintained since the barrier consists of a system of regularly spaced islands and tidal inlets. The reasoning is as follows. Transport of sand by large waves over the North Sea bottom becomes effective at a depth of about 20 m, with significant shoreward transport at perhaps 10 m. The 10 m depth contour in front of the islands, although generally convex, is composed of a number of separate concave sections. The most seaward extensions of those sections are the outer edges of submarine delta in front of the inlets. Supply of sand from deeper water is thus directed from the deltas to the center of every island, assuring its stability. Nearshore sand transport by smaller waves, chiefly inside the 5 m contour, subsequently carries sand through the tidal inlet into the tidal basin and on to the adjacent intertidal sand flats, from where the ebb flow transports it back to the outer edge of the delta (Fig. 6). The curvature of the 10 m depth contour depends on island length and size of the delta. These two, again, are related to the size of the tidal prims within the tidal basin. An island length of 20–30 km seems most efficient.

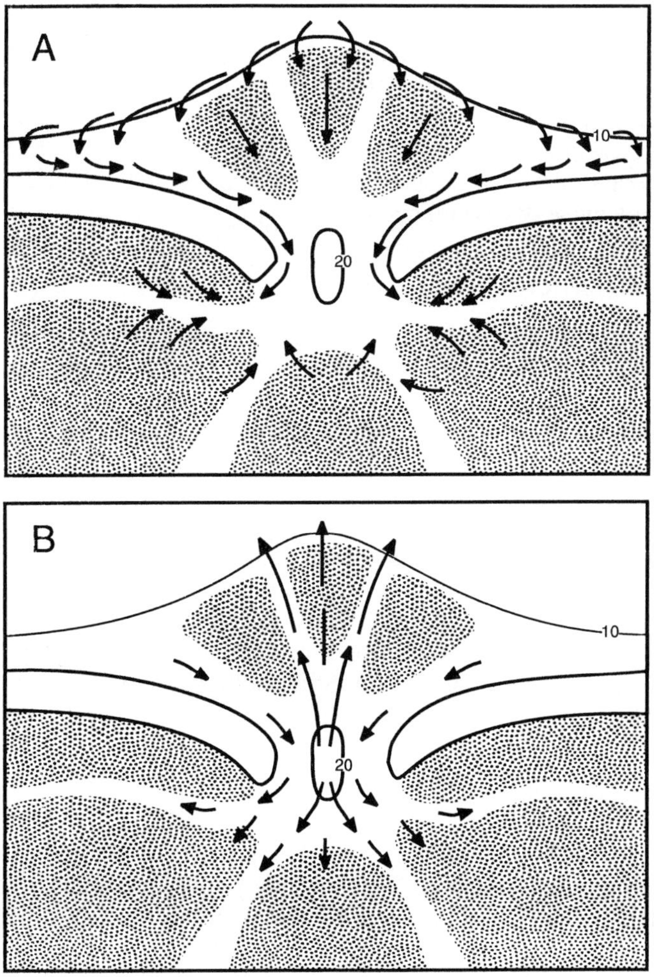

Fig. 6. Residual sand transport by waves (A) and tidal currents (B) in a Wadden Sea tidal inlet. After Postma (1982).

Ideally the above is a closed system. The magnitude of transport in the annual cycle may amount to several hundred thousand tons. Losses are nevertheless inevitable where prevailing winds and currents often carry sand in one direction. Bypassing of sand across a tidal inlet from one island to another occurs by a discontinuous displacement of sand banks over the inlet through a complex interaction of waves and currents. The total amount of sand stored in a submarine delta, therefore, can vary considerably. In case of delayed transport, beach erosion may be significant on the leeward side of the island.

This erosion manifests itself in a retreat of the dune front in the beach section where a passing sand bank has arrived. For example, an average longshore sand transport on the order of 0.5×10^6 m^3/year (as probably occurs along the Frisian islands of the Wadden Sea) might be a reasonable assumption. Over 10 years this could temporarily build up a submarine tidal delta with 5×10^6 m^3 of sediment, which is easily stored. It also will erode the dunes by the same amount. If this happens over half the length of the island, it could cause a temporary dune retreat of 50–100 meters depending on dune height. This loss, however, can be reversed in subsequent years, such that the barrier island beach will continually exhibit gains and losses on a cycle of several decades.

HUMAN INTERVENTION

Globally, erosion of barrier islands beaches presently seems to prevail over accretion (see Bird, this volume). The process, however, is gradual due to the very slow rise of sea level, about 10–15 cm in the last hundred years. More evident are the effects of human interference. Stabilization of beach sections by groynes, for example, decelerates longshore sand transport at the cost of unprotected sections. Sand excavation in tidal basins creates space that is partly filled by input of beach sand. In the western Dutch Wadden Sea, for example, 3.5×10^6 m^3 of sand are presently excavated and used inland each year, an amount that far surpasses the natural input into the basin, and this net loss must come at the expense of the barrier's front. Another reason for net erosion is the fact that the ocean supply of "new" fluvial sand has been diminished or completely stopped by the construction of dams and reservoirs upstream. Still another problem is that dunes can be destabilized by removal of vegetation and by fresh water extraction.

Future Sea-Level Rise

If sea-level rise were to accelerate 50–100 cm per century, it could become the single most important erosional factor in many low land areas, and could even threaten barrier island stability. Three principal threats are envisioned: beach and dune erosion, the deepening of lagoons, and increase in the size of inlets. Secondary effects would include changes in the tidal and wave regimes.

A rise of even 50 cm would cause tens of meters of shoreline retreat, assuming that the shape of the shore profile does not change significantly. The consequences for barrier island stability probably would not be much worse than that caused by present interruptions in sand delivery. However, if the barrier were small, washover events could become more frequent and displacements could be larger, leading to break-throughs.

SAND TRANSPORT AND DEPOSITION

The deepening of tidal lagoons and the increased size of inlets are coupled phenomena. Since lagoons are mostly very shallow, a one-meter rise of mean sea level could easily double the volume of the tidal lagoon. In this situation the main danger to the barrier island would be from sand starvation, which implies that most of the sand transported into the lagoon would not be returned with ebb flow to the outer edge of the tidal delta.

For the western Wadden Sea (Texel inlet) an additional 3×10^6 m^3/year of sand would be needed to keep pace with a 0.5 m sea-level rise, approximately equal to the amount now extracted commercially. The delta also would suffer through lack of sand supply, and a smaller delta itself could threaten barrier island stability. In this instance flood-tide sand transport through an inlet evidently must exceed that during the ebb. An exact measurement of this difference would give an indication of barrier island stability.

To appraise whether such a measurement would be possible, the Texel inlet can serve as an example. Sand transport back and forth through this inlet is restricted mainly to the weeks of spring tides, since during the rest of the lunar cycle tidal currents are insufficiently strong. The volume transport through the inlet is about 10^9 m^3 per tidal phase. For the section of the Wadden Sea fed by this inlet to follow a rise of 0.5 m per century, a simple calculation shows that at spring tide the flood current should contain 25 g/m^3 more sand than during ebb flow. Measurements of suspended sand around spring tide in the inlet are presently on the order of 100 g/m^3 (Fig. 7), so that a 25 g/m^3 difference between flood and ebb seems possible, and any increased elevation of the tidal basin would not lag far behind the rise of sea level. Artificial holes now created near the inlet by sand excavation are indeed rapidly filled with new sand.

Thus, it seems probable that sea-level rise would not lead to sand starvation, primarily because of erosion of the sand barrier. It is interesting to note in this connection that of the sand supplied artificially to the beaches outside the Texel inlet to compensate for present-day erosion, about 10^6 m^3 is eroded annually. The total amount of sand in an average barrier island is about 500×10^6 m^3, suggesting that total break-down of the barrier would happen in a few hundred years if sea level continues to rise by about 0.5 m per century. A faster rise at 1 m per century, of course, could be far more detrimental. Severe damage to the outer line of dunes would occur after several decades, and this damage would be in addition to that caused by factors mentioned earlier.

Lagoonal Muds

A large part of the deposits in the tidal basins consists of fine grained material, which must be taken into account when considering the effects of sea-level rise. Mud enters the lagoons via the rivers (from land) and via the inlets (from sea). Permanent deposition occurs in relatively quiet environments and is greatly enhanced by vegetation (see Fig. 4). Along the inner coastline of a lagoon, intertidal mud flats gradually merge into marshes that are only flooded during extreme high tides and/or storms. If sufficient mud were available, the trapping of mud by vertical growth of marsh and lagoonal plants could easily keep pace with sea-level rise. Most present-day marshes seem to grow upward at about 0.5 m per century, and intertidal mud flats seem to have kept pace. With a faster rise, the import of new mud to many areas probably will be a limiting factor. For example, mud deposition on marshes and tidal flats in the Dutch Wadden Sea is now about 3×10^6 m^3/year, of which most is trapped on mud flats; for the Texel inlet alone it is about 0.3×10^6 m^3/year. The transport of mud across the tidal inlets in the North Sea is about 10^7 m^3/year,

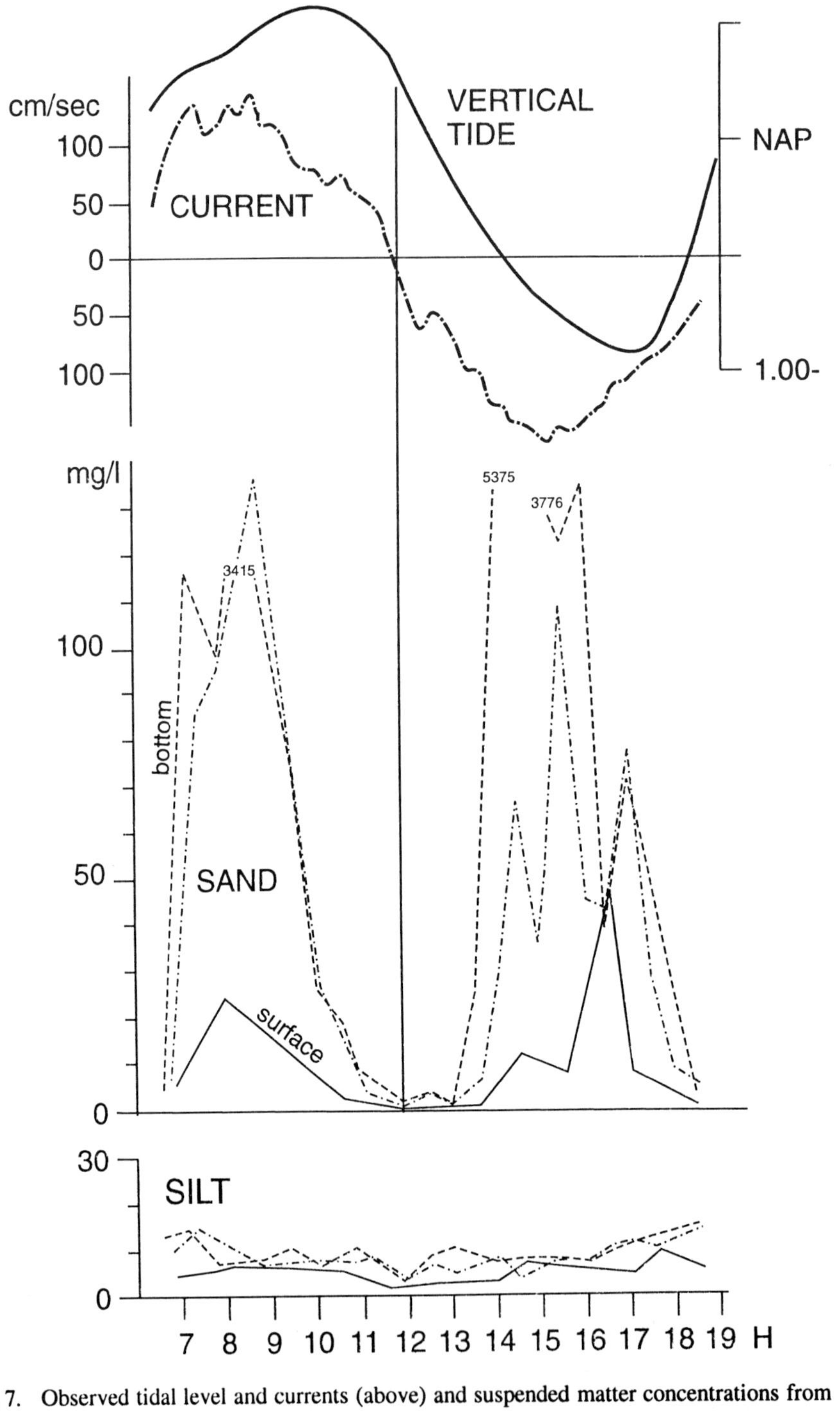

Fig. 7. Observed tidal level and currents (above) and suspended matter concentrations from three water depths in te Texel tidal inlet, spring tide 2 June 1949. Water depth is 33 m. After Postma (1954).

which, if all were trapped, would be sufficient to allow a 0.5 m per century shoaling of the mud flats.

Where mud supply is abundant, for example, where a silt-loaded river directly flows into a tidal basin, fine-grained deposits might occupy most or all extra space provided by the rise in sea level. Formation of new mud flats and marshes would then be enhanced further through the trapping of mud by underwater plants, such as seagrasses, benthic algae, etc. Ultimately, this could result in the transition from a sandy to a muddy environment, and barrier islands might remain largely intact.

Mitigating Measures

Measures to defend the barrier island system include the supply of additional sand to the tidal basins, the decrease of their surface area by partial reclamation, a damping of the tidal range, and promoting the growth of vegetation. From the point of view of the conservation of tidal basins as natural habitats, a combination of these measures might provide the best solution.

In the case of increased sand supply, the material must be collected from the open sea several km offshore, that is seaward of the barrier sand system, which is now common practice. In practice this would be in water depths where wave action is negligible. Where available, sand coarser than that in the tidal lagoon itself would be preferable to fine sand. Deposition in selected tidal channels some distance from the inlet would be most effective and certainly more effective than dumping the same amount on the beach. Input of a few million m^3 of sand per year in many cases would be sufficient.

Decrease of the surface area of a tidal lagoon by partial reclamation has long been a common practice to reclaim fertile land at the expense of natural marshes (Fig. 8). This activity may have to be intensified. Partial reclamation by building low sea dikes that allow periodic flooding might be preferred from the points of view of expense and nature conservation. Building similar dikes around selected intertidal sand flats might be an attractive alternative. If such flats were chosen near the tidal inlet, these also might function to decrease the vertical tidal range.

Promotion of vegetation growth is already used for mud fixation, but this method could be applied on larger scales. New types of eel grass and *Spartina*, for example, with wider environmental ranges, might be developed.

Conclusions

Even if the present slow rise of sea level were accelerated, which is still not a certain prediction, the future of barrier islands could be secured for a long time if appropriate measures are taken. Since every barrier island system is different, solutions will tend to be site-specific. They should be based not only on measurements of waves and tides., but as well on a profound knowledge of the geological history and present-day situation, as well as considerations of sediment budgets and movements.

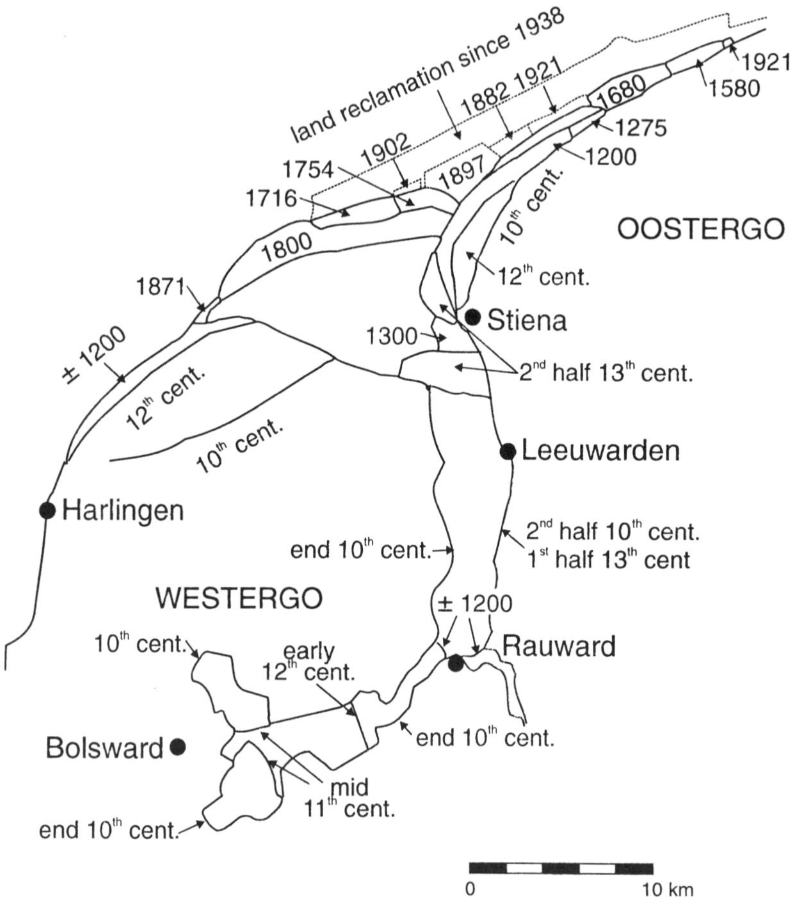

Fig. 8. Dike construction and land reclamation in the Frisian "Middelzee" and adjacent Wadden Sea since the 10th century (A.D.). After van der Spek (1994).

References

Jelgersma, S., 1979. In E. Oele, R.T.E. Schottenheim, and A.J. Wiggers (eds.), The Quaternary History of the North Sea. Acta Univ. Uppsala.

Postma, H., 1954. Hydrography of the Dutch Wadden Sea. *Arch. Nederl. Zoologie*, 10:1-106.

Postma, H., 1982. *Hydrography of the Wadden Sea*. Wadden Sea Working Group Rept. 2, 75 p.

van der Spek, A., 1994. *Large-scale evolution of Holocene tidal basins in The Netherlands*. Unpubl. Thesis, Univ. Utrecht.

Veenstra, H.J., 1980. Introduction to the geomorphology of Wadden Sea area. In: Wadden Sea Working Group, *Geomorphology of the Wadden Sea area*. Rept. 1.

Wadden Sea Working Group, 1980. *Geormophology of the Wadden Sea area*. Rept. 1, 135 p.

CHAPTER 15

Subsidence, Sea-Level Rise, and Wetland Loss in the Lower Mississippi River Delta

John T. Wells

ABSTRACT: Since late Cretaceous, depocenters with oscillating deltas and migrating shorelines have provided a fundamental geologic rhythm to the coast of Louisiana. Sites of deltaic sedimentation have shifted, sea level has risen and fallen by more than 100 m, and sequences of preserved deltas have been vertically stacked in the geologic record. This paper summarizes, in the form of a case history, recent changes in the modern Mississippi River Delta with special emphasis on the causes for geometrically increasing rates of wetland loss that have been experienced since the turn of the century. Rates of relative sea-level rise and discharge of freshwater down the main stem of the Mississippi River (north of Louisiana) appear to have been constant throughout the 1900s, indicating that the demise of the Mississippi Delta is probably a result of an inadequate sediment supply and an inefficient sediment delivery network. The combined effects of levees that prevent overbank flooding and funnel sediments to deep water, upstream dams that trap sediments in the Missouri and Arkansas River basins, and formation of a new delta lobe 150 km to the west have had a profound effect on sediment supply. This loss of sediment load is occurring as the Mississippi Delta is nearing the end of its natural 1000-yr life cycle, and has overwhelmed the ability of fragile wetlands, already in a state of delicate balance, to survive the combined effects of global sea-level rise and subsidence. Mitigation through creation of an extensive network of artificial diversions will slow the rate of delta deterioration but will not be able to rejuvenate a dying delta lobe.

Introduction

The pioneering research more than 50 years ago by R.J. Russell (1936) and later by H.N. Fisk (1944) called attention to the fact that the wetlands of Louisiana were undergoing rapid change and many were in a state of delicate balance. As early as 1949 O'Neil (1949) noted that "... subsidence and Gulf encroachment

J.D. Milliman and B.U. Haq (eds.), Sea-Level Rise and Coastal Subsidence, 281-311.

into inactive deltas are a product of years of delta evolution which has overpowered practically every plant species adapted to the coastal marshes". Although there was little cause for alarm in the 1940s, later studies consistently showed that the rate of land loss was increasing steadily and geometrically (Gagliano and Van Beek, 1970; Adams et al., 1976; Gagliano et al., 1981; Dozier et al., 1983). Estimates by Gagliano et al. (1981) placed the rate of land loss in Louisiana (as of 1980) at 102.1 km^2/yr, substantially higher than the 1967, 1946, and 1913 rates of 72.8, 40.9 and 17.4 km^2/yr, respectively. Stated another way, about 0.8% of Louisiana's wetlands are being lost every year through a combination of natural and human-induced causes (Baumann and Turner, 1990).

Although rates are highly variable, Gagliano et al. (1981) noted that particularly severe land loss problems existed in the four easternmost parishes (counties) of Lafourch, St. Bernard, Terrebone, and Plaquemines which had life expectancies, projected from 1980, of 205, 152, 102, and 52 years, respectively. The modern Mississippi Delta is located within the boundaries of Plaquemines Parish. Tracing its evolution from historic maps and charts, Wells et al. (1984) predicted that, if present trends continued, the Mississippi Delta would be partially under water in 17 years and completely under water in 34 years. A recent analysis by Britsch and Dunbar (1993) shows, however, that on a regional scale land loss rates for the entire coastal plain may have already reached a maximum and are now declining.

The causes for wetland loss are complex and varied (Walker et al., 1987). Much of the deterioration is a result of the complicated interaction of natural physical, chemical and biological processes as well as man's utilization of this environment (Table 1). These processes operate at different scales, both in time and space; the rates at which some changes occur are amenable to manipulation by man, such as in fluid extraction, whereas others, such as natural subsidence, are beyond our control. The net effect of the processes is an accretion deficit, a situation in which new land is not being built vertically or laterally at sufficient rates to compensate for the losses.

The purpose of this paper is to summarize, in the form of a case history, what we know about landscape changes in the lower Mississippi River Delta. The goal is to answer two fundamental questions: 1) why is the Mississippi delta, a distinctly prograding coastal feature, now deteriorating? and, 2) why have rates increased geometrically for most of this century? Information presented in the following paragraphs is derived from about 60 studies, conducted mostly within the last 25 years, and is focused on three aspects of the problem: sea-level rise and subsidence, loss of river-borne sediments, and deterioration of subdeltas.

Geological Setting and History

Processes of deltaic sedimentation in the northern Gulf of Mexico have been active since at least the end of the Cretaceous, approximately 65 million years ago (Woodbury et al., 1973). During this long period of geologic time, the edge of the continental shelf has moved steadily gulfward as a result of oscillating deltas and migrating shorelines (Lehner, 1969). Major shifts in the areas of maximum deltaic deposition, or depocenters, correlate quite well with Plio-Pleistocene shelf-

TABLE 1
Factors Contributing to Wetland Loss

Factors	Causes/Effects
Natural Processes	
1. Subsidence	Lowering of the marsh surface by consolidation and dewatering of sediments
2. Sea-Level Rise	Melting of glacial icecaps adding more water to the oceans
3. Process of Delta Switching	Upstream diversions leading to new depositional sites for sediments
4. Hurricanes	Storm surge and high salinities from offshore waters
5. Biological Degradation	Oxidation and decomposition of marsh peats; damage from muskrat and nutria
Human-Induced Processes	
1. Canal Dredging	Spoil bank barriers and intrusion of salt water
2. Fluid Extraction	Lowering of marsh elevation from depressurization of aquifers and hydrocarbon reservoirs
3. Levee Construction	Prevention of overbank flooding and processes of crevassing
4. Sediment Diversion	Trapping of sediment upstream by construction of dams and reservoirs
5. Boat Wakes	Bank erosion in dredged and natural canals

edge locations (Woodbury et al., 1973). The most recent episode of depocenter migration (the last 9 million years) was 320 km to the southwest, with each depocenter occupying a particular site for approximately 3 million years.

Sites of deltaic sedimentation also have shifted during the Holocene, but at much higher frequencies. During the last 7,500 yr six major lobes, with perhaps as many as 16 smaller lobes, have built the entire coastal plain of Louisiana from sequences of overlapping deltas (Kolb and Van Lopik, 1966; Frazier, 1967; Coleman, 1988). Each major lobe or complex remained active for ca 1,000–2,000 yr and covered an area of approximately 30,000 km^2 (Fig. 1). The abandonment and subsequent deterioration of each lobe was initiated by a course change in

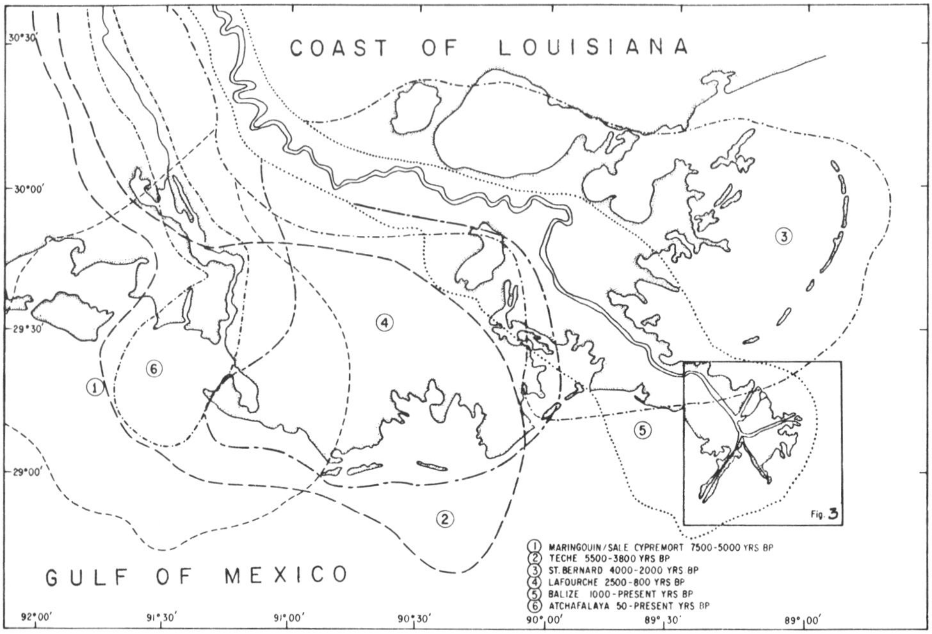

Fig. 1. Sites of deltaic sedimentation during the last 7,500 yr (Modified from Coleman, 1988; originally from Frazier, 1967). Modern Mississippi River Delta (Balize or birdsfoot delta) is enclosed by the box and is shown in more detail in Figure 3.

the main stem of the Mississippi River. Fisk (1961) determined that each main channel occupied a succession of favored distributaries and abandoned the others by plugging their channels with sandy sediments. Specifically, the loss of hydraulic gradient by gulfward progradation (Coleman, 1976) and the loss of efficiency by processes of channel bifurcation (Wells et al., 1984) were most likely responsible for changes in river course, sealing of old channels, and the shifting of depositional sites.

Each shift in location shown in Fig. 1 was accompanied by a transgressive depositional history that was driven by subsidence and marine reworking. The abandonment of a delta lobe and thus loss of sediment supply was followed by a 3-stage sequence (Fig. 2). The first stage is one in which the subaerial delta, deprived of sediment, begins to be inundated by marine waters. In stage 2, subsidence causes detachment of the mainland from the eroding headland and marine reworking forms a barrier island arc of transgressive sands, leading to stage 3, where the barrier islands become subaqueous shoals.

It is this natural cycle of constructional and destructional phases (Scruton, 1960) that allows for orderly repetition of depositional events and, eventually, the vertical stacking of newer deltas upon older deltas. Moreover, these natural deltaic sequences provide a fundamental rhythm that leads to alternating coastal erosion and progradation (Frazier, 1967; Scruton, 1960), governs the origin and dynamics of adjacent barrier islands (Penland and Boyd, 1981; Levin, 1993), and controls

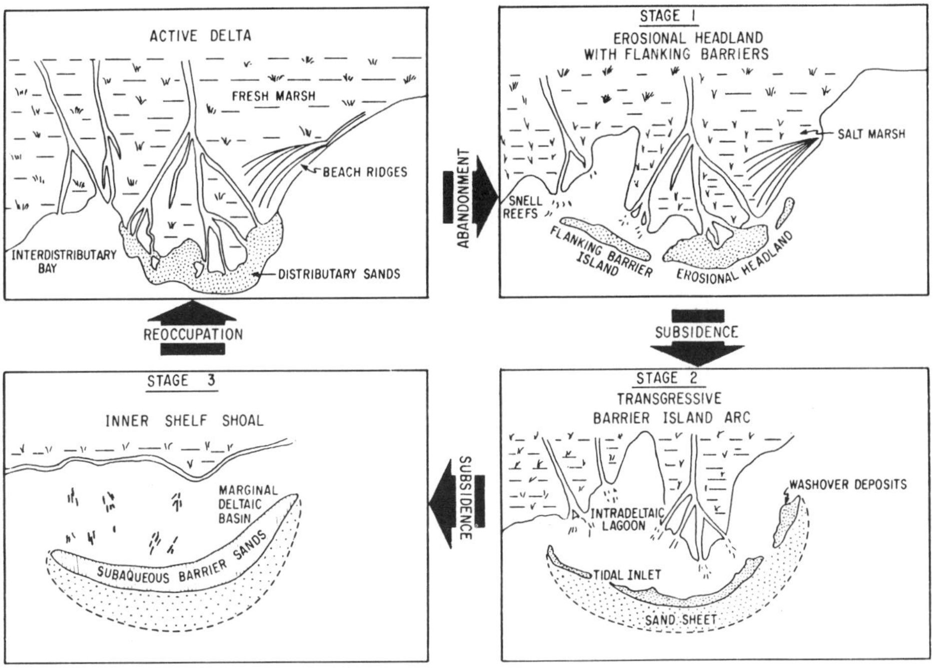

Fig. 2. Evolutionary stages in a delta cycle (Modified from Coleman, 1988; originally from Penland and Boyd, 1981).

the timing and extent of downdrift mudflat deposits that may be located more than 100 km to the west (Gould and McFarlan, 1959; Wells and Kemp, 1981; 1982).

The modern birdsfoot delta (800–1000 years old) is now at the end of its active delta phase and, without the intervention of man, would have already entered stage 1 of the transgressive cycle. Unlike previous Holocene lobes, the modern delta is small and elongate with distributaries that extend nearly to the edge of the continental shelf (Fig. 3). According to Fisk et al. (1954), the delta platform is made up of elongate fingers of sand underlying the distributaries. This skeletal framework of "bar finger sands", which have maximum thicknesses of 75–90 m and widths of 7–9 km, is surrounded by shallow-water sedimentation in the interdistributary troughs. However, calculations from Fisk's data show that only 18% of the birdsfoot delta below Head of Passes (Fig. 3) lies above mean sea level (Morgan, 1977). The remainder has been building subaqueously into progressively deeper water at a slow rate. Two of the three major passes are today within 20 km of the 200 m isobath and some 80% of the total discharge ends up on the outer continental shelf.

The suspended-sediment load of the Mississippi River, derived from a drainage basin that covers 41% of the United States, is 65% clay and 35% silt and very fine sand (Coleman, 1976; 1988). The present-day sediment load (including the Atchafalaya distributary) is estimated to be 210 × 10^6 t/yr from an average water discharge of 18,390 m^3/sec (Milliman and Meade, 1983). There is a gradual rise in discharge, beginning in January, peaking at about 57,900 m^3/sec in April–

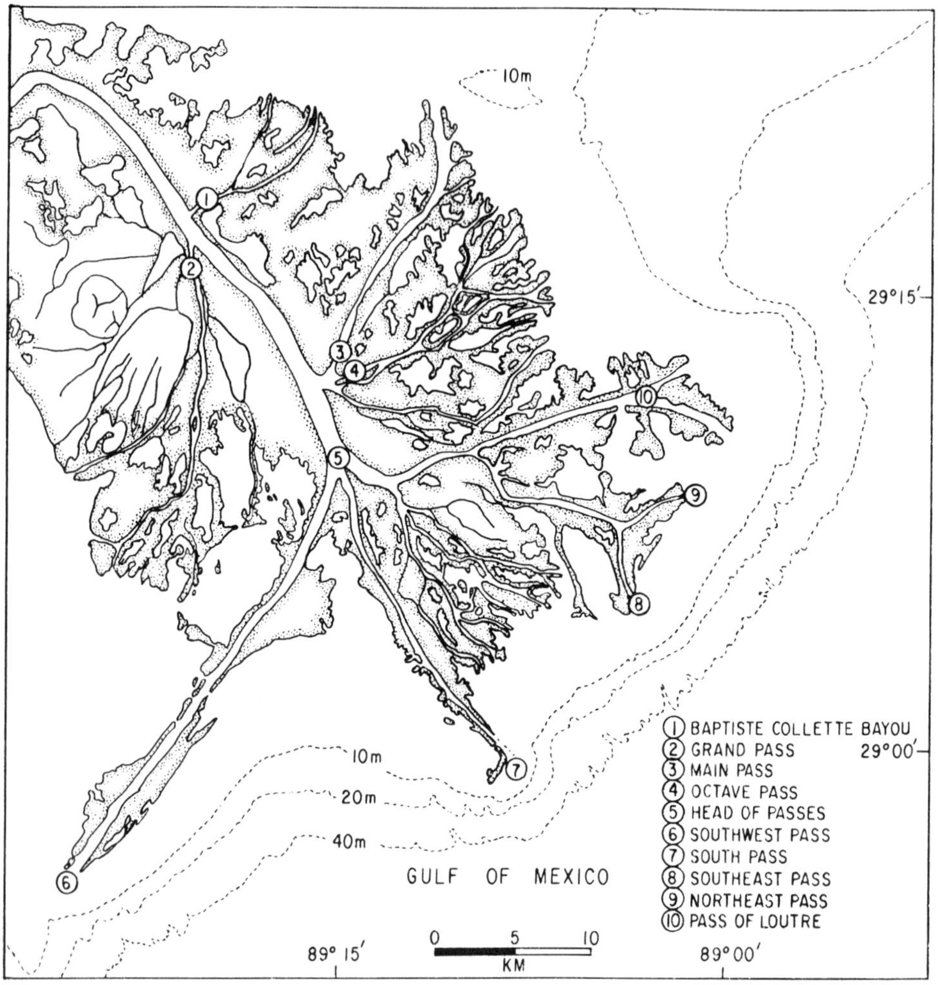

Fig. 3. Modern birdsfoot delta showing major distributaries (passes) and offshore bathymetry. Many of the small channels are nearly sealed and disperse sediments only during spring floods. See Figure 1 for regional setting.

May, followed by a gradual decrease in discharge to a low of about 2,830 m^3/sec during September-November. Hydraulic sorting of sediment at the river mouth and on the inner shelf has produced a characteristic seaward decrease in grain size (Scruton, 1960; Wright and Coleman, 1971); the high rates of coarse-sediment deposition in the form of distributary mouth bars and the low, but widespread, fine-sediment deposition farther offshore as prodelta clays are a result of deceleration and spreading of flow upon reaching the Gulf of Mexico.

As a receiving basin for sediments of the Mississippi River, the northern Gulf of Mexico can be characterized as a low energy environment. Tide range is approximately 0.5 m and wave energy, 0.034×10^7 ergs/sec per m of coast at the delta shoreline, is among the lowest of the world's major deltas (Wells and Coleman, 1984). During summer, winds are southeasterly, then shift to northeasterly in autumn. Longshore currents are typically less than 10 cm/sec (Wells et al., 1984) and cold fronts are the only regular forcing events that elevate water levels and increase the magnitude of nearshore currents (Moeller et al., 1993). Although tropical cyclones and hurricanes provide the most energetic conditions experienced on the shelf and at the shoreline, their statistical frequencies of occurrence are only once every 2.3 and 4.9 years, respectively (Wells et al., 1981).

This combination of low energy in the receiving basin and high sediment discharge from the river has made the modern Mississippi Delta one of the world's best examples of a fluvially-dominated delta. In contrast to former deltas that were characterized by many large distributaries, there are only three major channels in the modern birdsfoot delta: Southwest Pass, South Pass, and Pass a Loutre (Fig. 3). However, the modern delta is partitioned by numerous smaller active and abandoned distributary channels with natural levees that formed during flood stage when the river exceeded its banks and deposited part of its load near the channel margins.

Sediments carried into interdistributary basins by overbank flooding and through breaches in natural levees, referred to as crevasses, provide the essential shallow-water sedimentation marginal to and between the distributary channels that account for most of the total area in the subaerial delta. These sediments are covered by a luxuriant growth of freshwater marsh, except at the seaward shoreline fringes where more salt-tolerant intermediate marsh species have colonized. Marshes can maintain themselves during periods of relative sea-level rise by accumulating organic material and capturing inorganic mineral constituents (DeLaune et al., 1987). However, considerable sediment accumulation is required to offset the combined effects of subsidence and sea-level rise. Restricted sediment input reduces both vertical accretion and the resistance to hydrodynamic forces such as weaves and currents (DeLaune et al., 1990).

Because of the low elevations and gentle gradients, the Mississippi Delta is, and perhaps always has been, in a state of extremely delicate balance. One of the most interesting aspects of the wetland deterioration is the way in which it occurs: the Mississippi Delta is virtually falling apart from within (Fig. 4). Loss of deltaic wetlands is an internal process, not an inward and uniform transgressive process as might be expected along shorelines fronted by beaches and barrier islands. Although landloss rates are highly dependent on local factors, the typical pattern is for small, shallow ponds to form, then enlarge and merge, to produce expansive bodies of open water that can accommodate storm waves. Similar patterns of deterioration have been noted in Atlantic coast marshes (Stevenson et al., 1985).

Sea-Level Rise and Subsidence

Small changes in relative sea level have the potential for leading to large, irreversible changes in the subaerial Mississippi Delta. Relative or apparent sea-level variations in the northern Gulf of Mexico include the effects of global (eustatic) change in water elevation (Barnett, 1984), regional subsidence and compaction

Fig. 4. Typical pattern of marsh deterioration in the modern delta.

of local land surfaces (Holdahl and Morrison, 1974), freshwater runoff (Meade and Emery, 1971), variations in seawater density (Whitaker, 1971) and persistent winds (Behrens et al., 1977). Questions concerning sea level history are central to the questions of why the Mississippi Delta is presently deteriorating and why the rates has been increasing geometrically this century.

Relative rise in sea level is easy to measure but difficult to interpret. For the northern Gulf of Mexico, sea-level data obtained from coastal tide gauges (Suhayda, 1987) are complicated by subsidence and other local factors; the records include a combination of information about the local relative change in level of the Gulf as well as the land upon which the gauges are placed. Because of atmospheric effects, tide, and local runoff, monthly variations can exceed annual variations, making it difficult to detect the true sea-level signal (Turner, 1991). Further, some data sets are incomplete and documentation of the data is often insufficient for establishing their accuracy and precision. Despite the limitations, all records seem to show a secular trend with superimposed moderate fluctuations in values of mean annual sea level. Numerous independent analyses of eustatic sea level show nearly identical rates for global change: e.g., Gutenberg (1941), 1 mm/yr; Fairbridge and Krebs (1962), 1.2 mm/yr; Hicks (1978), 1.5 mm/yr; and, Gornitz et al. (1982), 1.2 mm/yr.

Several recent studies (Suhayda, 1987; Turner, 1987, 1991) have re-examined previously published tide gauge data (from Penland et al., 1986, 1987, and Barnett, 1984) for the northern Gulf of Mexico, with special reference to the Louisiana coast. Figures 5A and 5B show Suhayda's sea-level curve, together with smoothed annual values (5-yr running mean) for several coherent stations in the Gulf of Mexico.

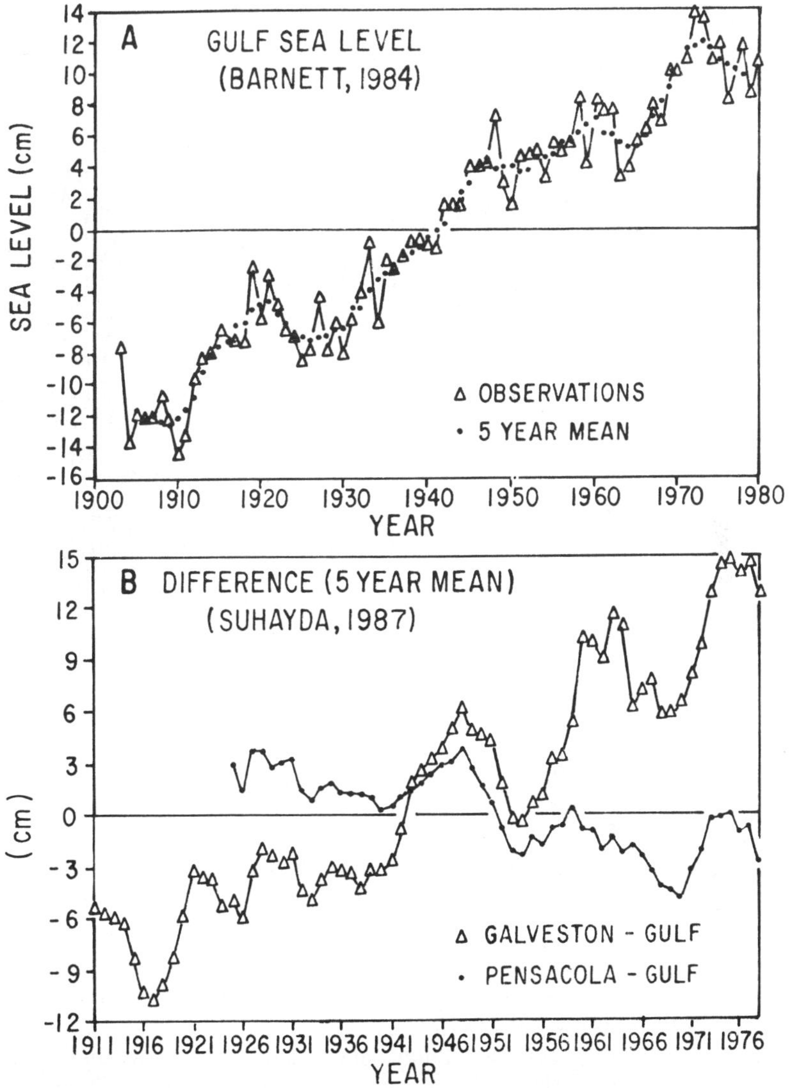

Fig. 5. (A) Sea level curve for the Gulf of Mexico using data from Suhayda (1987) and Barnett (1984). (B) Sea level curves for Galveston minus Gulf and Pensacola minus Gulf, showing the effects of subsidence at Galveston and local runoff and wind effects at Pensacola.

This curve does not include data from subsiding tide stations in Louisiana, but does include Galveston, Texas, a site itself undergoing rapid subsidence. In the past, Galveston water levels were used extensively as a surrogate for investigating long-term changes at other stations because of the record length (approximately 85 yr) and the coherence with other gauge records (Turner, 1991). When averaged over the entire record, the rate of rise in Figure 5A is about 2.3 mm/yr. Although there is little change in underlying rate of rise, there are periods of several years during which

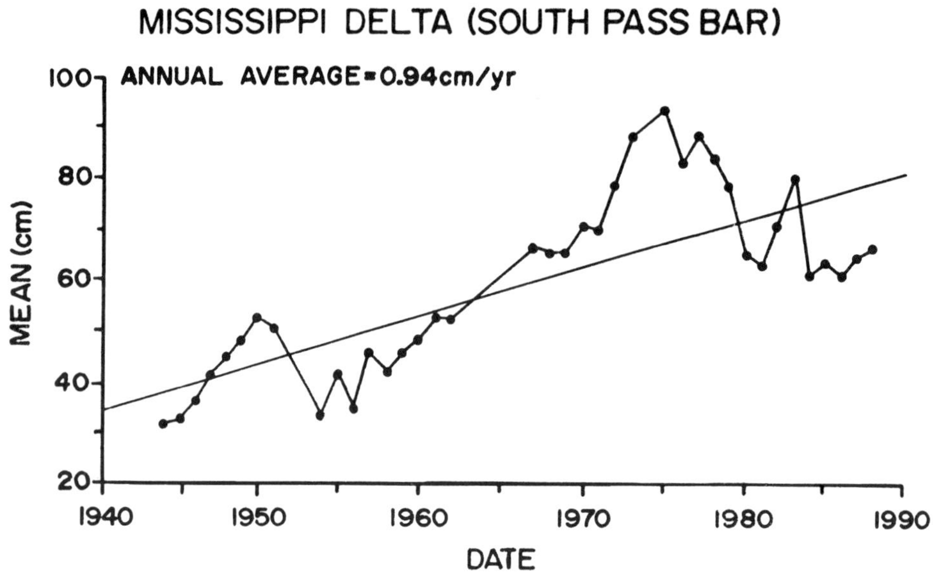

Fig. 6. Sea level curve for the lower Mississippi River Delta from U.S. Army Corps of Engineers data, approximately 5 km north of the Gulf of Mexico, for the 1944–1988 time period (from Penland and Ramsey, 1990).

the rate accelerates and during which the rate is negative.By subtracting the Gulf of Mexico annual values from tide records at Pensacola, Florida and Galveston, Texas, Suhayda (1987) has calculated (Fig. 5B) changes in land elevation and/or the influence of local hydrodynamic factors such as wind setup and freshwater runoff. The rationale for plotting differences between local curves and annual values of the Gulf "eustatic" curve was to provide reference stations to the east and west of the Louisiana coast. Whereas the Galveston minus Gulf record shows the effects of subsidence, the Pensacola minus Gulf record shows water mass and weather influences (see Meade and Emery, 1971; Marmer, 1952) since the Pensacola region is considered to be tectonically stable.

The most recent analyses of tide gauge data with reliable record quality indicate average rates of sea-level rise in Louisiana are about 0.90 cm/yr (Turner, 1991), with maximum rates of 1.04–1.19 cm/yr (Penland and Ramsey, 1990). A single 44-yr record for the Mississippi Delta (the only record with sufficient quality and duration) shows an average rate of 0.94 cm/yr (Fig. 6). Although short-term rates are variable, there is no evidence of an acceleration in the long-term rate of sea-level rise along the Louisiana coast in general, or specifically along the Mississippi Delta. Given the apparent long-term stability, average subsidence can be estimated by subtracting the average eustatic rate from the long-term relative rate. Using a eustatic rate of 0.1 cm/yr, average coastwide subsidence is on the order of 0.8 cm/yr and considered to be temporally stable.

Subsidence, however, is spatially quite variable. Sinking of land surfaces results from a combination of geosynclinal downwarping, sediment dewatering and

TABLE 2
Summary of subsidence and Sea-level Rise (Modified from Turner and Cahoon, 1987)

Process	Rate (cm/yr)	Duration (yr)	Impact
SUBSIDENCE			
Fluid Withdrawal	0–4	25	Local
Levee Subsidence	0–4	>100	Local
Geologic Consolidation	0.2–2.5	>100	Regional
Marsh Surface	0.05–1.0	<100	Regional Consolidation
WATER LEVEL			
Eustatic Sea Level	0.23	>100	Regional
Basin Water Level	−0.2–+0.2	20 cycle	Regional
Basin Water Level	−20–+20	1 (annual)	Regional
Basin Water Level	−2.0–+2.0	monthly	Regional

compaction, decomposition of organic material, and fluid withdrawal. Most scientists agree that there is a general trend of increasing subsidence south and southeast toward the Mississippi Delta (Swanson and Thurlow, 1973; Holdahl and Morrison, 1974; Trahan, 1982), and that local compaction is the most important short-term component in the subsidence records. The highest rates are where recent sedimentation rates are highest, e.g. deltas, and where water management has been most intense (Turner, 1991). Data in Figure 6 suggests that an average rate of subsidence for the Mississippi Delta is on the order of 0.84 cm/yr.

Suhayda (1987) examined the effects on subsidence from fluid withdrawal, including several sites in the Mississippi Delta region, by using models from Geertsma (1973) and Martin and Serdengecti (1984). He concluded that fluid withdrawal from oil and gas reservoirs had a localized and generally insignificant influence on subsidence, although it was possible for sinking to amount to as much as 80 cm directly above a reservoir over its life. Extensive growth faults that are associated with subsurface reservoirs may compartmentalize the land subsidence thus creating a phenomenon that is highly local rather than regional.

Table 2 is a summary of the time scales and magnitudes of the most important subsidence and water level processes that operate in the Mississippi Delta. Obviously, the relative importance of a particular process depends on the time scale and spatial scale of interest. Given the fundamental question of why deterioration of the delta has accelerated this century, the time scale of interest is the last 80 yr or so, and the spatial scale of interest is on the order of several thousand km^2. The conclusions, then, are 1) there is no evidence for an increase in rate of relative sea-level rise in the Gulf during the last 80 years and thus sea-level changes do not satisfactorily explain accelerating wetland loss, 2) between 50 and 96% of the total rise in relative sea level during the last 80 years (using Turner's 1991 analysis) can be attributed solely to combined subsidence effects, and 3) natural factors,

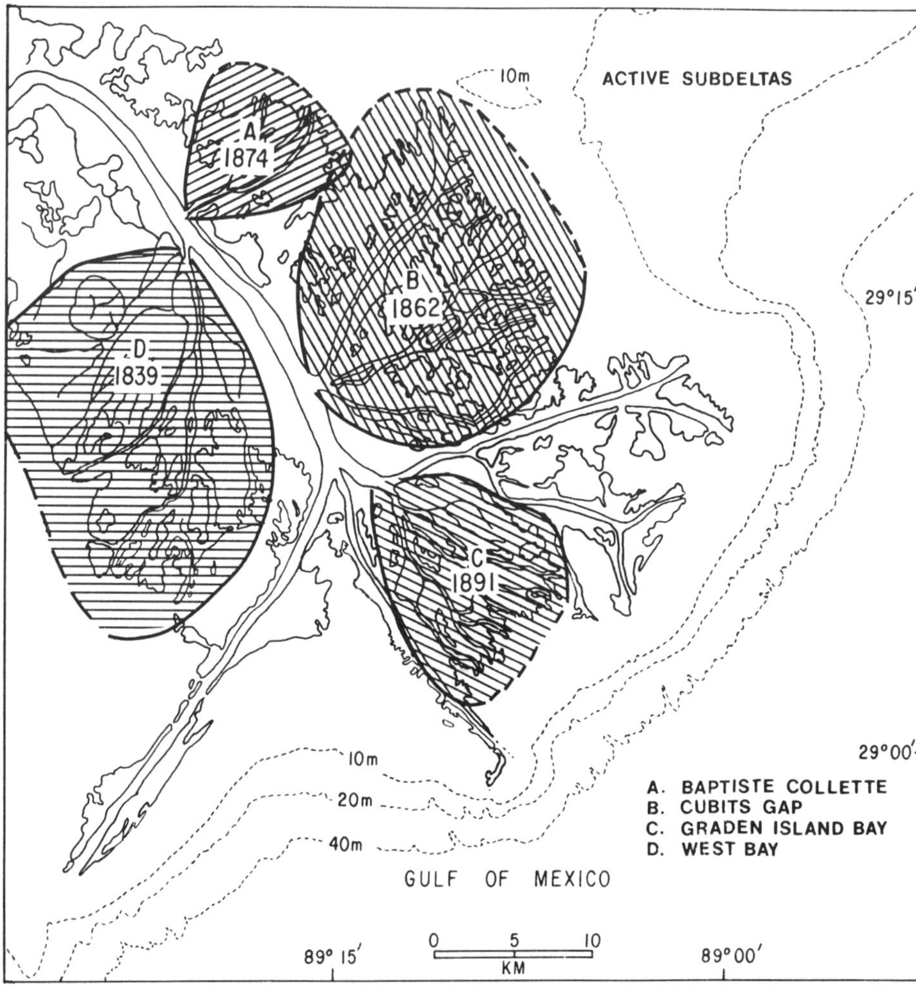

Fig. 7. Index map showing the four active subdeltas of the modern Mississippi Delta. Dates indicate years of the crevasse openings.

particularly compactional dewatering, are more significant at the time and space scales of interest than local changes caused by below-ground fluid withdrawal or subsidence from spoil bank levees.

Subdeltas and Wetland Loss

Whereas deep-water sedimentation has been responsible for construction of the deltaic platform, shallow-water sedimentation in the form of subdeltas has provided the subaerial expression (Fig. 7). Subdeltas are scaled-down versions of major delta lobes, both in time and space, and serve as excellent models for the larger lobes within which they occur. Just as each major delta lobe has a life cycle of 1000–

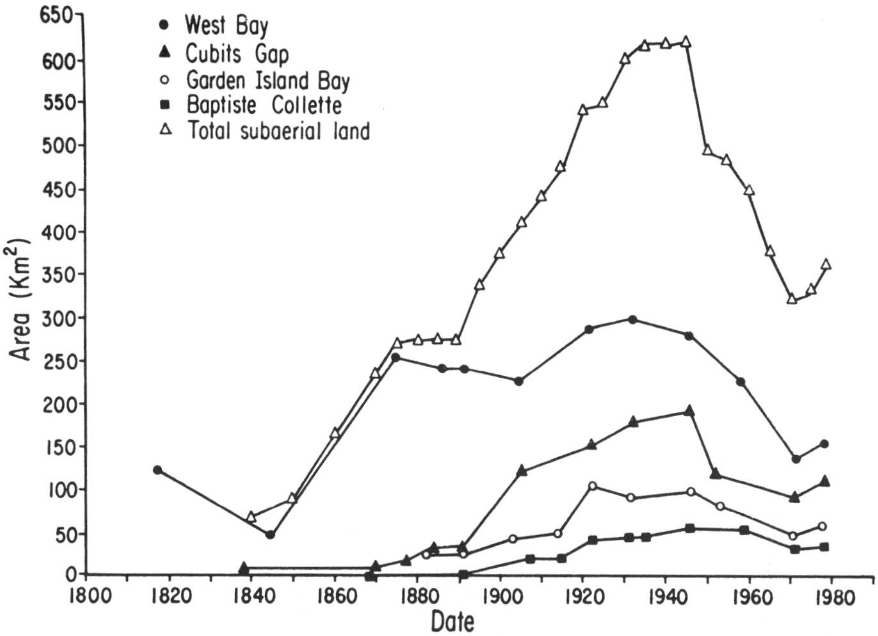

Fig. 8. Plot showing the composite and individual change in the Mississippi River subdeltas.

2000 yr, each subdelta has its own natural life cycle that also consists of growth and deterioration. In the modern birdsfoot delta, the rapid deterioration of marsh is controlled, at least in part, by subdelta processes that operate on time scales of 100–200 years.

A subdelta or bay-fill deposit is a sequence that forms initially from a break or crevasse in a major distributary natural levee during flood stage, enlarges as flow increases through successive floods, reaches a peak of maximum discharge and deposition, then wanes and becomes inactive (Coleman et al., 1969). As a result of high subsidence rates by compactional dewatering, an abandoned subdelta is gradually inundated by brackish water and reverts to an open-bay environment, thus completing one sedimentary cycle of infilling and abandonment.

Four subdeltas within the modern lobe of the Mississippi River Delta have been active since the first accurate survey on the lower delta was conducted in 1838 (Fig. 7). Three of the subdeltas lie above Head of Passes (Fig. 3) and together take, via flow through crevasse systems, approximately 18.4% of the flow that reaches the lower delta (Morgan, 1977). Using maps, charts, and aerial photographs, Wells and Coleman (1987) constructed curves for rates of change in land area, sediment volume, and linear progradation in each of the subdeltas.

Figure 8 shows on a single plot the cyclic nature of the four Mississippi River subdeltas; each cycle consisted of growth followed by deterioration. In fact, subdeltas seem to follow an orderly, almost predictable life cycle that lasts from 115 to 175 years. Average rates of growth ranged from a low of 0.8 km^2/yr in the Baptiste Collette subdelta, to a high of 2.7 km^2/yr in the West Bay subdelta.

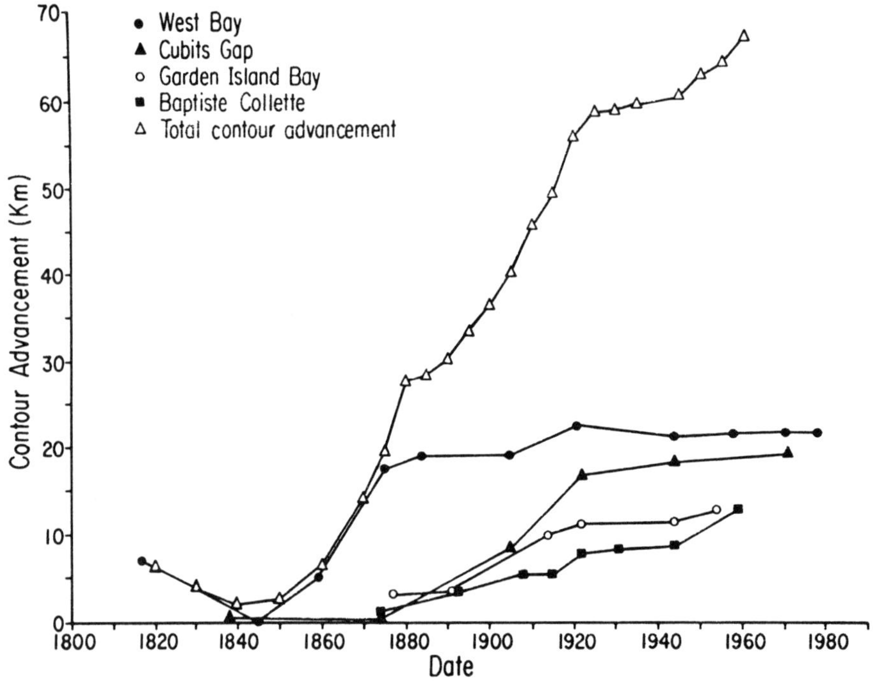

Fig. 9. Plot showing the composite and individual curves for linear progradation of the Mississippi River subdeltas.

Maximum rates of growth for the same two subdeltas, respectively, were 1.1 and 7.0 km^2/yr. A composite curve shows that the average rate at which subaerial land was added to the modern birdsfoot delta from all subdeltas combined was 5.2 km^2/yr.

Maximum and average rates of deterioration were on the same order as rates of growth, typically several km^2/yr. The exception was the Cubit's Gap subdelta which, during its maximum deterioration (12.9 km^2/yr in the late 1940s), accounted for nearly one-third of the total land loss in Louisiana. The most apparent, and certainly the most important, feature of subdelta deterioration was the abrupt onset of land loss that began in 1945, shown by the composite curve.

Figure 9 shows rates of progradation of the leading edge of each subdelta, together with a composite curve plotted at 10-yr intervals from 1820–1870 and 5-yr intervals from 1870–1960. Maximum rates were on the order of 0.5 km/yr. Nearly all of the composite curve contour advancement prior to 1880 was a result of growth of the large West Bay subdelta. After 1860, there was a 40-yr period of rapid, steady progradation of the subaerial delta front (composite curve) that tracked the rate of subaerial progradation shown in Figure 8. During their life cycles, individual subdeltas extended a maximum of 13–22 km across the bays into which they were building.

Unlike subaerial growth, extension of subdeltas through contour advancement never showed a decline. Rather, it simply leveled off, or, in the case of the composite

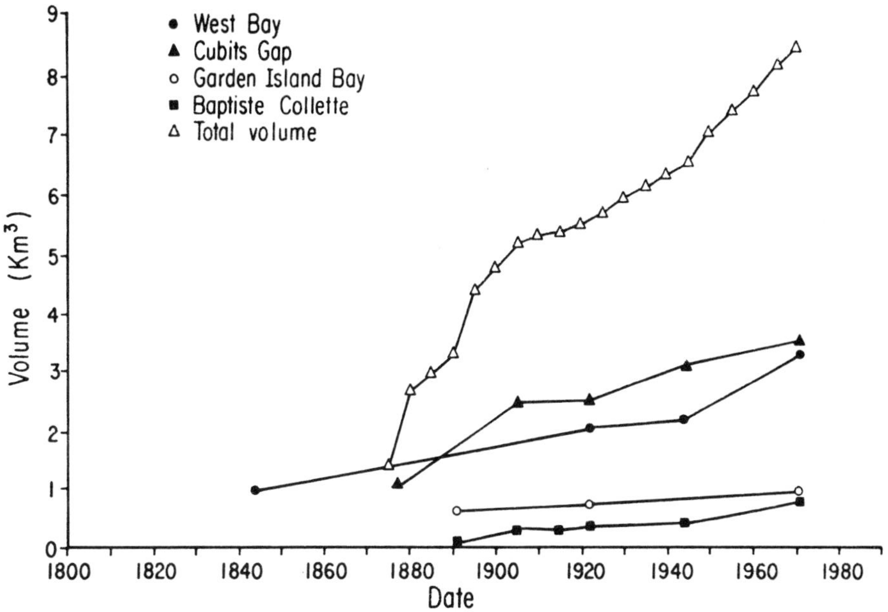

Fig. 10. Plot showing composite and individual curves for volume infilling of the Mississippi River subdeltas.

curve, continued to increase but at a slower rate. Thus even during the rapid decline in amount of subaerial land, the subaerial delta front maintained its position.

Figure 10 shows the rates of volume infilling, together with a composite curve for all subdeltas plotted at 5-yr intervals from 1875–1970. Over this 100-yr period, the Mississippi Delta acquired approximately 7 km^3 (7000×10^6 m^3) of sediment through the process of subdelta growth. Average rates of sediment accumulation in the subdeltas were in the range 4–26 $\times 10^6$ m^3/yr, and maximum rates were 4–45 $\times 10^6$ m^3/yr. Unlike subaerial growth and, to a lesser degree contour advancement, volume infilling began immediately upon opening of the crevasse. Although rates of volume increase for individual subdeltas appeared to remain quite steady throughout the life cycles, this could be an artifact of the limited available data (maximum of six data points per subdelta).

The receiving basins or bays into which the subdeltas built were as shallow as 2 m (Baptiste Collette) and as deep as 9 m (Cubit's Gap). The two larger subdeltas, Cubit's Gap and West Bay, had infilling rates higher by factors of 2–4 than the two smaller subdeltas, Baptiste Collette and Garden Island Bay. Similar to the contour advancement curves, the volume curves continued to increase throughout the period of subaerial deterioration. Thus, even though subaerial land was disappearing, sediments were carried through the crevasse openings and continued to accumulate in the subdeltas. The overall rate of infilling, however, decreased in the early 1900s (composite curve).

Figure 11 is an example, from selected surveys, of growth and deterioration in the Cubit's Gap subdelta. The patterns in Cubit's Gap were typical of all the

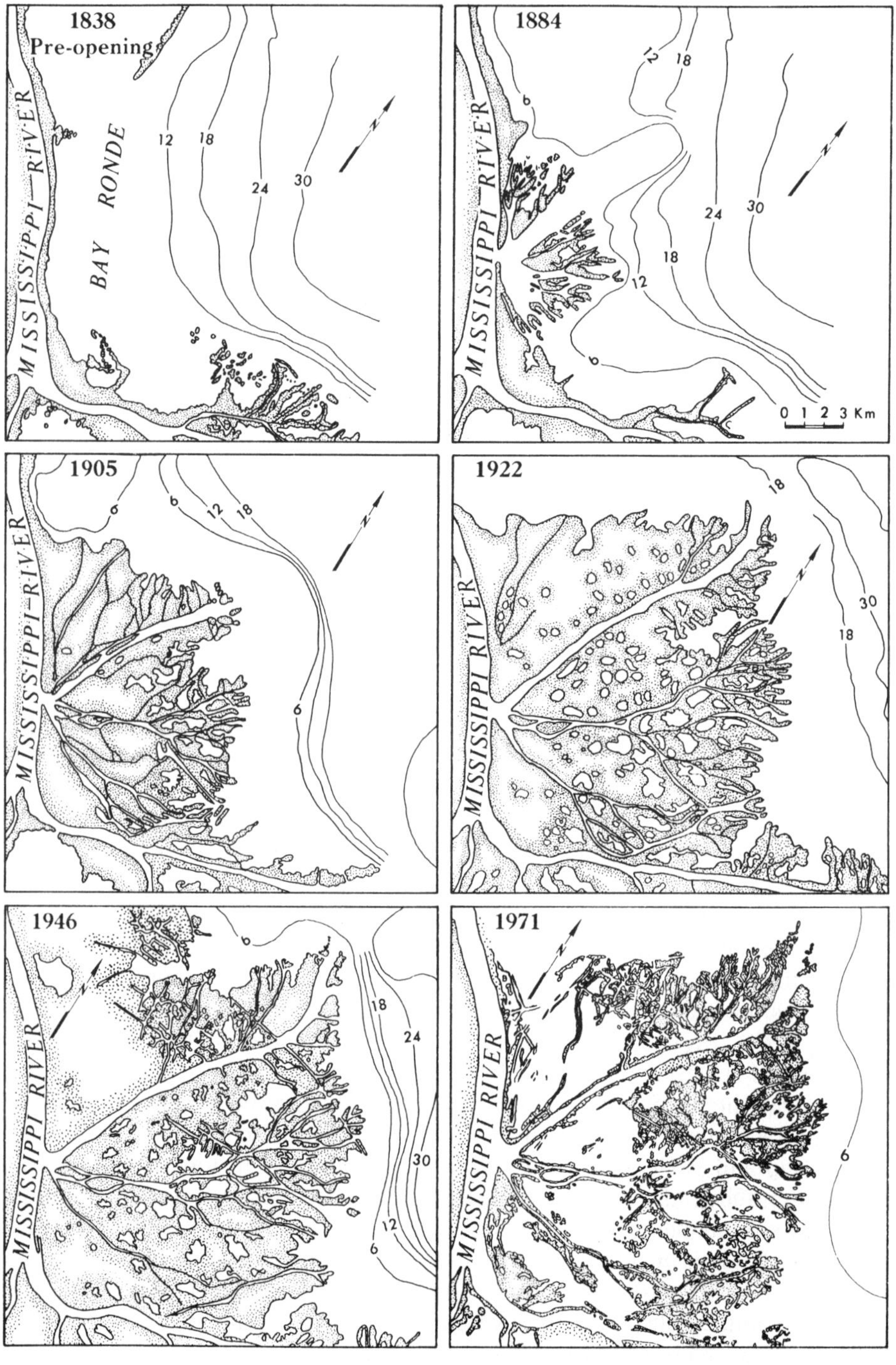

Fig. 11. Sequential development of the Cubits Gap subdelta, constructed from Coast and Geodetic Survey and U.S. Geological Survey charts.

Fig. 12. Crevasse channels extending across natural levee into an interdistributary bay. At point of bifurcation a mid-channel bar usually develops into a subaerial lobe.

subdeltas (Wells et al., 1984). A major channel leading from crevasse to bay usually bifurcated several times within the first decade (Fig. 12). Subaerial growth then accelerated through processes of channel extension and bifurcation (van Heerden, 1980; Wells et al., 1984). Mid-channel bars, the obstructions which cause flow to bifurcate, quickly became subaerial by vertical growth during floods. As delta lobes grew, the cross-sectional area of the channels decreased. Much of the channel system was eventually eliminated, setting the stage for deterioration. Deterioration occurs more rapidly in the older, proximal portions of the subdeltas, typically from small water bodies that revert to larger bays. Often the only subaerial evidence of former channels is the fringe of marsh that continues to grow on the topographically-high natural levees.

Figures 8–11 show that during the progradational phase of development, subdelta growth begins after crevasse development, but not at a constant rate. This is because of the uneven distribution of coarse sediment, which is deposited initially only in the vicinity of the break. Fine-grained sediments are carried farther into the bay and build up a platform for future progradation and channel development. At some point in the subaqueous infilling process, usually after 10–15 years (Wells, et al., 1984), major channels become established and a well-organized pattern of bifurcations becomes evident (Fig. 12). It is only after the development of a well-defined channel pattern, hence sediment delivery network, that subaerial growth begins to increase rapidly. For a period of 70–90 years (30 yr in the case of Garden Island Bay), the subdeltas maintain growth from deposition of coarse sediment by bed-load transport during low-flow periods and by overbank flooding involving suspended as well as bed-load transport during flood stage. As progradation and areal extent of the subdeltas increase rapidly, so does the growth of marsh, which forms a cap on the underlying sedimentary deposits.

The fact that the total volume of sediments in every subdelta has continued to increase despite the loss of subaerial land (Figs. 8, 10) is a direct result of subsidence. Assuming an average subsidence rate of 1.5 cm/yr and an average life of 150 years, then 2.25 m of subsidence will occur during the life of a subdelta. In some cases, this amount of subsidence could, in the absence of sedimentation, double the depth of the bay in which land is being built. Thus, sedimentation may not be sufficient to keep pace with subsidence, but it nevertheless continues to fill the subdelta receiving basins as they slowly revert to open water. The 1946 and 1971 maps of Cubit's Gap (Fig. 11) show that, even during the rapid decline in amount of subaerial land, the delta front is maintained and subsidence is more important than marine erosion in loss of wetlands. Figure 4 shows the complexity of vegetation patterns when a subdelta loses its land by processes other than erosion along a coherent seaward margin.

THE DETERIORATION PHASE

It appears that the demise of the Mississippi Delta is a result of simultaneous deterioration of its subdeltas. Whereas an increase in the number of subdelta channels leads to rapid subaerial growth, it also ultimately leads to a slower progradation rate and eventual loss of subaerial land. Examination of subdelta channel patterns suggests that sediment delivery becomes inefficient and deterioration of the subdeltas begins after the number of distributaries exceeds approximately 25. Often, however, a point of balance is reached wherein the subdelta is able to keep pace with subsidence and sea-level rise, despite the inability to deliver sediments efficiently to the subdelta land surfaces. The double peak in growth curves for West Bay and Garden Island Bay and the flat crest in the growth curve for Baptiste Collete (Fig. 8) probably represent such periods of attempted stabilization. Finally, the marsh surface can no longer maintain its growth rate to keep up with processes of deterioration, and the marsh cover will begin to open into many small lakes and bays.

Construction of artificial levees and extension of the entire birdsfoot delta to the edge of the continental shelf are usually cited as reasons for the loss of subaerial land. In an general sense, these are contributing factors to the poor health of the delta because they favor deep-water sedimentation; however, they are not the immediate reasons for the rapid loss of subaerial land which became so apparent after the 1945 peak in the composite curve (Fig. 8). Artificial levees end upstream of the four active subdeltas and thus do not prevent sediment from entering the subdelta system by way of overbank flooding (levees do, however, prevent renourishment to a vast area of south Louisiana). In the case of the modern delta, artificial levees may even serve as conduits to funnel water and sediments to the lower delta that would ordinarily be lost to overbank flooding farther upstream. Moreover, geologic reconstruction of the modern delta platform by Fisk et al. (1954) indicates that the

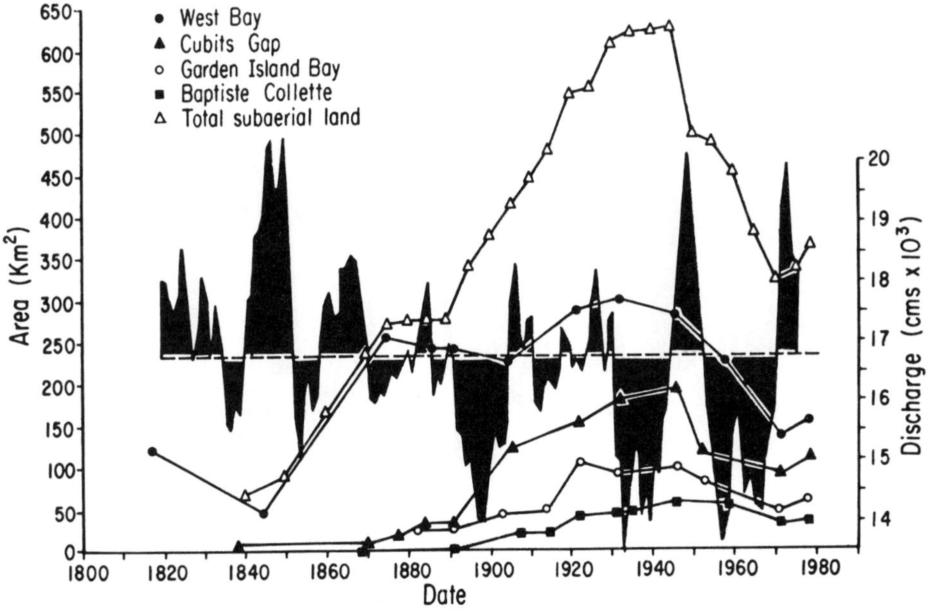

Fig. 13. Composite and individual subaerial growth curves showing their relationships to a seven-year moving average of Mississippi River discharge, 1817–1977, taken from a gauging station at Vicksburg, Mississippi.

delta was approaching the shelf edge and depositing part of its sediment load in deep water well before the abrupt deterioration began in 1945.

Numerous studies in the early 1980s attributed much of the landloss problem in Louisiana to excavation of oil and gas access canals (e.g. Turner et al., 1984). Dredging and its associated activities (direct and indirect impacts, respectively) were felt to represent significant development pressures on coastal wetlands by changing normal flooding patterns and allowing salt water penetration into previously fresh marshes. However, a recent analysis of long-term salinity records, including one in the Mississippi Delta, indicates that salinity trends are extremely weak and the magnitude of predicted changes in mean salinity are generally small enough to be non-detrimental to adjacent marsh plants (Wiseman and Swenson, 1987). According to a recent assessment by Turner and Cahoon (1987), direct and indirect impacts from oil and gas activities accounted for only 4–5% and 4–13%, respectively, of total wetland loss between 1955 and 1978.

Wells and Coleman (1987) attempted to explain the loss of subaerial land by relating long-term variations in river discharge to subdelta life cycles (Fig. 13). However, in only four instances were they able to find an explicable relationship between growth or deterioration and the subaerial composite curve. These were: (1) initial growth during a period of high water discharge from 1840–52, but nearly all of which was due to growth of the large West Bay subdelta; (2) a pause in the rapid growth between 1875 and 1890 that corresponded to mildly fluctuating (near average) discharge; (3) a period of stabilization between 1930 and 1945

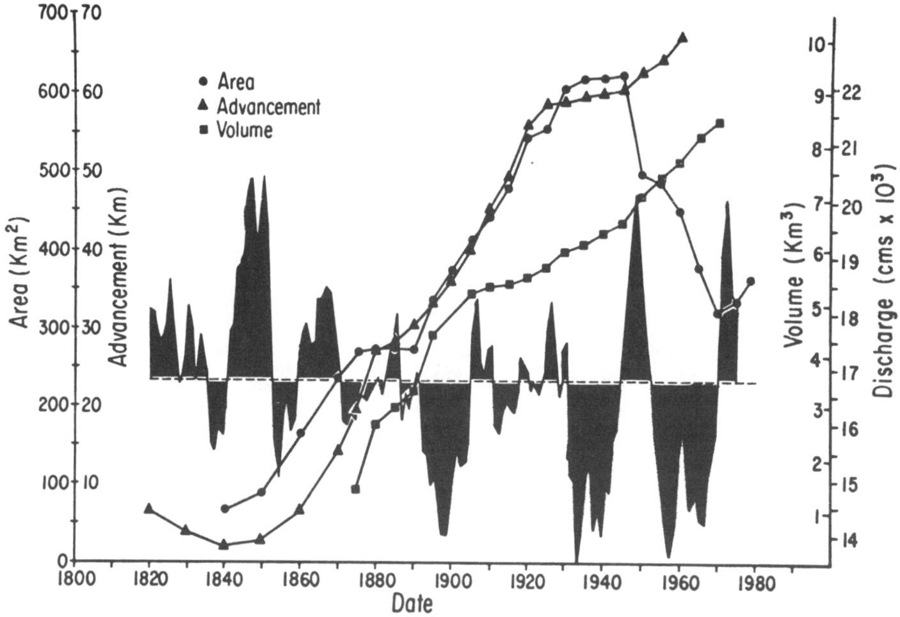

Fig. 14. Plot showing subaerial growth, linear progradation, and volume infilling relative to a seven-year moving average of Mississippi River discharge, 1817–1977, taken from a gauging station at Vicksburg, Mississippi.

that took place during very low discharge over the same time period; and, (4) a period of renewed growth between 1971 and 1978 that resulted from three years of extremely high discharge (1973–75). Further, the lack of correlation between river discharge and either of the other two composite curves, contour advancement and volume infilling (Fig. 14), leads to the conclusion that water discharge alone is not particularly important in subdelta dynamics and does not offer a satisfactory explanation for accelerating deterioration. Rather, there is now substantial evidence that loss of sediment load in the main stem of the river and the lack of an efficient sediment delivery network within the delta are largely responsible for the present high land loss rates.

The Sediment Problem

With respect to sediment loads, three significant events have occurred this century. First, the river has been leveed along virtually its entire length. Although construction of levees for flood protection began with the earliest permanent colonization in Louisiana, the most intense period of construction has been since 1900. Second, dams along the Missouri and Arkansas River basins have trapped much of the suspended sediment load that, prior to 1953, was delivered to the Mississippi Delta. Third, the Atchafalaya River, a major distributary of the Mississippi River, has been taking progressively more water and sediment discharge from the main stem of the river since the early 1900s. The combined effect of these activities has

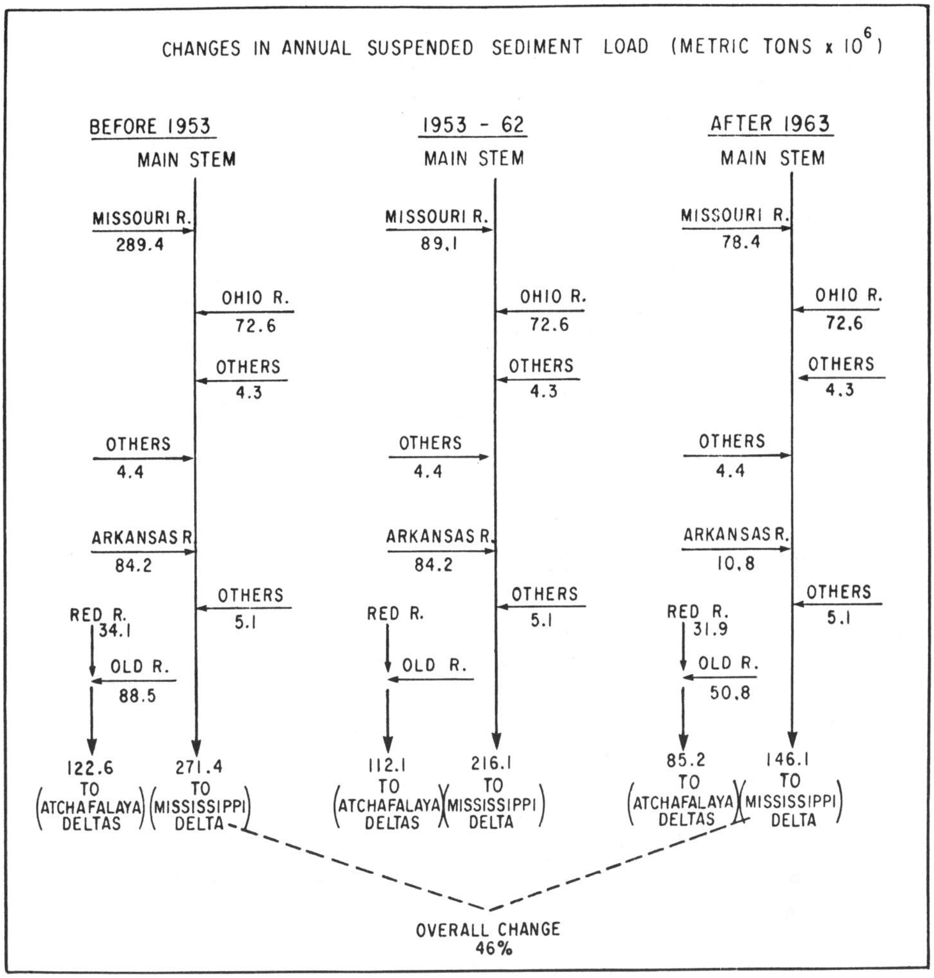

Fig. 15. Flow diagram showing the changes in suspended sediment load on the Mississippi River and its major tributaries (data from Keown et al., 1981). Note that values have been converted from short tons into metric tons.

been an estimated 46% decrease in suspended sediment load reaching the modern birdsfoot delta (Fig. 15).

Although levees confine sediments that, under natural conditions, would provide overbank material to the adjacent flood plain and wetland habitats, they are, as previously discussed, a minor contributing factor to wetland loss in the Mississippi Delta itself. However, other streambank protection measures such as revetments, meander bend cutoffs, and spur dikes have either eliminated scour and bank caving as sediment sources or trapped sediments upstream (Keown et al., 1981). Even the offshore jettying at the mouths of the major passes may indirectly increase the loss of sediments to deep water by funnelling sediment farther offshore. Instead of being reworked over wide distributary bars or back into the marsh, the sediments are more likely to be concentrated into small elongate bars offshore (Morgan, 1977).

Figure 15 shows diagrammatically the changes in suspended-sediment load that have occurred as a result of dam closures and partial diversion to the Atchafalaya River Delta. A series of multipurpose dams were constructed in the Missouri River Basin between 1953 and 1967 and on the Arkansas River main stem between 1963 and 1970 (Keown et al., 1981). Although river water that enters a reservoir (because of dam construction) is eventually released, much of the sediment is permanently trapped (Meade and Parker, 1985). Prior to placement of two dams in 1953, the Missouri River contributed approximately 290×10^6 metric tons/yr of sediment to the Mississippi River. The effect of these two dams, which was to diminish sediment loads immediately in this upper-basin tributary, could be observed all the way to the Gulf of Mexico (Meade and Parker, 1985). A second major loss of sediment was from construction of impoundments on the Arkansas River, beginning in 1963. In this system, the sediment load decreased from 84 to 11×10^6 metric tons/yr over a period of several years.

Perhaps the most significant event that has occurred on the Mississippi River during historic times is the attempted diversion and capture of main-stem flow at Old River, Louisiana (31° 00′). At the Old River diversion, part of the main stem sediment load is carried into the Red-Atchafalaya River system and the remainder continues to the Gulf of Mexico (Fig. 15). Prior to the completion of a control structure at Old River in 1963, volume of flow to the Atchafalaya River from the Mississippi River increased steadily from 13% in 1900 to nearly 30% in 1952 (Morgan et al., 1953). Flow through the original control structure and an auxiliary structure that was opened in 1987 is now limited to 30% of the combined flow from the Mississippi River and the Red River.

Loss of suspended sediment from the main stem to the Atchafalaya River, plus the loss of sediment from dams and other upstream engineering works, has profoundly changed the amount and distribution of available sediment. Between 1917 and 1960, several large shallow lakes in the Atchafalaya Basin served as natural traps for sediments that were being increasingly diverted through the Atchafalaya distributary. As these lakes approached a sediment-filled state by the mid-1900s, more sediment reached the coast and two subaerial deltas in the Atchafalaya Bay were formed in the early 1970s (Fig. 16; Roberts et al., 1980; Wells,1987). Together these deltas added 14×10^6 m^3 of new sediment to the bay each year (subaerially equal to several km^2/yr), most of which, prior to the 1900s, would have been transported to the modern birdsfoot delta. However, had man not intervened to prevent full-scale diversion into the Atchafalaya distributary through the Old River Diversion, the amount of sediment now reaching the modern Mississippi Delta (146×10^6 metric tons/yr) would be even less. One example of the loss in sediment load is reflected in the subdelta volume infilling curve (Fig. 10) which decreased from an infilling rate of 12.6×10^6 m^3/yr between 1875 and 1905 to only 5.4×10^6 m^3 between 1905 and 1970.

These changes in amount and distribution of available sediment have created a sediment disparity in the modern birdsfoot delta. Simply defined, the disparity is the difference between sediment accretion and relative sea-level rise; even though

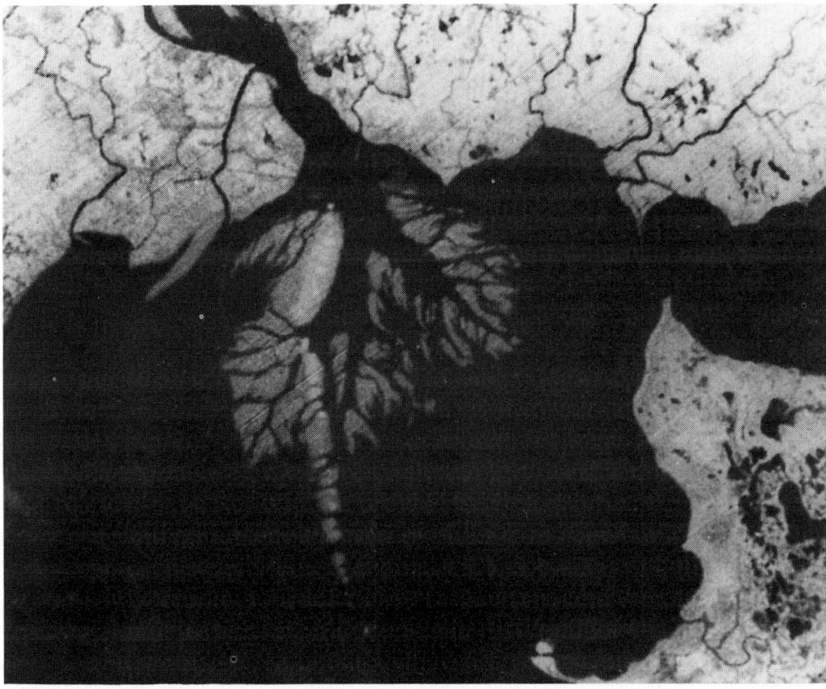

Fig. 16. Landsat image of the lower Atchafalaya River Delta (January 1976). Location of Atchafalaya Bay is shown by lobe No. 6 in Figure 1.

there is no evidence for a long-term change in rate of relative sea-level rise (Figs. 5 and 6), a disparity occurs when either the rate of sediment input decreases or the annual variation in basin water levels create biologically significant (20–25 yr) land/water differences. Both types of disparity occur in the delta.

Although the sediment disparity offers an explanation for long-term demise of the Mississippi River Delta, it does not address the geometrically-increasing land loss rates that have plagued the coastal marshes since the turn of the century, nor does it explain the abrupt onset of subaerial deterioration that began in 1945. The sharp reduction in sediment load resulting from dam construction did not occur until after the growth peak in 1945 (Fig. 8) and therefore cannot specifically be linked to rapid subdelta deterioration. It is more likely that the primary reason for such abrupt deterioration was the loss of critical stream gradient at a time when small channels were forming a complex and inefficient network for delivering the available sediments. Shallow-water, fluvially-dominated deltas cease to grow when they develop too many small, inefficient, bifurcating distributaries, rather than when the deltas reach a given size (Wells et al., 1984). In general, the longer and slower a delta grows, the more likely it is to reach a state of inefficiency with too many channels before reaching a size limited by receiving basin area (Wells, 1987).

The possibility also exists that the delicate balance of the mid-1900s was upset by a change in the magnitude-frequency relations of large floods on the river. For example, the magnitude of the 2-percent-probability flood between 1920 and 1949 was only about 2000 m^3/sec (Upper Mississippi Valley) whereas it was greater than 3000 m^3/sec for two other periods of approximately the same length (1867–1895 and 1950–1980; Meade et al., 1990). Large floods are significant because of the non-linear relationship between water discharge and sediment discharge. Most of the sediment is discharged in only a few days to weeks each year and, therefore, large floods play a disproportionately large role in long-term transport of sediments (Meade, et al., 1990).

If deforestation for agricultural purposes has allowed an abnormally high input of sediments to the Mississippi River during the 19th century, it could be argued that the decrease in sediment load this century has simply returned the river to its normal condition. This assertion might suggest that the subdeltas grew only because of an abnormal sediment load. However, previously cited geological evidence indicates that this is not the case. Before the influence of man, major and minor delta lobes were part of a natural rhythm that produced a highly dynamic coastal plain with sequences of vertically stacked deltas. The influence of man has not changed the underlying processes, but only the rates at which the processes occur.

Future Mitigation: Artifical Diversions

The present transformation from marsh to open bay can be slowed only if there is a reintroduction and redistribution of riverine sediments. It is interesting to note that freshwater diversions have been advocated in Louisiana since the turn of the century as a way of altering salinities and controlling floodwaters. Today the greatest long-term benefit of diverting freshwater is probably in aiding subdelta growth.

In 1973, Gagliano et al. (1973) proposed massive manmade diversions to create new subdelta lobes, increase upper deltaic plain aggradation, and control salinity in the estuarine regions. Seven potential diversion sites were selected within and just upstream of the modern birdsfoot delta. They determined that a 70% sediment retention rate for subdelta deposition would build 32 km^2 of new land per year, replacing about 75% of the net annual loss in coastal wetlands (41 km^2/yr). Since their study, however, land loss rates for the coast (as a whole) have increased to over 100 km^2/yr, and between 10 and 25 new potential diversion sites in the delta plain have been identified (Moger and Faust, 1991; van Heerden, 1993).

In 1985, the State of Louisiana in conjunction with the U.S. Army Corps of Engineers, excavated the first controlled diversions for the sole purpose of creating subdeltas. The Garden Island Bay subdelta (Figs. 7, 17), which utilizes only 4% of total river flow, was selected as a test field site. The maximum rate of infilling in Garden Island Bay was between 1914 and 1922 when subaerial land grew at 6.9 km^2/yr (Wells et al., 1984). The highest point on the growth curve was reached in 1922, only 31 years after the crevasse opened; for the next 24 years (1922–1946) an overall balance between sedimentation and relative sea-level rise was roughly maintained.

Since this was a pilot field project, no data were available to use in the engineering design, and predictions for future growth were made primarily by using the

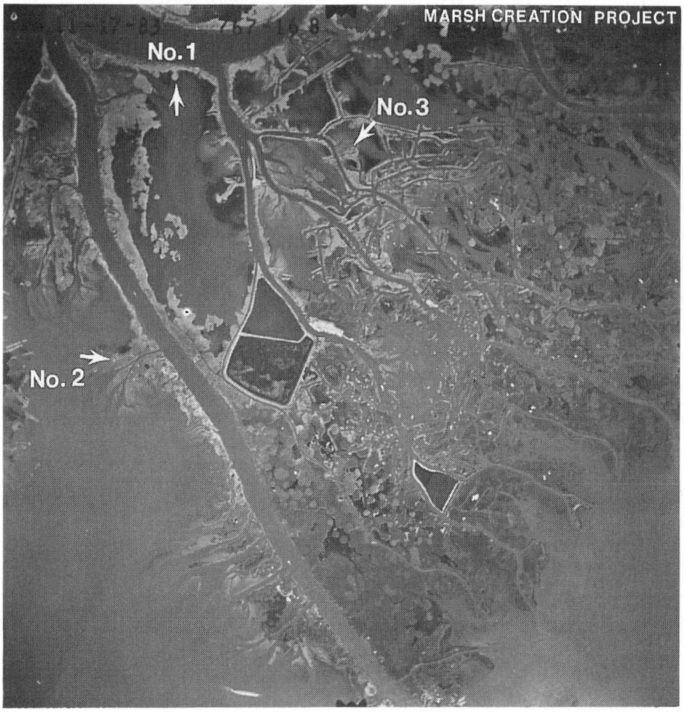

Fig. 17. Aerial photograph of the Garden Island Bay subdelta showing the location of three artificial diversion sites.

four active subdeltas as a model. Figures 18 and 19 show the volume computations and subaerial configurations that were used in this projection (Howard, Needles, Tammen & Bergendoff, 1985). It was assumed that the rate of subaerial growth would be controlled by 1) volume of sediment passing through the artificial cut into the interdistributary bay, 2) percentage of sediment retained in the immediate vicinity (1–2 km radius) of the cut (as opposed to being carried out of the system), and 3) volume of infilling necessary for sediments to reach the elevation of mean sea level and become subaerial. The mass of sediment transported through each diversion was taken as a product of concentration of suspended sediment, cross-sectional area of the diversion, and average current speed through the opening. From the computations it was concluded that these small subdeltas would add a total of approximately 1 km^2/yr of subaerial land to the larger Garden Island Bay subdelta. A follow-up survey conducted in 1991 (Table 3) showed that a total of approximately 2.5 km^2 of subaerial land had been added over a 5-yr period, considerably less than predicted (Louisiana Department of Natural Resources, 1993).

At best, it is reasonable to be somewhat cautious about the future of artificial diversions in the Mississippi Delta. Although diversions clearly offer the best (and perhaps only) hope of slowing the rate of wetland deterioration, they probably do

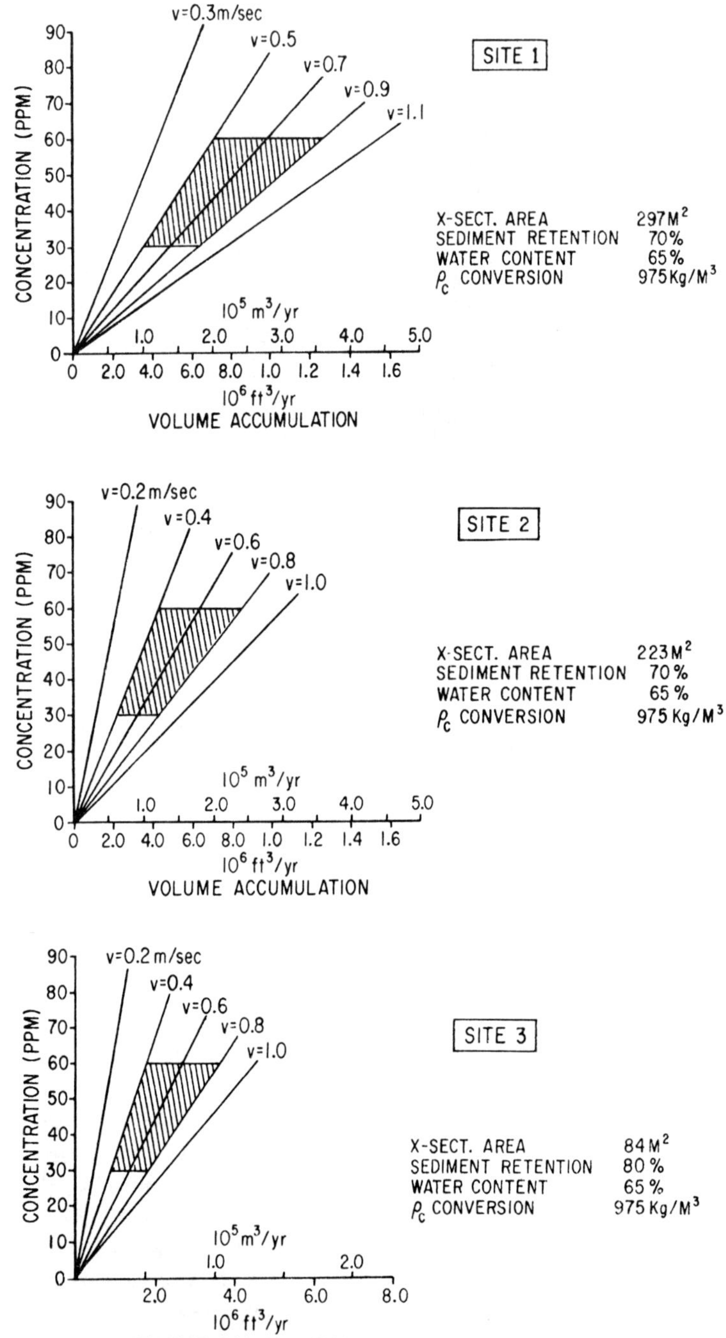

Fig. 18. Rates of predicted volume infilling for the three diversion sites as a function of sediment concentration and current speed.

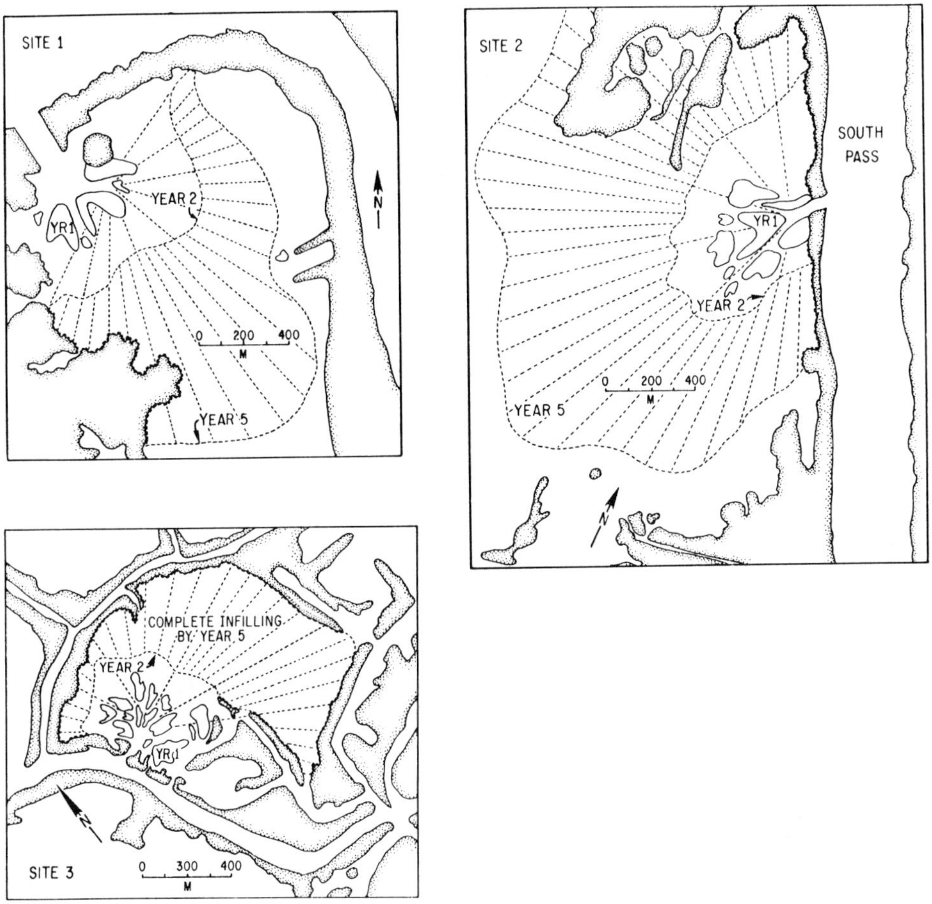

Fig. 19. Projected extent and configuration of new land at three diversion sites. Locations shown in Figure 17.

not offer a long-term solution. In the case of the modern birdsfoot delta, diversions are an attempt to rejuvenate a dying delta lobe. The basic ingredients for a healthy substrate are an efficient sediment delivery network in addition to an adequate sediment supply. Both of these are now missing in the delta (studies have shown that sufficient sediment is available in the river to create substantial new coastal wetlands (van Heerden, 1993), but the sediment must be effectively captured and distributed). Unfortunately, even if sediment load was restored to its previous level, another 200–300 years might be required for the newest major delta lobe, the Atchafalaya Delta, to grow at rates sufficient to offset in southcentral Louisiana the land loss occurring in the modern delta of southeastern Louisiana.

TABLE 3
Summary of crevasse growth patterns and receiving bay characteristics (from Louisiana Dept. Natural Resources, 1993)

Characteristics	Site 1	Site 2	Site 3
Volume of Growth (m^3) 1988–1991	22,258	951	12,024
Linear growth (m) 1986–1991	1,676	3,505	1,219
Subaerial growth (km^2) 1986–1991	0.69	1.51	0.46
Receiving bay area (km^2)	12.86	36.44	1.29
Receiving bay depth (cm)	60	40	30

References

Adams, R.D., Barrett, B.B., Blackmon, J.H., Gane, B.W., and McIntire, W.G., 1976. Barataria Basin: Geologic processes and framework. Center for Wetland Resources, Louisiana State Univ., Sea Grant Rept. LSU-T-76-006, 117 pp.

Barnett, T.P., 1984. The estimation of "global" sea level change: a problem of uniqueness. *Jour. Geophys. Res.*, 89, 7980-7988.

Baumann, R.H. and Turner, R.E., 1990. Direct impacts of outer continental shelf activities on wetland loss in the central Gulf of Mexico. *Environ. Geol. Water Res.,*, 15, 189-198.

Behrens, E.W., Watson, R.L., and Mason, C., 1977. Hydraulics and dynamics of New Corpus Christi Pass, Texas: a case history, 1972-1973. U.S. Army Corps of Engineers, C.E.R.C., GITI Rept. 8.

Britsch, L.D. and Dunbar, J.B., 1993. Land loss rates: Louisiana coastal plain. *J. Coastal Res.*, 9, 324-338.

Coleman, J.M., 1976. *Deltas: processes of deposition and models for exploration*. Int. Human Res. Dev. Comm. Publ., Boston,102 pp.

Coleman, J.M., 1988. Dynamic changes and processes in the Mississippi River delta. *Geol. Soc. Am.,Bull.*, 100, 999-1015.

Coleman, J.M., Gagliano, S.M., and Morgan, J.P., 1969. Mississippi River subdeltas, natural models of deltaic sedimentation. Louisiana State Univ., Coastal Studies Inst. Ser. Bull. 3, 23-27.

DeLaune, R.D., Smith, C.J., Patrick, W.H., and Roberts, H.H., 1987. Rejuve-nated marsh and bay-bottom accretion on the rapidly subsiding coastal plain of U.S. Gulf coast: A second order effect of the emerging Atchafalaya delta. *Est. Coastal Shelf Sci.*, 25, 1-9.

DeLaune, R.D., Pezeshki, S.R., Pardue, J.H., Whitcomb, J.H., and Patrick, W.H., 1990. Some influences of sediment addition to a deteriorating salt marsh in the Mississippi River deltaic plain: a pilot study. *J. Coastal Res.*, 6, 181-188.

Dozier, M.D., Gosselink, J.G., Sasser, C.E., and Hill, J.M., 1983. Wetland change in southwestern Barataria Basin, Louisiana, 1945-1980. Louisiana State Univ., Center for Wetland Res., LSU-CEL-83-11, 102 pp.

Fairbridge, R.W. and Krebs, O.A., 1962. Sea-level and the southern oscillation. *Geophys. J.*, 6, 532-545.

Fisk, H.N., 1944. Geological investigation of the alluvial valley of the lower Mississippi River. U.S. Army Engr., Miss. River Comm., Vicksburg, 78 pp.

Fisk, H.N., 1961. Bar-finger sands of the Mississippi Delta. In: Peterson, J.A. and Osmond, J.C. (eds.), Geometry of Sandstone Bodies. *Am. Assoc. Petrol. Geol., Spec. Publ.*, 29-52.

Fisk, H.N., McFarlan, E., Jr., Kolb, C.R., and Wilbert, L.J., Jr., 1954. Sedimen-tary framework of the modern Mississippi Delta. *J. Sed. Pet.*, 24, 79-99.

Frazier, D.E., 1967. Recent deltaic deposits of the Mississippi River, their development and chronology. *Gulf Coast Assoc. Geol. Soc., Trans.*, 17, 287-315.

Gagliano, S.M. and Van Beek, J.L., 1970. Geologic and geomorphic aspects of deltaic processes, Mississippi Delta system. Louisiana State Univ., Center for Wetland Resources, Rept. 1, 144 pp.

Gagliano, S.M., Light, P., and Becker, R.E., 1973. Controlled diversions in the Mississippi Delta system: an approach to environmental management. Louisiana State Univ., Center for Wetland Resources, Rept. 8, 146 pp.

Gagliano, S.M., Meyer-Arendt, K.J., and Wicker, K.M., 1981. Land loss in the Mississippi River deltaic plain. *Gulf Coast Assoc. Geol. Soc., Trans.*, 31, 295-300.

Geertsma, J., 1973. Land subsidence above compacting oil and gas reservoirs. *J. Petrol. Tech.*, June, 734-744.

Gornitz, V., Lebedeff, S., and Hansen, J., 1982. Global sea-level trend in the past century. *Science*, 215, 1611-1614.

Gould, H.R. and McFarlan, E., Jr., 1959. Geologic history of the chenier plain, southwest Louisiana. *Gulf Cost Assoc. Geol. Soc., Trans.*, 9, 1-10.

Gutenberg, B., 1941. Changes in sea-level, postglacial uplift, and mobility of the earth's interior. *Geol. Soc. Am. Bull.*, 52:721-772.

Hicks, S.D., 1978. An average geopotential sea-level series for the United States. *J. Geophys. Res.*, 83, 1377-1379.

Holdahl, S.R. and Morrison, N.L., 1974. Regional investigations of verticacrustal movements in the U.S., using precise relevelings and mareograph data. In: R. Green (ed.), Recent Crustal Movements and Associated Seismic and Volcanic Activity. *Tectonophysics*, 23, 373-390.

Keown, M.P., Dardeau, E.A., and Causey, E.M., 1981. Characterization of the suspended-sediment regime and bed-material gradation of the Mississippi River basin. Environmental Lab., Potamology Program, U.S. Army Corps of Engr., Vicksburg, 2 Vol.

Kolb, C.R. and Van Lopik, J.R., 1966. Depositional environments of the Mississippi River deltaic plain, southeastern Louisiana. In: Shirley, M.L. and Ragsdale, J.A. (eds.), *Deltas*. Houston Geol. Soc., 17-62.

Lehner, P., 1969. Salt tectonics and Pleistocene stratigraphy on continental slope of northern Gulf of Mexico. *Am. Assoc. Petrol. Geol., Bull.*, 53, 2431-2479.

Levin, D.R., 1993. Tidal inlet evolution in the Mississippi River delta plain. *J. Coastal Res.*, 9, 462-480. Louisiana Department of Natural Resources, 1993. Accretion and hydrologic analyses of three existing crevasse splay marsh creation projects at the Mississippi Delta. Final Rept., 1986 Artificial Crevasing Project

Marmer, H.A., 1952. Changes in sea level determined from tide observation. Proc., 2nd Conf. Coastal Engr., Berkeley, p. 62-67.

Martin, J.C. and Serdengecti, S., 1984. Subsidence over oil and gas fields. *Rev. Engr. Geol.*, 6, 23-34.

Meade, R.H. and Emery, K.O., 1971. Sea-level as affected by river runoff, eastern United States. *Science*, 173, 425-427.

Meade, R.H. and Parker, R.S., 1985. Sediment in rivers of the United States. National Water Supply Summary. U.S. Geol. Survey Water-Supply Paper 2275, p. 49-60.

Meade, R.H., Yuzyk, T.R., and Day, T.J., 1990. Movement and storage of sediment in rivers of the United States and Canada. In: M.G. Wolman and H.C. Riggs (eds.), *Surface Water Hydrology*, Geol. Soc. Am., p. 255-280.

Milliman, J.D. and Meade, R.H., 1983. World-wide delivery of river sediment to the oceans. *J. Geol.*, 91, 1-21.

Moeller, C.C., Huh, O.K., Roberts, H.H., Gumley, L.E., and Menzel, W.P., 1993. Response of Louisiana Coastal Environments to a cold front passage. *J. Coastal Res.*, 9, 434-447.

Moger, J. and Faust, K.J., 1991. Sediment diversion as a form of wetland creation in the Mississippi Delta. in: H.S. Bolton (ed.), Coastal Wetlands. Am. Soc. Civil Eng., New York, p. 157-163.

Morgan, D.J., 1977. The Mississippi River Delta, legal-geomorphologic evaluation of historic shoreline changes. Louisiana State Univ., *GeoScience and Man*, v. 16, 196 pp.

Morgan, J.P., Van Lopik, J.R., and Nichols, L.G., 1953. Occurrence and development of mudflats along the western Louisiana coast. Louisiana State Univ., Coastal Studies Inst., Tech. Rept. 2, 34 p.

O'Neil, T., 1949. The muskrat in the Louisiana coastal marshes. Louisiana Dept. Wildlife and Fisheries, New Orleans, 152 p

Penland, S. and Boyd, R., 1981. Shoreline changes on the Louisiana barrier coast. IEEE Oceans, 81, 209-219.

Penland, S. and Ramsey, K.E., 1990. Relative sea-level rise in Louisiana and the Gulf of Mexico: 1908-1988. *J. Coastal Res.*, 6, 323-342.

Penland, S., Ramsey, K.E., McBride, R.A., Moslow, T.F., and Westphal, K.A., 1986. Relative sea level rise and subsidence in Louisiana and the Gulf of Mexico. Louisiana Geol.l Survey, Coastal Geol. Publ. Series, Tech. Rept.

Penland, S., Ramsey, K.E., McBride, R.A., Mestayer, J.T., and Westphal, K.A., 1987. Relative sea-level rise, delta plain development, subsidence, and wetland sedimentation in the Teche and Lafourche delta complexes. Terrebonne Parish, Louisiana. Final Rept., Louisiana Geol. Survey, Coastal Geology Program, 236 p.

Roberts, H.H., Adams, R.D., and Cunningham, R.H.W., 1980. Evolution of sand-dominant subaerial phase, Atchafalaya Delta, Louisiana. *Am. Assoc. Petrol. Geol., Bull.*, 64, 264-279.

Russell, R.J., 1936. Lower Mississippi River delta: reports on the geology of Plaquemines and St. Bernard Parishes. Louisiana Dept. Conservation Geol. Bull., v. 8, 199 pp.

Scruton, P.C., 1960. Delta building and the deltaic sequence. In: F.P. Shepard (ed.), *Recent Sediments, northwest Gulf of Mexico*. Am. Assoc. Petrol. Geol., p. 82-102.

Stevenson, J., Kearney, M., and Pendleton, E., 1985. Sedimentation and erosion in Chesapeake Bay brackish marsh system. *Marine Geol.*, 67, 213-235.

Suhayda, J.N., 1987. Subsidence and sea level. In: R.E. Turner and and D.R. Cahoon (eds.), Causes of Wetland Loss in the Coastal Central Gulf of Mexico, U.S. Dept. Interior, Minerals Management Service, v. II, 187-202.

Swanson, R.L. and Thurlow, C.I., 1973. Recent subsidence rates along the Texas and Louisiana coasts as determined from tide measurements. *J. Geophys. Res.*, 78:2665-2671.

Trahan, D.B., 1982. Monitoring local subsidence in areas of potential geopressured fluid withdrawal, southwestern Louisiana. *Gulf Cost Assoc. Geol. Soc., Trans.*, 32, 231-236.

Turner, R.E., 1987. Tide gage records, geological subsidence, and sedimenta-tion patterns in the northern Gulf of Mexico marshes. In: R.E. Turner and D.R. Cahoon (eds.), Causes of Wetland Loss in the Coastal Central Gulf of Mexico, U.S. Dept. Interior, Minerals Management Service, v. II, 203-211.

Turner, R.E., 1991. Tide gauge records, water level rise, and subsidence in the northern Gulf of Mexico. Estuaries, 14, 139-147.

Turner, R.E. and Cahoon, D.R., 1987. Causes of wetland loss in the coastal central Gulf of Mexico. U.S. Dept. Interior, Minerals Management Service, v. I. 32 pp.

Turner, R.E., McKee, K.L., Sikora, W.B., Sikora, J.P., Mendelssohn, I.A., Swenson, E.M., Neill, C., Leibowitz, S.G., and Pedrazini, F., 1984. The impact and mitigation of man-made canals in coastal Louisiana. Water Sci. Tech., 16, 497-504.

van Heerden, I.L., 1980. Sedimentary responses during flood and non-flood conditions, new Atchafalaya delta, Louisiana. Louisiana State Univ., Unpubl. M.S. Thesis, 76 pp.

van Heerden, I.L., 1993. Conceptual review of potential sediment input sites into Barataria Basin including Myrtle Grove, for the purpose of restoring and/or creating coastal wetlands. Contract Rept. for Minerals Management Service, 16 pp.

Walker, H.J., Coleman, J.M., Roberts, H.H., and Tye, R.S., 1987. Wetland loss in Louisiana. *Geography Annales*, 69:189-200.

Wells, J.T., 1987. Effects of sea-level rise on deltaic sedimentation in south-central Louisiana. In: O.H. Pilkey, Nummedal, D., and Howard, J.D. (eds.), *Soc. Econ. Paleont. Mineral. Spec. Publ.* 41:157-166.

Wells, J.T. and Kemp, G.P., 1981. Atchafalaya mud stream and recent mudflat progradation: Louisiana chenier plain. *Gulf Coast Assoc. Geol. Soc., Trans.*, 31:409-416.

Wells, J.T. and Kemp, G.P. 1982. Mudflat and marsh progradation along Louisiana's chenier plain: a natural reversal in coastal erosion. In: Boesch, D.F. (ed.), Proc., Conf. on Coastal Erosion and Wetland Modification in Louisiana: Causes, Consequences, and Options. U.S. Fish and Wildlife Service, Rept. FWS/OBS-82/59, p. 39-51.

Wells, J.T. and Coleman, J.M., 1984. Deltaic morphology and sedimentology, with special reference to the Indus River Delta. In: B. U. Haq and J.D. Milliman (eds.), *Marine Geology and Oceanography of Arabian Sea and Coastal Pakistan.* Hutchinson and Ross, Stroudsburg, PA, p. 85-100.

Wells, J.T. and Coleman, J.M., 1987. Wetland loss and the subdelta life cycle. *Estuarine, Coastal and Shelf Sci.*, 25, 111-125.

Wells, J.T., Crout, R.T., and Kemp, G.P., 1981. An assessment of coastal processes, dredged-sediment transport, and biological effects ofdredging, coast of Louisiana. Louisiana State Univ., Coastal Studies Inst., Tech. Rept.No. 312, 36 pp.

Wells, J.T., Chinburg, S.J., and Coleman, J.M., 1984. The Atchafalaya River Delta: generic analysis of delta development. Waterways Experiment Station, Hydraulics Lab., Tech. Rept. HL-82-5, Vicksburg, 89 pp.

Whitaker, R.E., 1971. Seasonal variations of steric and recorded sea level of the Gulf of Mexico. Office of Naval Research Contract Rept. N00014-68-A-0308-0002, 110 p.

Wiseman, W. J., Jr. and Swenson, E.M., 1987. Long-term salinity trends in Louisiana estuaries. In: R.E. Turner and D.R. Cahoon (eds.), Causes of Wetland Loss in the Coastal Central Gulf of Mexico. U.S. Dept. Interior, Minerals Management Service, v. II, 101-121.

Woodbury, H.O., Murray, I.B., Jr., Pickford, P.J., and Akers, W.H., 1973. Pliocene and Pleistocene depocenters, outer continental shelf, Louisiana and Texas. *Am. Assoc. Petrol. Geol., Bull.*, 57, 2428-2439.

Wright, L.D. and Coleman, J.M., 1971. Effluent expansion and interfacial mixing in the presence of a salt wedge, Mississippi River delta. *J. Geophys. Res.*, 76, 8649-8661.

CHAPTER 16

Economizing Human Responses to Subsidence and Rising Sea Level

James M. Broadus†

ABSTRACT: Potential economic impact and required human adjustments due to land subsidence and rising sea level in low-lying coastal areas are explained in order to help guide public response, and to anticipate the ways in which people will affect and be affected by these problems. A form of benefit-cost analysis to estimate the expected cost of an event in the absence of mitigation is an economic impact assessment. For example, rough estimates of the scale of potential impact of land subsidence and sea-level rise in Bangladesh and Egypt are reported on the basis of inundation scenarios up to the year 2050. Using strong assumptions about economic growth rates, land rent as a segment of the national product, intertemporal discount rate, and rate of inundation, a method is formulated for extending this characterization to an even cruder estimate of potential economic loss. This is a certainty-equivalent value confined to land inundation. Aside from obvious sources of imprecision, it underestimates several important phenomena: lost capital structures, increased exposure to storm damage and interior flooding, and such secondary effects as saline intrusion, crowding and factor reallocation costs. Even so, it is probably an overestimate because it is not based on probability, assumes a linear rate of land loss, and, most importantly, ignores cost reductions arising from human responses. Five economic topics involved in such responses receive special attention. These are: (1) the advantages of incremental responses to gradual change; (2) principles for managing uncertainty; (3) the 'retrofit' problem and capital durability; (4) economic discounting of future values; and (5) the nature and implications of common property, spillover, transboundary and informational effects.

Introduction

The combined effects of eustatic sea-level rise and land subsidence can create problems for low-lying coastal areas that are more serious than either alone. In Bangladesh and Egypt, for example, it is estimated that the two, combined, could inundate nearly 20% of habitable land by 2050 (Milliman et al., 1989). To help guide public response to such problems, and to anticipate the ways in which people

J.D. Milliman and B.U. Haq (eds.), Sea-Level Rise and Coastal Subsidence, 313-325.

TABLE 1
Bangladesh 2050

	Best Case	Worst Case	Catastrophic Case
Total Relative Rise (cm)	13	144	209
Global	(13)	(79)	(79)
Local Subsidence	(0)	(65)	(130)
% Loss Habitable Land	n.a.	16	18
% Population in Exposed Area	n.a.	13	15
% Current GDP Originating in Area	n.a.	10	13

Source: Milliman, et al. (1989)

will affect and be affected by them, it is useful to examine potential economic impacts and human adjustments.

Economic impact assessments are a form of benefit-cost analysis. Generically, benefit-cost analysis is simply the counting up of estimated benefits and costs associated with some event or action in order to compare them to each other or with estimates for an alternative (Smith, 1986). Typically, the estimated benefits and costs accrue from different sources and at different levels over time, and it is important to try to include them all but not to count a factor more than once. A premise of benefit-cost analysis is that economic decision makers tend to act rationally, that they act in such a way as to do the best they can with the information and other resources available. The goal in dealing with prospective actions is to identify the one with either the greatest net benefits or the least net cost. For potential events such as accelerated relative local sea-level rise, the goal may be to estimate the expected cost of the event in the absence of mitigation. Thus, given the cost of mitigation investment, it can be determined whether or not mitigation is logically feasibly.

Most attempts to assess the economic impact of projected relative sea-level rise employ a kind of "coloring book" approach. First, a scenario is selected that describes a hypothetical rise, often within a specified time period. The scenarios are usually selected from a range of projections (Hoffman et al., 1986) and may be adapted to account for local land subsidence (Milliman et al., 1989; Barth and Titus, 1984). The area subject to inundation from such a rise is then identified with topographic information and "colored in" on a map (Figs. 1 and 2). Expected effects other than inundation (such as saline intrusion, storm exposure, and structural damage) also can be identified to complete the scientific description of the physical "dose-response" function. Benefit-cost techniques are then applied to estimate the potential loss in economic terms.

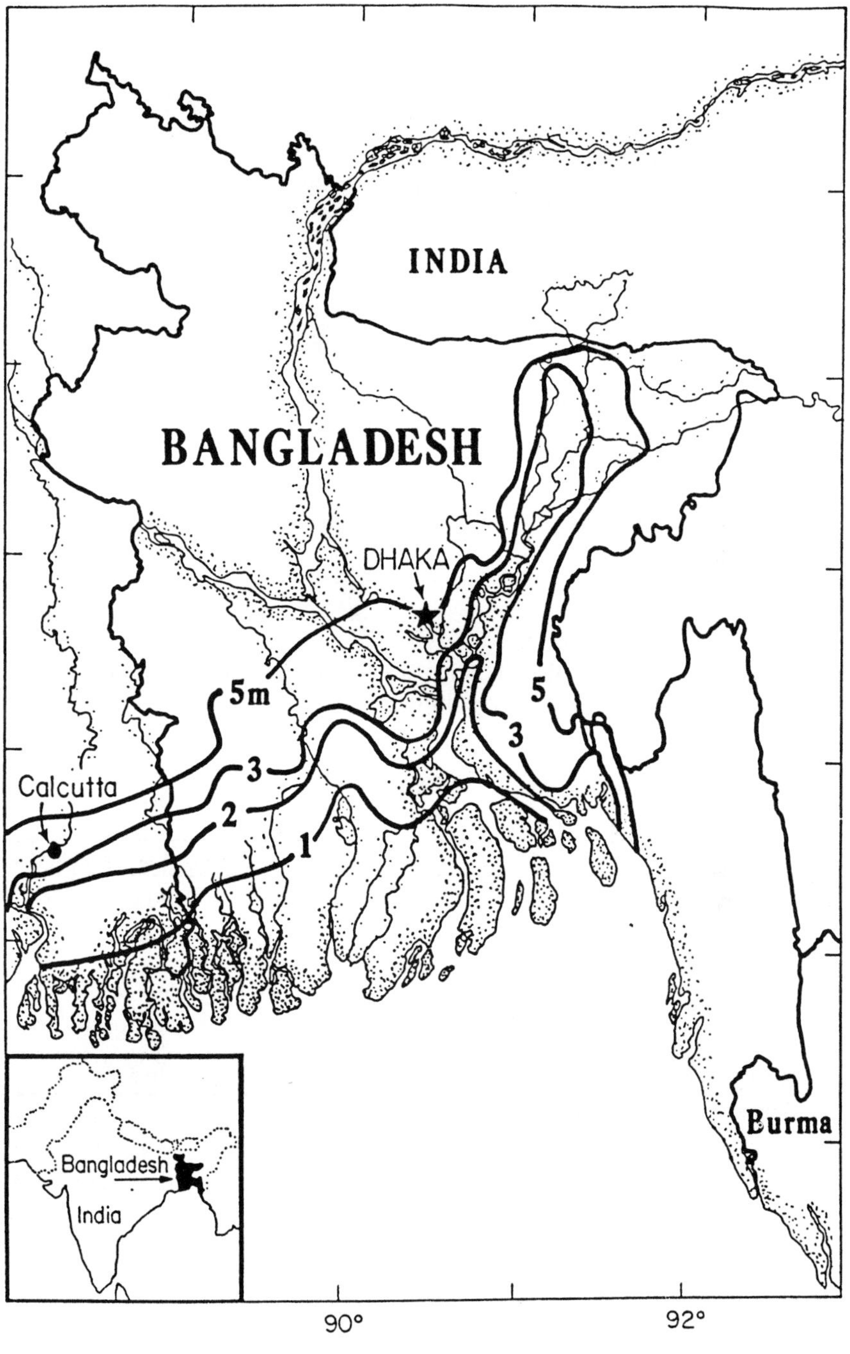

Fig. 1. Inundation scenarios for Bangladesh showing approximate 1 m, 2 m, 3 m and 5 m contours. (after Milliman et al., 1989)

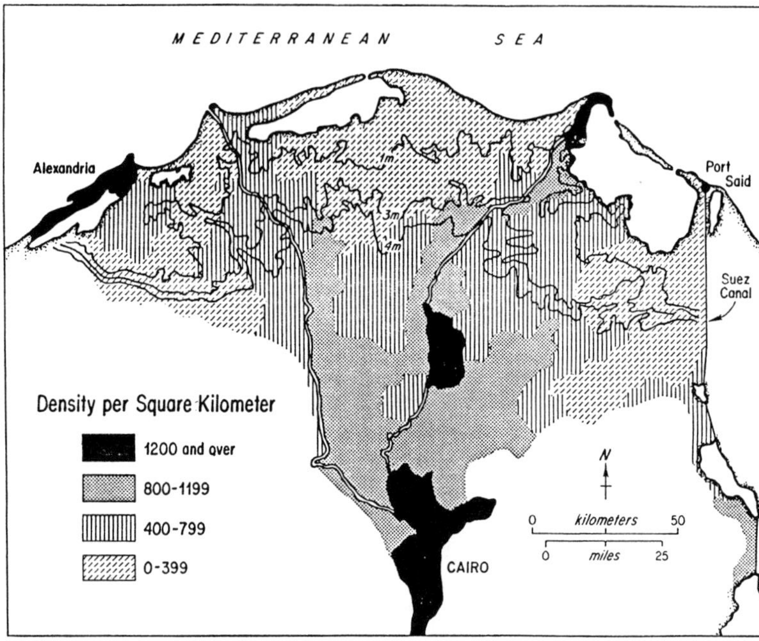

Fig. 2. Inundation scenarios for the Nile Delta, Egypt, superimposing approximate contours for 1 m, 3 m and 4 m relative sea-level rises on population density. (from Milliman et al., 1989)

TABLE 2
Egypt 2050

	Best Case	Worst Case	Catastrophic Case
Total Relative Rise (cm)	13	101	144
Global	(13)	(79)	(79)
Local Subsidence	(0)	(22)	(65)
% Loss Habitable Land	n.a.	15	19
% Population in Exposed Area	n.a.	14	16
% Current GDP Originating in Area	n.a.	14	16.5

Source: Milliman, et al. (1989)

Example Assessments for Bangladesh and Egypt

Impact estimates of this sort have been attempted on a national scale for Bangladesh and Egypt (Broadus et al., 1986; Milliman et al., 1989). Areas subject to future inundation were delineated for various scenarios up to the year 2050. Then, demographic information, land use patterns, and national income accounts were used to estimate the current scale of economic activity taking place within the potentially affected areas (Tables 1 and 2). It was found that given a 144–209 cm rise, 16–18%

of habitable land and 13–15% of current population were estimated to occupy the area subject to inundation in Bangladesh, with 15–19% of habitable land and 14–16% of population in the potentially affected area of Egypt. Up to 13% of current gross domestic product (GDP) is estimated to originate in the potentially affected area of Bangladesh and up to 16% of current GDP in Egypt.

This result poses serious economic damage to Bangladesh, a densely populated nation of about 110 million people. Eighty percent of its 144,000 km^2 area is made up of the complex Bengal Delta. The country's population is widely distributed and heavily dependent on agriculture used for basic subsistence. Agricultural production comprises approximately half the nation's gross domestic product. Net cropped/agricultural areas are by far the major use of land in Bangladesh. Based on the areal distribution of rice and jute production, 5–8% of the nation's agricultural output (some 7–11% of its crops) appears to originate in the area seaward of the 1 m contour. About 10% of the nation's current GDP is estimated to originate in the area exposed to the "worst case" rise of 144 cm and about 13% in the area involved in the "catastrophically worst case" of a 209 cm relative rise.

Of Egypt's 1,000,000 km^2 area, only about 35,000 km^2, or 3.5%, is cultivated and settled. The population of 53 million is clustered about the banks of the Nile River and throughout the river's densely inhabited delta, most of which is devoted to intensive multi-crop agriculture and to urban and industrial uses. The major city of Alexandria, in the northwest, has been exposed since antiquity to the forces of the sea and historically was reached by a causeway extending to its walled enclosure. The major existing natural defenses against coastal attack are the series of sand dunes along the delta's coast and the increasingly brackish lakes that are also a major source of the nation's approximately 100,000 tons of annual fish catch.

The area affected by the worst case 1 m relative sea-level rise in Egypt represents approximately 15% of the nation's arable land and, based on the assumption that population is evenly distributed within each governorate, contains approximately 14% of the nation's population. The area affected by a 144 cm rise in relative sea level represents inundation of about 19% of the nation's arable land and the forcible displacement of about 16% of Egypt's population. Assuming that agricultural output in the delta is distributed as arable area, and assuming that all other sectors of economic activity are distributed as population, approximately 14% of Egypt's GDP originates in the area likely to be affected by a 1 m increase in relative sea level, with about 16% of the GDP in the area affected by the 144 cm rise. (If, as in historical times, Alexandria's elevation were to spare it from an encroaching sea, then the 1 m figures should be adjusted downward. Specifically, less than 7% of the nation's current population is then in the potentially affected area and closer to 6% of current GDP.)

An Extension of Crude Loss Estimates

Although techniques have been developed and employed elsewhere to measure the monetary value of potential net economic losses associated with physical scenarios for relative sea-level rise (Barth and Titus, 1984), the approach to impact assessment reported above is not intended to estimate such future damages. It is rather an entirely mechanical approach intended only to portray the current scale of economic activities and population in potentially affected areas, as a first approximation of the economic stakes involved.

By imposing an additional sequence of strong assumptions, however, it is possible to extend the analysis to generate even coarser estimates of potential economic losses associated with the inundation scenarios. Obviously crude, and at best approximate, this value provides another concise estimate of the potential scale of economic impact, expressed in terms of net present value. It is useful primarily for early examination of the potential problem and should be read with caution.

The method used for the extension can be described simply. It begins with gross domestic product (GDP_0) which is known. Assume a time horizon (T) with a constant economic growth rate (r), for cumulative GDP over the full time horizon without any sea-level effects:

$$GDP_0 \sum_{t=0}^{T} (1+r)^t .$$

Weighting this cumulative value by agriculture's land-rent share (a) of GDP gives cumulative agricultural land rent:

$$aGDP_0 \sum_{t=0}^{T} (1+r)^t .$$

The last parcel lost to inundation over the time horizon will produce its full share of output right up until the last year of the time horizon. The annual rent value of the first parcel lost, on the other hand, accrues as an additional economic loss every year over the time horizon. The actual rate of land loss would most likely be nonlinear, but the simplest case is a rent-loss that is a linear function of time, $l(t)$, giving a cumulative loss of land rent over the time horizon:

$$L = aGDP_0 \sum_{t=0}^{T} l(t)(1+r)^t .$$

Since all of this rent loss from relative sea-level rise would occur in the future — much of it many years from now — it is helpful to express the full stream of losses in discounted present value terms. The social rate of discount is the factor by which future costs and benefits are reduced to make them comparable with present costs and benefits. With social rate of discount, d, the net present value of the cumulative loss (of agricultural land rent) becomes:

$$NPVL = aGDP_0 \sum_{t=0}^{T} l(t)(1+r)^t (1+d)^t .$$

With the unlikely, but arguably realistic and clearly convenient assumption that the economic growth of the economy just equals the social discount rate, $r = d$, then the "back of the envelope" expression for estimated loss simplifies to:

$$NPVL = aGDP_0 \sum_{t=0}^{T} l(t) .$$

Empirically, we know that agriculture accounts for about one half of GDP in Bangladesh and for about 15% in Egypt. Further, we reasonably assume that

TABLE 3
NPVL Estimates for Bangladesh and Egypt, 2050

Bangladesh		
Scenario Rise (m)	144	209
% Loss Habitable Land	16	18
NPVL* (billion takas)	26	29
NPVL/GDP_0 * (%)	6	7
* GDP_0 = 418 billion takas, $T = 61$, $a = 0.0125$.		
Egypt		
Scenario Rise (cm)	101	144
% Loss Habitable Land	15	19
NPVL* (million Egyptian pounds)	840	1060
NPVL/GDP_0 * (%)	3.5	4.5
* GDP_0 = 24 billion LE, $T = 61$, $a = 0.00375/0.05$		

the cost-price margin in agriculture (a surplus equivalent to an accounting profit) is 0.05, of which 50% is rent earned by the productive qualities of the land itself (with the other half accounted for by returns to capital and risk-bearing). Hence, for our purposes, a is set at 0.0125 (0.5×0.05×0.5) for Bangladesh and 0.00375 (0.15×0.05×0.5) for Egypt. Examination of reported land-use practice in Bangladesh suggests that the overwhelming bulk of land in the affected area that is arable and not waste or forest reserve, is cropped. So it is reasonable to treat all rent-bearing land as agricultural. For Egypt, however, as much as half the rent-bearing land in the delta may be nonagricultural, so a is divided by 0.5 (= 0.0075) to account for the loss of nonagricultural land rents there. The results of the exercise, with parameter values used, are reported in Table 3. Subject to several strong caveats and assumptions, we might thus expect that the present discounted value of the economic loss in Bangladesh and Egypt by 2050 of an unchecked rise in relative sea level would be on the order of 6–7% of current GDP in Bangladesh and 3–5% of current GDP in Egypt.

Qualitative Considerations

Caveats obviously apply, and strong ones at that. This method for extending current economic data to projections of potential future losses omits some important sources of potential loss but also abstracts from some key mitigating factors. Again, it is assumed that human response is largely passive in the face of rising local sea levels. The method results in a certainty-equivalent value confined to land inundation and, aside from obvious sources of imprecision, underestimates lost capital structures, increased exposure to storm damage and interior flooding, and such secondary effects as saline intrusion, crowding, and factor reallocation costs. Even so, it is probably an overestimate since it is not likelihood-weighted, assumes a linear rate of land loss, and, most importantly, ignores cost reductions arising from human responses.

Confining estimated losses to lost land rents doesn't take into account immobile physical capital (houses, wharfs, factories, bridges, etc.) that might also be lost to inundation. Where data on property values are available, and where market institutions are not severely distorted, the value of capital services will be reflected in reported property values. When only the estimated value of land rents is included, however, as above, then the omission of physical structures can be important. Even so, it should be recognized that only the *remaining* lifetime value of lost capital should be counted, not the original or replacement value; so that the abandonment of largely depreciated capital need not weigh very heavily in the equations.

For some settings, the omission of losses from increased future exposure to storms and flooding may be an even larger source of underestimation in NPVL. While coastal storms may not be a major concern in Egypt, exposure of the population to storm surge is an extremely grave consideration in Bangladesh. Severe cyclonic storms attack the country at a frequency of approximately 1.5 per year, with storm surge reaching as far as 160 km up the river courses in recent experience (Bird and Schwartz, 1985). With increasing population densities in the future, through the combined effects of natural population increase and loss of territory to coastal transgression, the exposure to severe storm surges, and associated costs, will almost certainly increase.

Another source of economic loss not captured in NPVL is the intrusion of saltwater into freshwater resources. This is already a serious problem for both Bangladesh and Egypt and can be expected to spread farther inland with relative local sea-level rise. Current estimates in the Bangladesh National Water Plan suggest that saltwater intrusion may now extend seasonally over 150 km inland in the western districts and along the Meghna River. If future damming or diversion of the Bengal rivers effectively cuts off new sediments from maintaining the delta, and if global rise in sea level over the next century is severe, the intrusion of saltwater could encroach on Dhaka.

Both Bangladesh and the Nile Delta are densely populated. Land loss from inundation and population displacement from threatened areas can only tend to increase population density and impose the additional costs in social services and management so often associated with crowding. The aggregate cost of relocation is itself a "passive response" cost not captured in NPVL. Clearly, even if the assumption of passive adaptation is accepted, some adjustment costs must be expected. If a nation loses its low-lying coastal rice production to sea-level rise, for example, this will alter the allocation of resources in the economy in ways that extend beyond the loss of rice production. The resources formerly applied to rice (including materials, mobile capital and labor) will tend to be redirected into other activities that eventually will help replace at least part of the economic benefits lost with the rice fields. How quickly and how completely this economic adaptation will occur clearly depends on the suddenness of the initial loss, on the organization of the economy, and on the availability of other opportunities for productive employment of the displaced resources. Some readjustment costs are inevitable, however, such as the penalty for industries shifting resources into their next best production.

All these additional sources of cost would be included in a refined, data-intensive impact assessment, and all would tend to increase the NPVL estimates. Three other qualitative refinements, however, including the benefits side of human responses, would tend to have the opposite effect.

PROBABILITY-WEIGHTED EXPECTED VALUES

The NPVL method above assumes that the hypothetical scenarios are realized (certainty-equivalence). If, in fact, we believe there is only a 10% chance of the worst case in Bangladesh by 2050, for example, then the actual expected NPVL should be only 10% of the reported 26 billion taka estimate, or 2.6 billion takas. But that implicitly assumes there is a 90% chance of no sea-level rise at all. What is needed, instead, is a probability distribution for each scenario so the expected loss associated with each can be weighted by its estimated likelihood of occurrence. If the likelihood of the best case outcome outweighs that of the worst case scenarios, then the worst case NPVL estimates overstate the true expected loss.

NONLINEAR INUNDATION RATE

If inundation in a scenario approaches its ultimate percentage coverage of total land area at a logarithmic or exponential rate over the time horizon, rather than linearly as assumed in the NPVL example, then the present value loss would be smaller since a larger proportion of the eventual land loss would occur later in the period, preserving the output of most land for longer periods. Such a nonlinear rate of rise could generate a loss less than half the size of the linear case.

ACCOUNTING FOR HUMAN RESPONSES AND ADAPTATION

Impacts will depend on human responses (Schelling, 1983; Bird, 1986; Bruun, 1986; Gibbs, 1986; Marine Board, 1987). Unless the rate of inundation is very gradual, it is ridiculous to imagine that people would passively accept such losses. Responses to imminent or observed rising sea level include: *mitigation* (reduce or reverse the causes), *defense* (dikes, seawalls, floodgates, etc.), *abandonment* (building setbacks, or simply using the land until it floods and then leaving), and *adaptation* and *development of alternative uses* (raising houses on stilts, reoccupying inundated areas with wharf towns or houseboats, converting rice fields to shrimp ponds, etc.). In this regard, it is worth emphasizing that humans tend to be quite competent at incremental adaptation, risk management, and technological advance.

The reasonable expectation of rational decision-making leads to a clear and strong prediction about human responses. That prediction is that possible courses of action will be selected to reduce the expected costs and maximize the expected benefits. Of course, mistakes may be made, but, in general, human actions can be expected to lower the net cost of sea-level rise. If they did not, they would most likely not occur. Factoring in the benefits of future human actions is one of the most pressing challenges for economic impact assessment.

Economic Considerations in Human Responses

Understanding human responses and the calculations on which they are based is important not only for descriptive and predictive purposes, but for prescriptive use as well. Both to anticipate human responses and to help guide them, several economic considerations warrant special notice. Five are discussed briefly.

Advantages of An Incremental Approach

Economists tend to stress that people have a strong talent for adjusting to change, and that their adjustments are most effective if incremental (Schelling, 1983). This view often also holds that decision makers closest to the facts of each choice are likely to devise the most appropriate incremental responses (McFadden, 1984). In a prescriptive sense, this view tends to be non-interventionist and favor letting things sort themselves out as the facts emerge.

Responding to Uncertainty

Because the timing, magnitude and physical effects of sea-level rise are still highly uncertain, potential near-term responses are more in the nature of risk management. This will tend to put a premium on information-seeking research efforts and on institutional arrangements that permit diversification and pooling of risks. Economic theory supports several rules of thumb for dealing with uncertainty that are also time honored as simple common sense:

- *Diversify* (risk pooling). Simply put, don't put all your eggs in one basket. Total risk can be reduced by maintaining a portfolio of alternatives.

- *Play it safe*. Conservative choices that err on the side of safety are reasonable in the face of uncertainty. Essentially, this means assigning more value to overestimates of future hazards than to underestimates.

- *Insurance* (risk spreading). By allowing a large number of individuals to share their risks, the total exposure of each can be reduced. A number of institutional arrangements, including governmental programs, can achieve this.

- *Procrastinate* (play for time). If waiting is not too expensive or the overall risk is low, then uncertainty can be reduced by waiting to see how things develop. Also, technology improves as time passes, reducing the cost of some plans and making others feasible.

- *Get more information*. Probably the most direct way to reduce uncertainty, ana often the most cost effective, is simply to invest in information. Conduct research, study, and learn.

- *Reduce fixed commitments.* Uncertainty puts a premium on flexibility, and flexibility is enhanced by a *minimum* of fixed commitments.

Capital Mobility, Durability, and the Retrofit Problem

Economic impacts are sensitive to the amount and values of durable *fixed capital*. Some physical capital in exposed areas can be moved, and some will have become worthless or nearly worthless before inundation. Therefore, it is really the value of long-lived or durable fixed capital (such as power plants, industrial factories, waste treatment facilities, highways, port infrastructure, etc.) that must be reckoned into estimates of future losses. The further the useful lifetime of such installations extends into the period of expected inundation, the more costly the potential impact. At the same time, extra front-end investments made in anticipation of uncertain sea-level rise can reduce the lifetime cost of such facilities if the rise does occur (Gibbs, 1986). Anticipating the rise can thus afford great savings compared to the cost of retrofit or investments in defensive reaction (Wilcoxen, 1987). However, if the rise does not occur or is more gradual than anticipated, the extra investment will have been wasted. The choice depends on the likelihood of the rise and on the ratio of anticipatory investment to retrofit cost or abandonment. In practice, it should be remembered that the future cost of retrofit may be overestimated unless the advantages of waiting (which yields better information and technological advances) are taken into account.

Discounting Future Values

It makes good economic sense to apply some discount rate to future values, because people do tend to value a current payment or benefit more heavily than a nominally equal payment at a future date. To do otherwise would be to ignore the additional earnings that could be gained from the current payment (e.g., through investing it and gaining compounded interest) in the time before the future payment comes through. Selecting the appropriate discount rate to apply in practice, however, is quite difficult (Lind et al., 1982). It involves strong judgments about the preferences of society and encounters serious ethical complications when inter-generational effects are at stake. Higher discount rates depress the estimated cost of future sea-level impacts. Lower discount rates make them appear larger. Relatively small changes in discount rate can exert a remarkable influence on the present value of future events (Table 4).

Common Property, External Influence, Transboundary Effects and Information

The value of cooperative study and planning and the responsibility of governments and other collective institutions to anticipate and help engineer effective responses should be emphasized. Collective intervention may be essential when common property is involved and when external economic pressures lead private decision makers to ignore the effects of their choices on others. Effects that cross boundaries,

TABLE 4
Gibbs' Sea-Level Impact Estimates ($ millions, 1980); Showing Effects of Discount Rate

	Discount Rate	
	3%	10%
Charleston, SC, 2075		
87.6 cm case	1250	20
159.2 cm case	1910	50
Galveston, TX, 2075		
92.4 cm case	555	10
164.5 cm case	965	20

Source: Gibbs (1984)

whether property lines or national frontiers, are obvious examples. More subtle external effects inhibit private investment in information, virtually assuring that too little will be spent on research and dissemination of information without governmental sponsorship.

Summary

Low-lying coastal areas are subject to substantial future financial costs due to the combined effects of land subsidence and global sea-level rise. The magnitude of these costs, however, will greatly depend on how well the problems are anticipated and on the nature of human responses. Although difficult to predict effects and responses decades in the future, crude economic estimates can be derived to help determine the potential scale of the problem and guide public investment decisions. Because these decisions will often be made in the face of great uncertainty, several guiding principles may be useful: diversify, play it safe, spread risk, play for time, get more information, and reduce fixed commitments.

Acknowledgements

The author is indebted to John Milliman who conceived the project for which this work was developed and whose original conjecture is reflected throughout. Financial support from the Pew Charitable Trusts, the U.S. Environmental Protection Agency, and SCOPE is gratefully acknowledged. Thanks also go to M. Alam, D. Aubrey, E. Bird, H. Brammer, S. Edwards, S. Raper, S. Sabhasri, M. El Sayed, and A. Solow for encouragement and constructive criticism. Research assistance was provided by Frank Gable.

References

Barth, M.C. and J.G. Titus (eds.), 1984. *Greenhouse Effect and Sea Level Rise*. Van Nostrand Reinhold Company, New York.

Bird, E.C.F., 1986. Potential Effects of Sea Level Rise on the Coasts of Australia, Africa and Asia. In: J.G. Titus (ed.), Effects of Changes in Stratospheric Ozone and Global Climate, Volume 4: Sea Level Rise. U.S. E.P.A., Washington D.C

Bird, E.C.F. and M.L. Schwartz (Eds.), 1985. *The World's Coastline*. Van Nostrand Reinhold, Stroudsburg (PA).

Broadus, J.M., J. Milliman, S. Edwards, D. Aubrey, and Frank Gable, 1986. Rising Sea Level and Damming of Rivers: Possible Effects in Egypt and Bangladesh. In J.G. Titus (ed.), Effects of Changes in Stratospheric Ozone and Global Climate, Vol. 4: Sea Level Rise, U.S. E.P.A., 165-189.

Bruun, P., 1986. Worldwide Impact of Sea Level Rise on Shorelines. In J.G. Titus (ed.), Effects of Changes in Stratospheric Ozone and Global Climate, Volume 4: Sea Level Rise. U.S. E.P.A., Washington D.C

Gibbs, M.J., 1986. Planning for Sea Level Rise Under Uncertainty: A Case Study of Charleston, South Carolina. In J.G. Titus (ed.), Effects of Changes in Stratospheric Ozone and Global Climate, Volume 4: Sea Level Rise. U.S. E.P.A.,Washington D.C.

Gibbs, M.J., 1984. Economic Analysis of Sea Level Rise: Methods and Results. In M.C. Barth and J.G. Titus (eds.), *Greenhouse Effect and Sea Level Rise*. Van Nostrand Reinhold Company, New York.

Hoffman, J.S., J. Wells, and J.G. Titus, 1986. Future Global Warming and Sea Level Rise. In: C. Sigbjarnarson (ed.), Iceland Coastal and River Symposium. National Energy Authority, Reykjavik.

Lind, R.C., K.J. Arrow, G.R. Corey, P. Dasgupta, A.K. Sen, T. Stauffer, J.E. Stiglitz, J.A. Stockfisch, and R. Wilson, 1982. Discounting for Time and Risk in Energy Policy. Resources for the Future, Washington, D.C.

Marine Board, National Research Council, 1987. *Responding to Changes in Sea Level: Engineering Implications*. National Academy Press, Washington, D.C.

McFadden, D., 1984. Welfare Analysis of Incomplete Adjustment to Climatic Change. In: V.K. Smith and A.D. Witte (eds.), *Advances in Applied Micro-Economics*, 3, 133-149.

Milliman, J.D., J.M. Broadus and F. Gable, 1989. Environmental and Economic Impact of Rising Sea Level and Subsiding Deltas: The Nile and Bengal Examples. *Ambio* 18, 340-345.

Schelling, T.C., 1983. Climatic Change: Implications for Welfare and Policy. in Changing Climate: Report of the Carbon Dioxide Assessment Committee. National Academy Press: Washington, D.C., 449-482.

Smith, V.C. (ed.), 1986. *Advances in Applied Micro-Economics*, Vol. IV: Risk, Uncertainty and the Valuation of Benefits and Costs. JAI Press.

Titus, J.G. (ed.), 1986. Effects of Changes in Stratospheric Ozone and Global Climate, Volume 4: Sea Level Rise. U.S. Environmental Protection Agency, Washington, D.C.

Wilcoxen, P.J., 1986. Coastal Erosion and Sea Level Rise: Implications for Ocean Beach and San Francisco's Westside Transport Project. *Coastal Zone Management J.*, 14, 173-191.

CHAPTER 17

Man-Made Projects and Relative Sea-Level Rise

J.G. de Ronde

ABSTRACT: Coastal areas in many countries could suffer the combined effects of sea-level rise and land subsidence. In this paper the consequences of local relative sea-level rise on man-made projects will be discussed. Some general principles are presented and the example of the the Dutch situation are discussed. The main concerns for The Netherlands will be: strengthening of our dunes and dikes against the threat of overtopping during storm surges and the adaptation of our water management system to ensure adequate drainage and the ability to cope with salt water intrusion. A method to study these combined problems and some of the first results are also shown

General Considerations of Impacts

HYDRAULIC CONDITIONS

Tidal Range – In continental shelf areas and in shallow seas a relative rise of the sea level will cause an increase in water depth, which in turn will cause a change in the tidal propagation (in amphidromic systems) and in tidal ranges in coastal areas. In most cases the increase (or decrease) of the tidal range will be relatively small compared to the rise of the sea level (see below). Only in a very shallow sea or in a basin that is almost in resonance with a tidal frequency will sea-level rise have an important effect by decreasing the bottom friction in a significant way or by bringing the basin into resonance.

Storm Surge – If winds remain the constant, a sea-level rise will cause a slight decrease in surge height because of a small increase of water depth which increases the return-flow. On the other hand, climatic change might increase storm frequency.

Wave Height – If bottom topography does not change, the water depth will increase and bottom dissipation will decrease. This means that wave heights will

J.D. Milliman and B.U. Haq (eds.), Sea-Level Rise and Coastal Subsidence, 327-342.

increase. In most cases however, the morphology of the area changes when sea level rises, and before anything can be said about the wave height, morphological changes must be considered.

Transport Rates – In estuaries and in estuarine systems with barrier islands, the ebb and flood flow through tidal inlets will change with sea-level rise, and thereby sediment transport will change. If bottom topography remains the same, the flow and sediment transport will increase. Again, because morphological change is expected, constant bottom topography is unlikely, increasing the potential importance of morphological changes.

MORPHOLOGY

Sandy Coasts without Defences – In most cases, a sandy coast with dunes will retreat when sea level rises. Depending on the grain size of the sand and the gradient of the shore face, the dunes will retread 40 to 400 m (or more) per meter sea-level rise.

Sandy Coasts with Defences – Where there are coastal defences such as sea walls and dikes, a relative sea-level rise will diminish the beach in front of it. Sand depletion will occur along the remaining beach. If the beach is absent, the shore face must be protected or the defence system eventually will be destroyed or flooded.

Estuaries – Changes in the morphology of an estuarine system resulting from sea-level rise is determined mainly by the change of the cross profile of the tidal inlet. The tidal volume is linearly related to the tidal discharge through the tidal inlet, which in turn is related to the cross-sectional area of the tidal inlet. Therefore, if the tidal volume increases, the tidal discharge will increase and the cross-sectional area of the tidal inlet will increase as well. The time needed to reach a new balance is dependent on the rate of sediment input. In case of a sea-level rise one can distinguish three scenarios:

a) Tidal volume remains constant. This will happen if there are few or no intertidal areas. The cross-sectional area will increase and tidal velocities will decrease. Decreased velocities may promote sedimentation, which will continue until a new balance has been reached between tidal volume and the cross-sectional area of the tidal inlet.

b) Tidal volume increases, but at a rate less than the increase of the cross-sectional area of the tidal inlet. This situation will be comparable with situation a) but sedimentation will be less extensive.

c) Tidal volume increases more rapidly than the cross-sectional area of the tidal inlet. This situation occurs when the intertidal area is relatively large. In this case, erosion will occur, especially in the tidal channels.

The most vulnerable areas will be those with a small tidal range, where the existing sedimentation rate is small (or negative, i.e. erosion) and where scenario c) occurs. Here the intertidal zone will diminish, especially if the estuary is surrounded by dikes. In an area without dikes or relatively steep embankments, the intertidal zone and higher zones can move upwards along the slope, but only if formerly dry areas become inundated.

Safety – Dikes, sea-walls, barriers, sluices and sometimes dunes will need to be strengthened as local sea-level rises, particularly if the same standard of safety is to be maintained. This is not only the case for coastal defences but also for safety structures along the lower courses of rivers.

Water management – Due to sea-level rise, salt intrusion will increase in rivers. In polders below sea level, the salt intrusion via ground water may become a major problem. Clearly pumps and drainage systems must be adapted for higher sea levels, particularly in inner distributary channels. The greatest adaptations may be needed in those areas where no safety and/or drainage systems are presently required.

Environment – When estuaries change and dunes retreat, safety and drainage systems must be built or strengthened. The consequences for the environment, e.g., natural preservation, bird and fish, etc., can be considerable and therefore must be examined carefully.

A Preliminary Study for The Netherlands

At Rijkswaterstaat (part of the Ministry of Public Works and Transport) preliminary studies have estimated the most important aspects of the impact of sea-level rise. At first, the studies were based on an extreme rise of 5 m. This unrealistic hypothesis (at least for the next century) was chosen to get a better understanding of the major mechanisms involved. At present the sea-level rise in The Netherlands is 15 to 20 cm per century (de Ronde, 1983). In the following section some results are discussed, starting with the impact on the tidal system in the North Sea, followed by impacts on wave propagation, morphology along the Dutch coast and the environment. Finally, costs are estimated.

Impact on the Tidal System in the North Sea

With a 2-dimensional computer model of the North Sea and part of the continental shelf (Verboom et. al., 1989; see Fig. 1) calculations were made for present conditions and for 2.5 m and 5 m higher water levels as well as with a 2.5 m lowered level. In case of a higher sea level (and therefore greater water depth) the tidal wave propagate faster and tidal amplitudes (for water levels as well as currents) increase. Furthermore, the locations of the amphidromic points change (Fig. 2), thereby increasing or decreasing local amplitudes. In and along the North Sea these two effects cause in most cases an increase in the tidal amplitude. Only

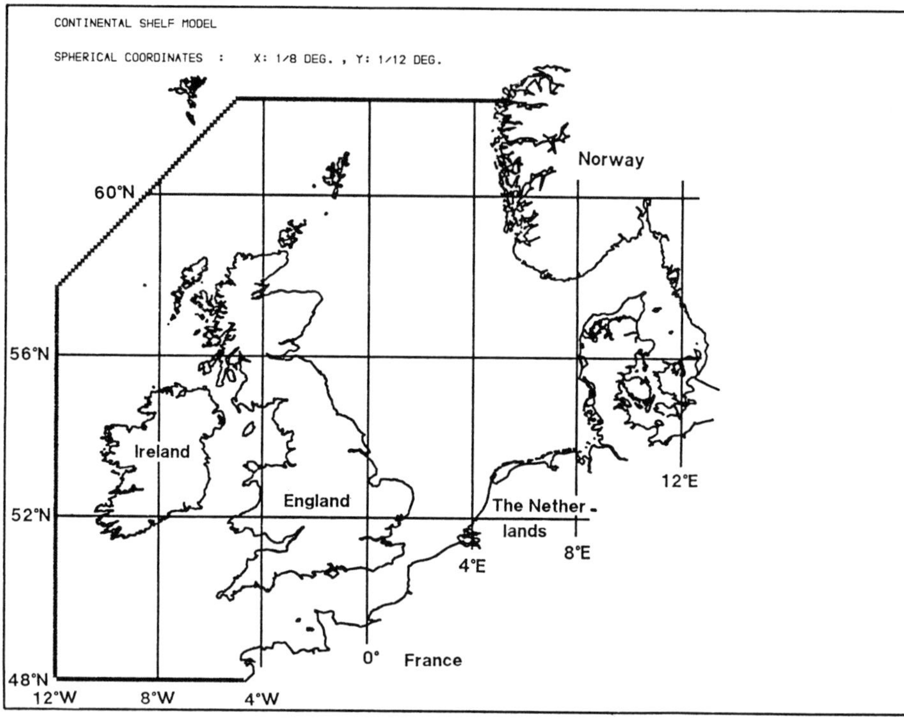

Fig. 1. Outlines of the continental shelf model.

locations near Cromer-Den Helder will show a decrease, as they are nearer the southern amphidromic point (Fig. 3).

In general one can say that in the North Sea a higher sea level will cause an increase in tidal motion. The levels will rise not only because of the sea-level rise but also due to the related (but smaller) rise in tidal amplitude. For instance, a sea-level rise of 2.5 m gives a rise of the high water levels between 2.4 and 2.7 m (the smallest rise is found in Cromer, the greatest rise in southend).

A higher sea level has hardly any influence on the height of a storm surge itself. A much more relevant question concerning storm surges is the possible increase or decrease in the number or strength of storms caused by climatic change. Overall one might say that a decreasing temperature gradient between the Equator and the Poles would cause fewer depressions, but that does not mean that this will happen in the North Sea.

Another feature potentially important is a change in residual transport, which could impact the morphology as well as the ecology. Especially in an area like the Wadden Sea this can be important. Model results indeed show a change of residual transports if sea level rises by 1 m (Fig. 4).

One must be careful with the quantitative use of these results because of the still inaccurate prediction of residual transport by these models. One also must

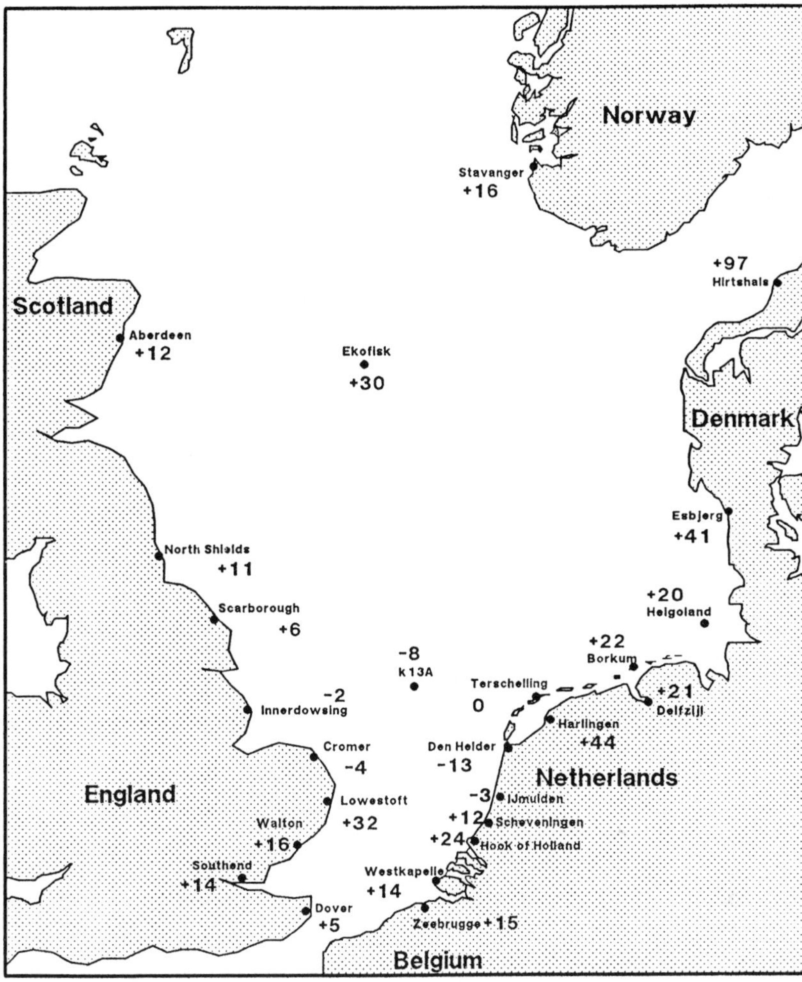

Fig. 2. Change in semidiurnal amphidrome in the southern North Sea, with sea-level rise of 5 m.

bear in mind that the bottom topography in this model was held constant, which may not happen in the real world.

IMPACT ON WAVES AND SWELL

A higher sea level and greater water depths will cause less dissipation of waves and swell. Especially in shallow water, wave heights will increase. For instance, calculations of wave height near the Haringvliet sluices, located behind a shallow sand ridge, the Hinderplaat, show an increase from 90 to 210 cm in wave height when sea level rises 5 m. An important assumption is that sea-bed morphology will not change. If the sea bed shoals as fast as the sea level rises, wave height

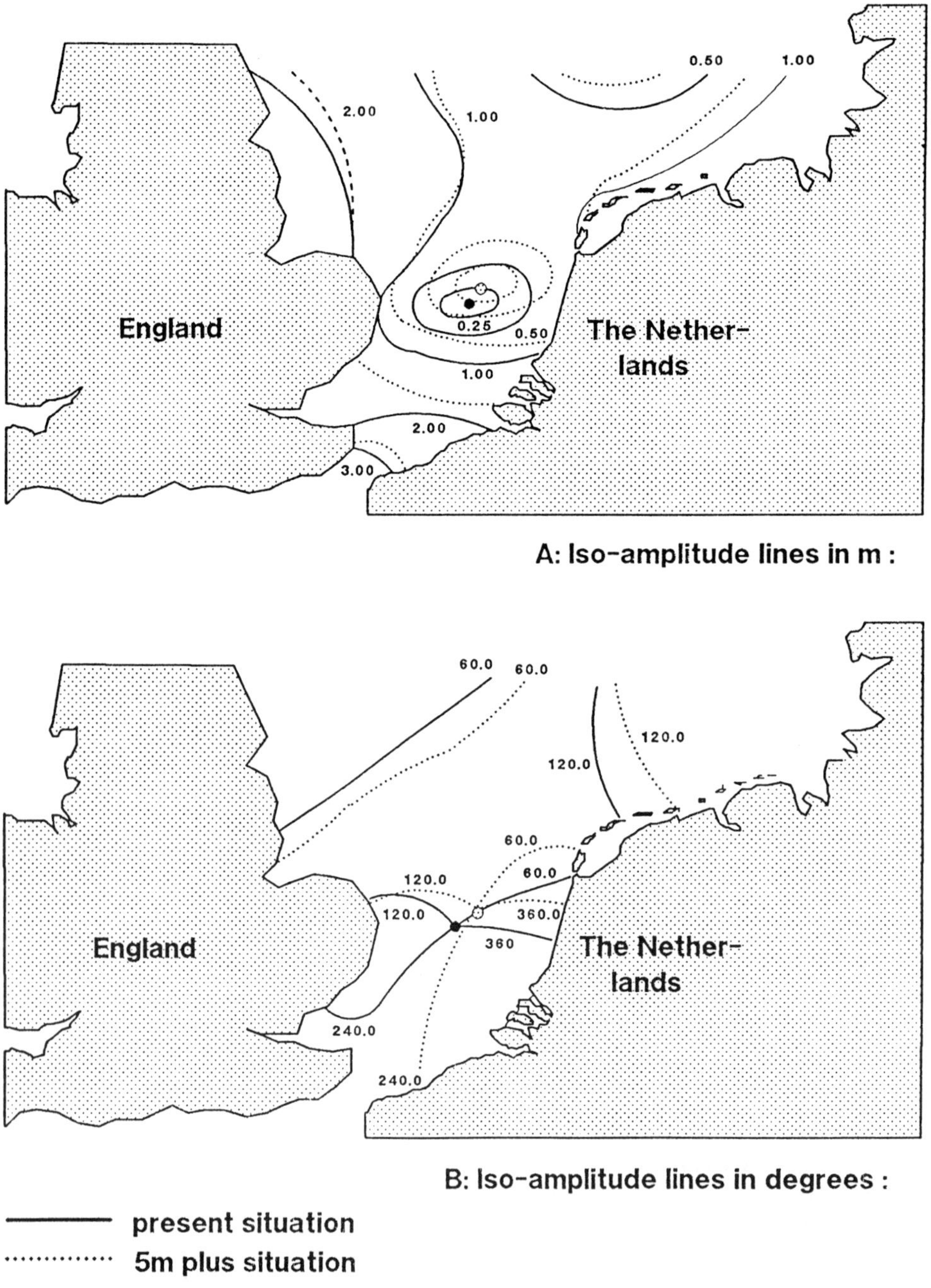

Fig. 3. Percent change in tidal amplitude, with sea-level rise of 5 m.

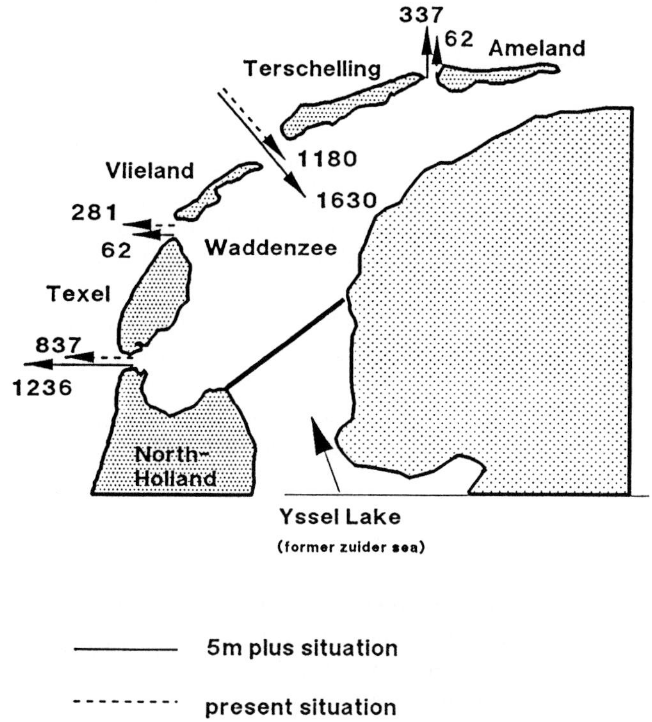

Fig. 4. Change in residual transport in the Wadden Sea, with sea-level rise of 1 m.

probably would not change significantly. The truth probably lies somewhere in between and depends on the change of the sea-bed morphology.

IMPACT ON MORPHOLOGY

Generally speaking, the Dutch coast can be divided in three different parts. The delta area in the south with the Eastern and Western Scheldt; the " closed" coast between Hook of Holland and Den Helder, without any inlets; and the Wadden Sea area.

Closed Coast – The so called "closed" coast, along which the coastal defence consists mainly of dunes, is the easiest to deal with. Roughly speaking the present coast profile is stable. A sea-level rise will destroy this balance, and the coastline will retreat. Present gradient of the beach slope is about 1:60, so a rise of 1 m results in a retreat (or inundation) of 60 m. This is complicated somewhat by grain size of the dune sand, which is finer than the grain size of the beach sand. Because of this, the expected gradient of a beach constructed from dune sand will be smaller and the retreat of the coastline will be greater (up to 80 to 150 m). In case of such a retreat, the dunes may not be strong enough to withstand the projected storm surge (frequency 10^{-4}) and the dunes will have to be strengthened. For large parts of the

Dutch coast, measures like sand depletion will be necessary. For the coastal area protected by dikes (or other defensive structures) the only solution is strengthening these structures.

The Wadden Sea Area – The morphology of the Wadden Sea area is very complex, and even in the present condition it is difficult to predict its future morphology if sea level does not change. One additional problem is that the Wadden Sea area has not attained equilibrium since the closure of the Zuider Sea in 1932.

The Wadden Sea area presently shows a transport of sand from the coastal zone (into the Wadden Sea) of 30×10^6 m^3/year ($\pm 15 \times 10^6$ m^3). If sea level rises 1 m in 100 year and the sea bed were to shoal by a similar amount, then 28×10^6 m^3/year of sediment would be required, which seems possible. In more detail, the Wadden Sea can be divided into 10 smaller tidal systems, each with its own tidal inlet. That allows us to use the three scenario's mentioned earlier.

Only in the tidal system with its inlet south of Texel (25% of the total Wadden area) is scenario (a) is applicable, and so sedimentation will occur. Scenario (c) is only applicable to the tidal system with its inlet west of Ameland (10% of the total area), and here there will be erosion. For the other tidal systems scenario (b) is applicable, indicating that sedimentation probably will not be sufficient in 2 or 3 tidal systems to keep pace with the sea-level rise.

It must stressed once more that the morphology of the Wadden Sea area is very complex and the models used here is very simple. For example: from the total amount of sedimentation of 30×10^6 m^3/year, one can account for the origin of only 25%. This comes from the coast of North-Holland and the Wadden islands, Texel and Vlieland, and from the North Sea.

Finally, one can conclude that even at the present rate of sea-level rise we don't know exactly what will happen to the morphology of the Wadden Sea. Simple models and calculations show that, in case of a sea-level rise of 1 m, sedimentation probably would be enough (at least in much of the Wadden Sea) for sea-floor shoaling to keep up with sea-level rise. So the intertidal area could remain about the same, which is very important for the ecology of the area (see below).

The Delta Area – The Eastern Scheldt as well as the Western Scheldt will have a greater increase of tidal volumes than the cross-sectional areas of the tidal inlets (situation c). This means that erosion will occur and that the intertidal area probably will decrease. Also, it must be stressed that the processes involved are very complex.

IMPACT ON SALT INTRUSION

There are two major ways in which salt water can intrude into coastal areas in The Netherlands. One way is inflow of sea water via the river channels. The Rotterdam Waterway is probably the most relevant for this process. With the advent of larger ships, the Rotterdam Waterway had to be deepened several times. The ensuing salt

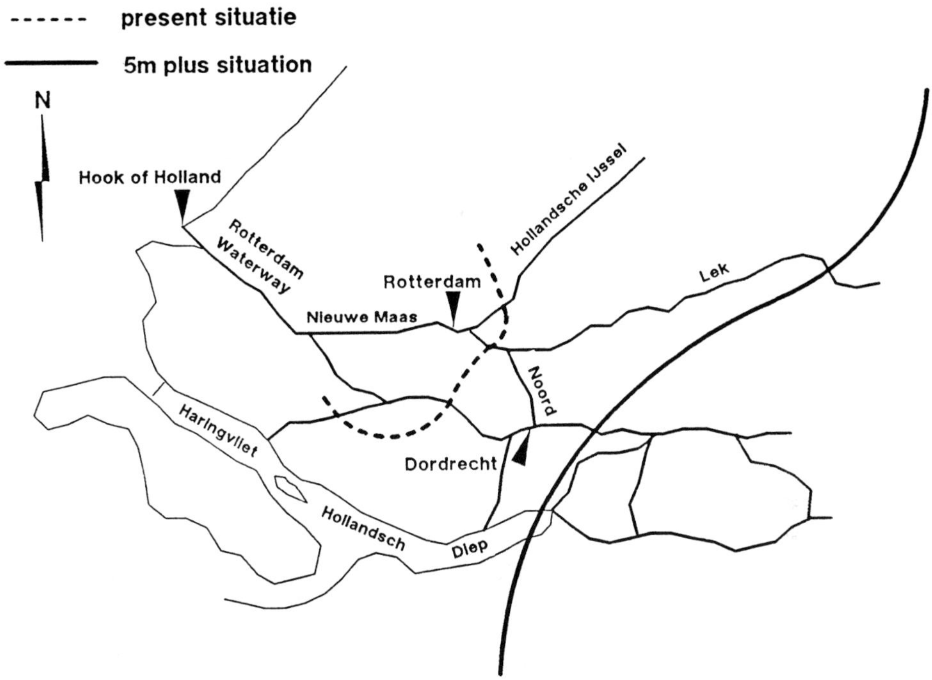

Fig. 5. Change in the salt intrusion in the Rhine estuary, with sea-level rise of 5 m.

intrusion has caused problems for both agricultural and drinking water uses. When Europoort, the new harbour area, was completed, further deepening of the Rotterdam Waterway was no longer necessary, but rising sea-level could again increase the salt intrusion. Figures 5 and 6 show the results of model calculations where sea level was raised over 5 m and bottom topography held constant. Big areas, like the Haringvliet, important for fresh water supply, under these circumstances could be intruded by salt water. This salt water will remain in the deeper parts of the Haringvliet, and it will be extremely difficult to remedy the situation. Of course, if the river bottom shoals as fast as sea level rises, the rate of salt intrusion could remain the same as at present.

The other way in which salt water intrudes is via ground-water flow. Large areas of The Netherlands are below sea level, which causes subsoil seepage to these lower parts. Because the deep ground-water layers in the western part of The Netherlands are brackish, this upward seepage is also brackish. In many polders the seepage rate is more then 0.25 mm per day. If nothing is done to minimize this intrusion, grassland and crops would be damaged.

The solution is to flush and rinse the polders and canals with fresh water. In very dry summers, however, insufficient fresh water is available for this purpose and crop damage is unavoidable. The amount of seepage is linearly related to the difference in sea level and ground-water level in the polders. It is also linearly

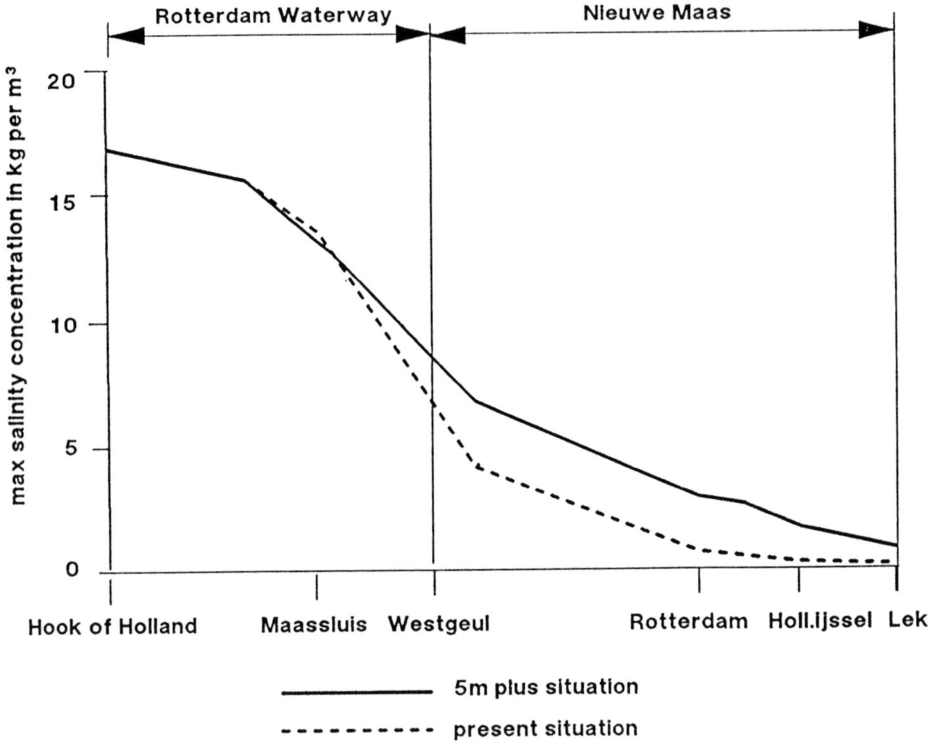

Fig. 6. Change in maximum salinity concentration during a tidal cycle in the Rhine estuary, with sea-level rise of 5 m (see Fig. 5 for locations).

related to the distance between polders and sea and the permeability of the subsoil. A higher sea level causes a higher gradient and increased seepage. Moreover, in certain areas with no present day seepage, a sea-level rise will initiate seepage. Simple calculations show an increase of the total seepage of 200 to 300% when sea level rises 5 m (Fig. 7). The amount of fresh water needed for flushing the polders will then increase by about 50%. In dry summers this extra amount of water is not available.

Impact on Water Management

In addition to salt-water intrusion, another important factor is the drainage of the surplus water. This is especially the case during winter when precipitation exceeds evaporation. For areas well below sea level there is only one solution for drainage, namely pumping. With a higher sea level, all pumping stations would have to be rebuilt because the surplus water must be pumped to a greater height, necessitating stronger pumps.

In The Netherlands some areas drain naturally by gravity. The solution here seems to build new pumping stations when sea-level rises, particularly for smaller

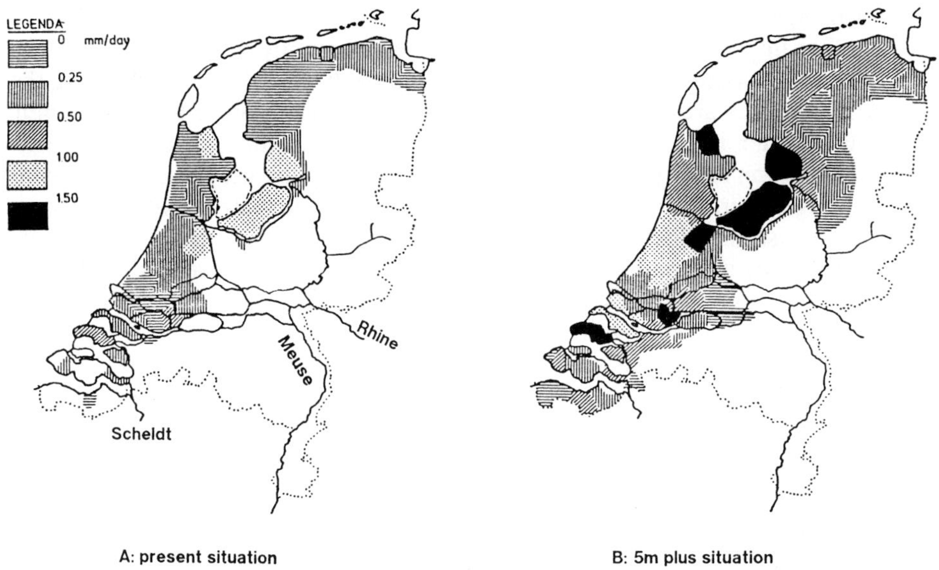

Fig. 7. Change in seepage of salt water through the subsoil in The Netherlands, with sea-level rise of 5 m.

areas. However in extensive areas drained by gravity with a large discharge, this solution might not be the best, for example, in the IJssel Lake and its drainage area (15,000 km^2). The IJssel River accounts for about 10% of the Rhine River discharge into the lake. IJssel Lake, itself, with an area of 1200 km^2, drains into the Wadden Sea through two sluices in the enclosing dam. Water level of the lake is maintained 40 cm below mean sea level during winter and 20 cm below mean sea level during summer, thereby allowing natural drainage during low tide. In a recent study, it was found that as much as a sea-level rise of 50 cm would allow the lake to drain in a natural way without necessitating the raising the of the lake's water level. Any additional increase of the sea level, however, would necessitate the building large pumping stations (total capacity of 300×10^6 m^3/day) to keep the water level below 1.30 m above mean sea level during extreme discharges of the Rhine. (The design discharge of the Rhine River is 16,500 m^3/sec, of which about 1650 m^3/sec goes through the IJssel River into the IJssel Lake). A possible better solution would be to raise the water level of the IJssel Lake by an amount equal to sea-level rise. The consequence of course, would be that all the dikes around the IJssel Lake and partly along the River IJssel would be elevated also. Moreover, a great number of pumping stations would need to be built for areas draining into IJssel Lake.

A rise in temperature, of course, means climate change as well as sea-level rise. An increase in winter precipitation, for example, would affect not only the drainage system, but also would mean more extreme river discharges of the Rhine and Meuse, necessitating higher dikes along the rivers. If the summertime evaporation increased and precipitation decreased, fresh water might be in more or less continual shortage.

Another important aspect is the drinking water supply. A major part of The Netherlands' drinking water supply comes from fresh water lenses underlying the dunes. The lenses are surrounded by saline water, and in order to extract large amounts of water, infiltration of fresh (river) water is necessary, as renewal by surplus precipitation is not sufficient. In effect, these lenses are used as storage basins for drinking water. Moreover, by filtering the river water, its quality improves. A sea-level rise therefore may distort the equilibrium of these lenses, especially where the dunes are small. Possible solutions would be increased infiltration of fresh water, extraction of less drinking water, or periodic sand replenishment on the landwarf side of the dunes. If sea level rises, the phreatic water level of the fresh water lense will rise as well. When the water level reaches the surface of the soil, it acts as an artesian well, and a further sea-level rise might destroy (a part of) the lens. In such a case sand replenishment on the landward side of the dunes becomes necessary.

ECONOMIC IMPACT

Assuming a sea-level rise of 1 m within 100 years, we get the following ecological impacts in four areas where the impact of sea-level rise would be concentrated:

The Closed Coast – Total dune area will diminish, a strip of land, 80 and 150 m in width, will disappear. Assuming that the dunes will be strengthened (where needed) with sand replenishment (not by dikes), the ecological impact in this area may not be too severe. It is assumed that beach and coastal gradients will be maintained with naturally or artificially.

The Wadden Sea Area – The intertidal area is very important to the Dutch estuarine ecology. For some bird species this is the only place where they can feed. A decrease of the intertidal area almost certainly will result in a smaller bird populations per species as well as a smaller number of species. In the case of the Wadden Sea, the assumed sea-level rise probably will hardly have any effect on the intertidal area, so the impact will be small. As mentioned previously, the residual transports in the Wadden Sea may change, which could affect the ecological system, although it is still very difficult to predict exactly how. The impact of temperature rise is another question that can not be answered meaningfully at the present time.

The Delta Area – In the Eastern and Western Scheldt, the intertidal area is likely to decrease, which could have the same negative consequences mentioned above. The bird species concerned will be mainly migratory birds (e.g. stilts).

The River Area – In much of The Netherlands' river system, both mean and extreme water levels will rise. The river forelands, which nowadays are only flooded occasionally during the winter, will be flooded more frequently, perhaps most of the time. As a result the characteristic ecology of these systems probably will change or disappear.

ECONOMIC IMPLICATIONS

Even with a 1-m rise of sea level, major resources will be needed for the direct defense against higher water levels in the coastal area and along the rivers. The strengthening of dikes, dunes, beaches and shore facilities are estimated to cost about 6 billion guilders. For comparison, the present amount spent for the maintenance of the coast is 60 million guilders per year. Interestingly, a storm surge barrier in the Rotterdam Waterway is under study at present. The main reason for this project is the high cost to strengthen the existing dikes along the lower river channels. The threat of accelerating sea-level rise also should be taken into account.

Although the possible measures and decisions regarding a 1-m rise in sea level might be more complex in the case of the water management, the amount of money involved may be less. An estimated 3 billion guilders will be needed for pumps and changes in the infrastructure, and the annual fuel costs could increase by about 10 million guilders per year. Furthermore, harbours, locks, bridges, etc. will need to be adapted to the new conditions (cost could be roughly 1 billion guilders per year).

The total costs if sea level rises 1 m, therefore, can be estimated at about 10 billion guilders.

Policy Analysis

Of course one must not only look at the impacts when discussing sea-level rise, but also the possible remedial measures and the impacts of these measures. One must be cognizant of best measures and when these should be taken. A policy analysis might follow a scheme like the one proposed in Figure 8.

One starts at looking at the impacts with the so-called T_0 alternative (no measures taken). First, study the changes in hydraulic conditions; these will affect the morphology. Together they provide the consequences for safety, water and environmental management, as well as costs. With this knowledge, alternative measures and construction plans can be designed. Given a certain measure or set of measures, the impacts on hydraulic conditions, morphology and related topics must be studied. When this has been accomplished, the different alternatives can be evaluated and compared.

At present two such studies have been undertaken. One is a specific policy analysis of the coastal zone (coastal defense study) to find the best way to deal with present-day erosion and problems associated with a future rise in sea level. In the second one, Rijkswaterstaat and Delft Hydraulics are studying The Netherlands as a whole, as part of the ISOS project (Impacts of Sea Level Rise On Society). The ISOS-study is a joint project of UNEP, Delft Hydraulics and the Dutch government (Rijkswaterstaat). The ISOS case study addresses not only the impacts of sea-

TABLE 1
Key scenarios (ISOS study)

Scenario 2090			
	PARAMETERS		
	Favorable Scenario (A)	Mean Scenarion (B)	Unfavorable Scenario (C)
Mean Sea level rise	+35 cm	+60 cm	+85 cm
Mean Rise of High Water	+40 cm	+65 cm	+90 cm
Mean Rise of Low Water	+30 cm	+55 cm	+80 cm
Wind force	–10%	0	+10%
Wind direction	–10°	0	+10°
Mean Rise of Design level	–20 cm	+65 cm	+150 cm
Precipitation			
Summer	+20%	+10%	0
Winter	0	+10%	+20%
Evapotranspiration			
Summer	0	+10%	+20%
Winter	0	+10%	+20%
River Discharge			
Summer	+10%	0	–10%
Winter	0	+10%	+20%

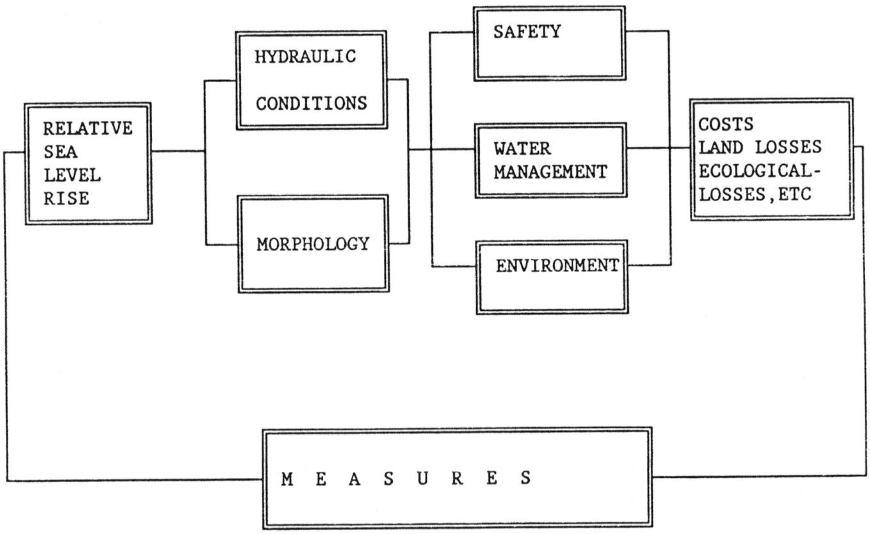

Fig. 8. Policy analysis plan.

TABLE 2
Necessary dike crest increments for the different scenarios

	Incr. of Crest due to a higher design level	Incr. of Crest due to a highjer wave attack	Total Increment of Crest
SCENARIO A			
SLR = 60 cm/100 year	60 cm	0–50 cm	60–110 cm
No increase of wind force			
SCENARIO B			
SLR = 60 cm/100 year	150 cm	40–110 cm	190–260 cm
10% increase wind force			
SCENARIO C			
SLR = 0 cm/100 year	80 cm	0–50 cm	80–135 cm
10% increase wind force			

level rise, but also the impacts of changing storm frequencies and of changing hydrological parameters, such as river discharge, precipitation and evaporation.

One of the main problems in these kinds of studies is the great uncertainty in predicting sea-level rise and other hydrological changes. One way of dealing with these uncertainties is to assess the sensitivity of different scenarios by following the policy analysis scheme in Figure 8. In Table 1 the main scenarios used in the ISOS study are given. Most of the scenario's deal only with different rates of relative sea-level rise. Some also deal with changes in storms and river discharge but one is less sure of these changes, and in some cases even the direction of change.

With these scenarios we hope to answer questions such as:

* Depending on the rate of acceleration of sea-level rise, *what* measures should be taken and *when* should these measures be initiated?

* If storm frequency and intensity as well as river discharge are also increasing, how important are they? In other words, should we not be equally worried about other changes as well?

Some preliminary answers from previous and on-going studies show that:

* Costs of maintaining the dune coast of The Netherlands will increase in the next decade when the rate of relative sea-level rise is predicted to increase from the present 20 cm per century to 60 cm per century.

* In case of a 60-cm rise in sea level (scenario A of Table 2), the crest level of the Dutch sea dikes have to be increased by 60 to 110 cm (depending on

wave attack and depth of the shore face), while in case of scenario C, with no extra relative sea-level rise but with an extra change in the windforce of +10 %, these numbers are 80 to 135 cm, respectively.

It follows that for the Dutch coast, a 10× increase in wind force has about the same importance as a 60 cm rise of the relative sea level. In this author's opinion, a sea-level rise of 60 cm during the next 100 years has a probability of 50% or more, while a 10% increase in the wind force will have a probability of about 25%.

References

De Ronde, J.G. , 1983. Changes of relative mean sea level and of mean tidal amplitude along the Dutch coast. In, A.R. Ritsema and A. Gürpinar (eds.), Seismicity and Seismic Risk in the Offshore North Sea Area, Proc. NATO Advanced Research Workshop Utrecht, June 1-4, 1982.

Verboom G.K., Ronde J.G. de, Dijk, R.P. van, 1992. A fine grid tidal flow and storm surge model of the North Sea. *Cont. Shelf Res.*, 12, 213.

CHAPTER 18

Impact of Sea Level Rise on Flood Control in Bangkok and Vicinity

Sanga Sabhasri and Ksemsan Suwarnarat

ABSTRACT: Long years of flood damage in Bangkok, Thailand have shown that flooding is not only a natural occurrence but also has resulted from urbanization and the utilization of natural resources. The steady rise in the mean sea water level, caused mainly by land subsidence, poses a threat for investment, operation costs and the safety level of the flood control system. The prevailing flood control scheme relies mostly on the protection from the rising estuarine and sea levels, and the estimated annual pumping costs for the Bangkok may reach US$ 20 million per meter rise.

Clearly more attention must be paid to this phenomenon and the ways and means to mitigate its effects. The exchange of experience and knowledge obtained in other countries may help minimize the problems and costs of "trial and error" in the developing countries such as Thailand.

Introduction

Many engineering schemes to protect the city of Bangkok and the vicinity from flood have been prepared since 1960. The location of the metropolitan area at the deltaic plain of the Chao Praya River estuary has made the city and the surroundings particularly vulnerable to flood. Most of the deltaic plain lies between 0–2.5 m above mean sea level (Fig. 1) with the exception of some relatively small areas (and narrow natural levees) that are 2.5 m to 5 above mean sea level. Much of the city area and the suburbs, which now support a population more than 5 million people, originally were as low as mean sea level. The city roads and most of the urbanisation in the vicinity are on land fill sites only slightly above the maximum tidal levels that reach about +2 m above MSL.

The need for Bangkok to find effective measures to contain or control the persisting flood damages (see Table 1) is most pressing. As demonstrated in 1983, a slowly rising yet persisting flood in this low lying area could inflict more than 6 billion Baht (approximately US$ 250 million) in a single event. For example,

J.D. Milliman and B.U. Haq (eds.), Sea-Level Rise and Coastal Subsidence, 343-356.

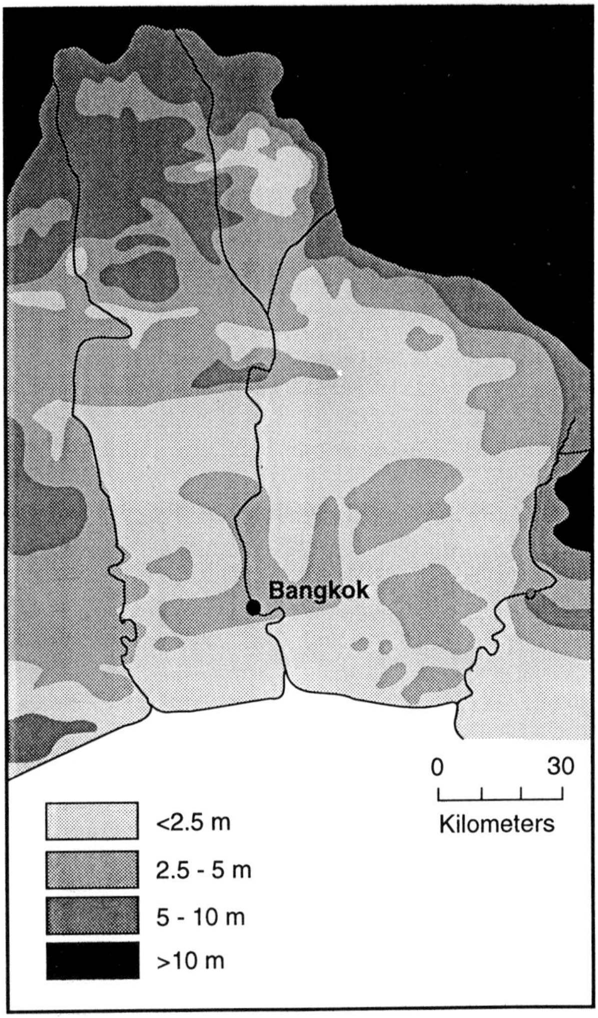

Fig. 1. General elevations of the deltaic plain of the Chao Praya River.

TABLE 1
Recorded flood damage in Bangkok and vicinity, 1975–1983.

Year	Damage (billion Baht)	Damage/unit area (million Baht/km^2)
1975	1.1	0.7
1978	0.28	0.18
1982	1.1	1.8
1983	6.6	4.2

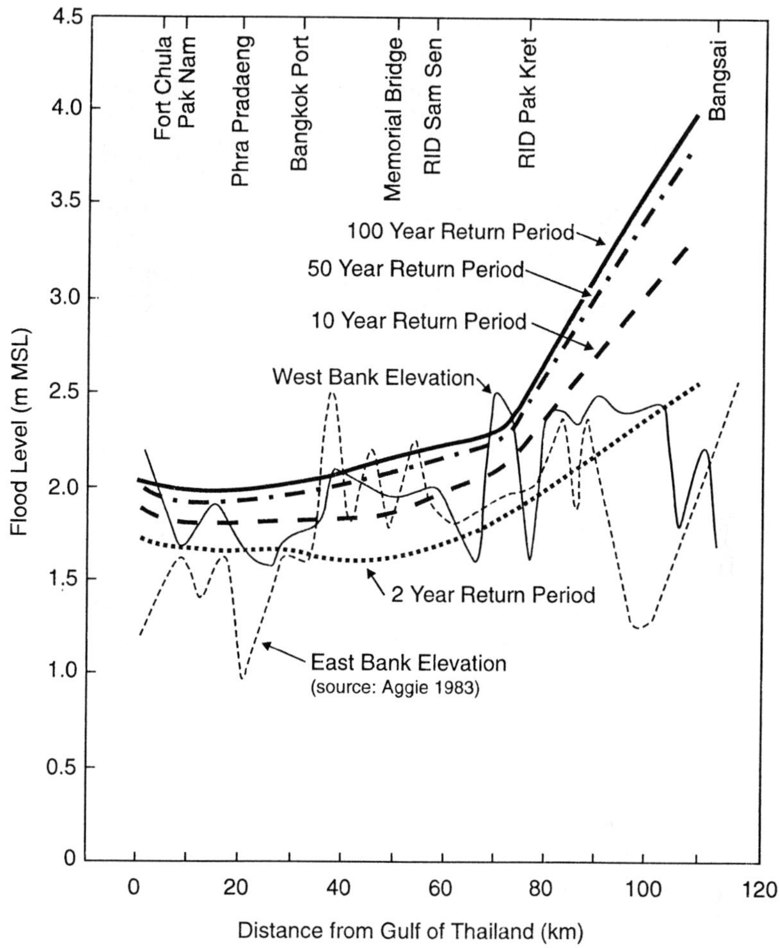

Fig. 2. Maximum river levels compared with bank elevations of the Chao Praya River at various return periods.

during the rainy season of 1988, Thailand was affected by flooding in 52 provinces and 283 districts, involving 62,000 families, and resulting in 19 deaths. Although the total budget spent for flood relief in various provinces was 16 million Baht, additional requests for flood relief funds exceeded 47 million Baht. During the same period, Bangkok and vicinity was hit by a tropical depression that caused inundation of 20 to 40 cm on many low-lying roads. Despite a much improved pumping capacity, many areas of the city were flooded.

One can conclude that there is an inherent risk of flooding in the central area of Thailand, especially in Bangkok and vicinity. Furthermore, the present provision of flood control is still not adequate to ensure safety against annual recurrences. Damage in Bangkok and vicinity is estimated to be on the order of 1 to 6 billion Baht annually, depending on the type and extent of the flooding.

Bangkok flooding is caused by:

1) High rate of river flow, with a river stage of about +2.5 m MSL and flow rate of 3600 m^3/sec. As average ground level is at MSL, over-spills affect the low-lying parts of the city.

2) Direct rainfall of up to 280 mm/day into the low-lying and subsiding urban areas.

3) High tide in the estuary, which can rise up to +2.1 m MSL (Fig. 2). It is important to note that tidal surges can raise levels 2500 mm/day compared to the so-called "50-year rainfall" of 280 mm/day or a river flow of 3500 m^3/sec compared with the high flow of 3600 m^3/sec for a "100-year event" (TAC, 1986).

Flood Control Plans

Flood-loss mitigation plans before 1980 consisted of structural measures, with river dikes and gates between the tidal river and the branching canals to protect the urban areas from tidal rise, and improved drainage and pumping systems to provide adequate discharge. For example, The Greater Bangkok Plan (Litchfield, 1960) provided a framework for development of a large number of regional services and facilities. The study area included the Bangkok-Thonburi metropolitan area, roughly the present Bangkok city and suburbs, and west of the Chao Praya River or the Thonburi City core area.

The recommendations included the following:

1) Flood protection against high river levels would be accomplished by raising roadways along the Chao Praya River.

2) To prevent flooding of the surrounding areas, two perimeter canals would be constructed and with the excavated material a dike would be constructed on the outer bank of these canals.

3) Improvement of storm-water drainage in the protected area would be facilitated by improvement of khlongs (local canals), by the excavation of the perimeter khlongs, and by the construction of storm sewers and pumps. It was proposed that the area be subdivided into 15 drainage districts, each of which would be provided with a central pumping station for evacuation of storm-water runoff to the river and main khlongs by means of sewers.

4) It was considered infeasible to construct a drainage system that would provide full protection against inundation from excessive rainfall. The degree of protection of different drainage districts would have to be a compromise between limited resources and impact factors such as damage, public health and inconvenience. For commercial and industrial areas protection strategy against five-year storms would be adopted; residential and agricultural areas would be protected to a lesser degree.

POLDER DIKES

In 1968, The Master Plan for Sewerage, Drainage and Flood protection systems for Bangkok and Thonburi by Camp, Dresser and McKee (CDM, 1968) was completed. The major recommended solution for flood control and drainage was

TABLE 2
Major investment cost of polder projects in Bangkok.

Projects	DIKES		PUMPING STATIONS		
	length (km)	cost (million Baht)	number	cost (million Baht)	Area (km^2)
City core	15.7	222	5	218	96
Eastern-suburbs	6.2	40.3	6	130	260
Thonburi	108	261	12	133	340
Smuthprakarn	52	501	13	104	220

to divide the area into a number of units (polders) that would be protected separately against over-spills from the river and also the overflow from the Tachin and Pasak Rivers.

Each polder would be protected from external flood waters by a system of surrounding flood protection barriers consisting of a combination of earth embankments, perimeter highways and concrete walls. Each polder would be bound either by the river or by a multipurpose khlong. The storm-water run-off from each polder would be discharged into these multipurpose khlongs by gravity or by pumping in the event of high exterior water levels.

The general practice of the polder system design is to provide protection dikes with crest levels above the maximum design flood. The free board to encounter the waves and ripples caused by winds and the water traffic is approximately 0.3 m in the case of the Chao Praya River. The bank elevations of the river (Fig. 1) indicates that the dike height of the east and west banks of the river would vary from 0.8 to 1.3 m. The dikes and the pumping stations are the major parts of investment cost of such a system (see Table 2).

The river dikes may have to reach up to km 70 at the border of the Bangkok metropolitan area plus another 100 km of sea-walls (TISTR, 1988) at the mouth of the river. The estimated investment cost of 50,000 Baht/m of dike would bring the total investment to 12,000 million Baht. As of 1990, maximum level of water in the estuary had risen about 0.5 m since 1970 (see Figs. 3 and 4), indicating that flood protection facilities will have to be increasingly higher and more expensive. As a consequence, maintaining multipurpose khlongs at high level between the polders as proposed by the CDM plan becomes increasingly expensive. In recent plans, which largely adhere to the polder principle of the CDM plan, flood protection would be provided along the river and inland khlongs would be protected against high river stages by control gates.

To prevent adverse effects for the area east of the barrier, drainage south to the sea would be improved by dredging a large number of khlongs. Furthermore, a 1–2 km wide green zone would be established outside the barrier. In this zone, urban development would be discouraged to provide a flood by-pass route. The Green Belt Project, which was a royally initiated flood prevention scheme, completed

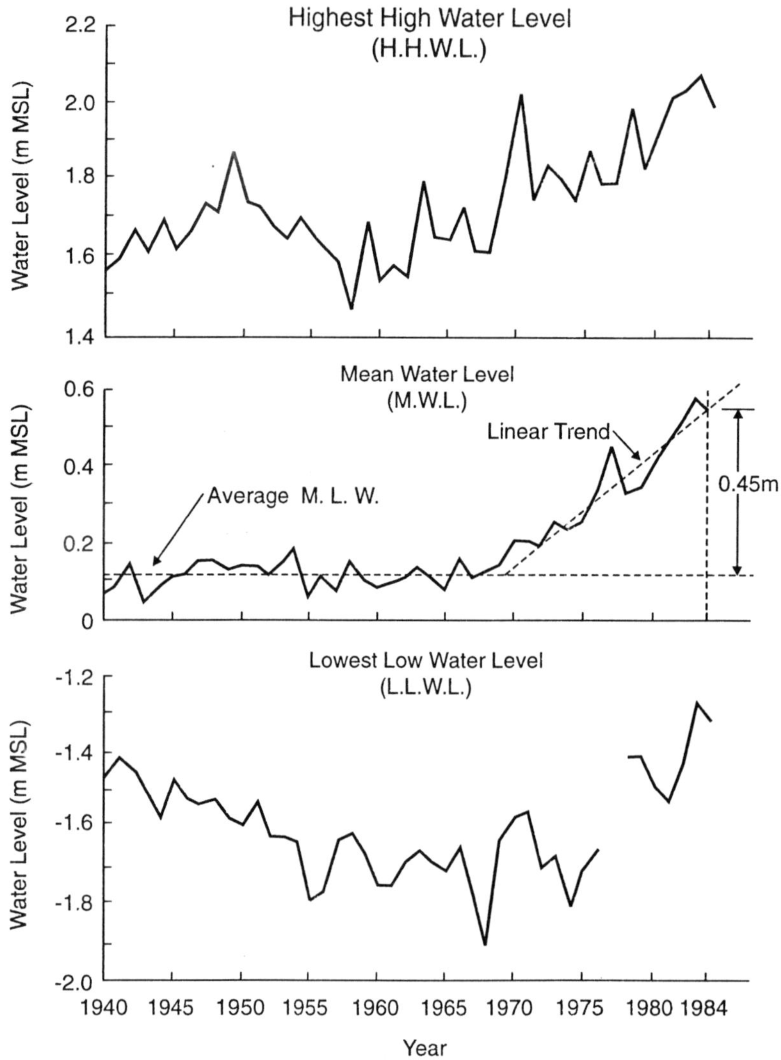

Source: Port Authority of Thailand

Fig. 3. Mean annual water levels at Pak Nam (on the Gulf of Thailand).

the polder approach on the east bank of the Chao Praya by connecting the eastern dikes into a single defense line to protect the city against the influx of the eastern overland runoff.

In 1983/1985 NEDECO (The Netherlands), NECCO and LM/SPAN (Thailand) performed general studies, feasibility studies and specific studies under the Bangkok Flood Control and Drainage Project (city core). The total cost of the

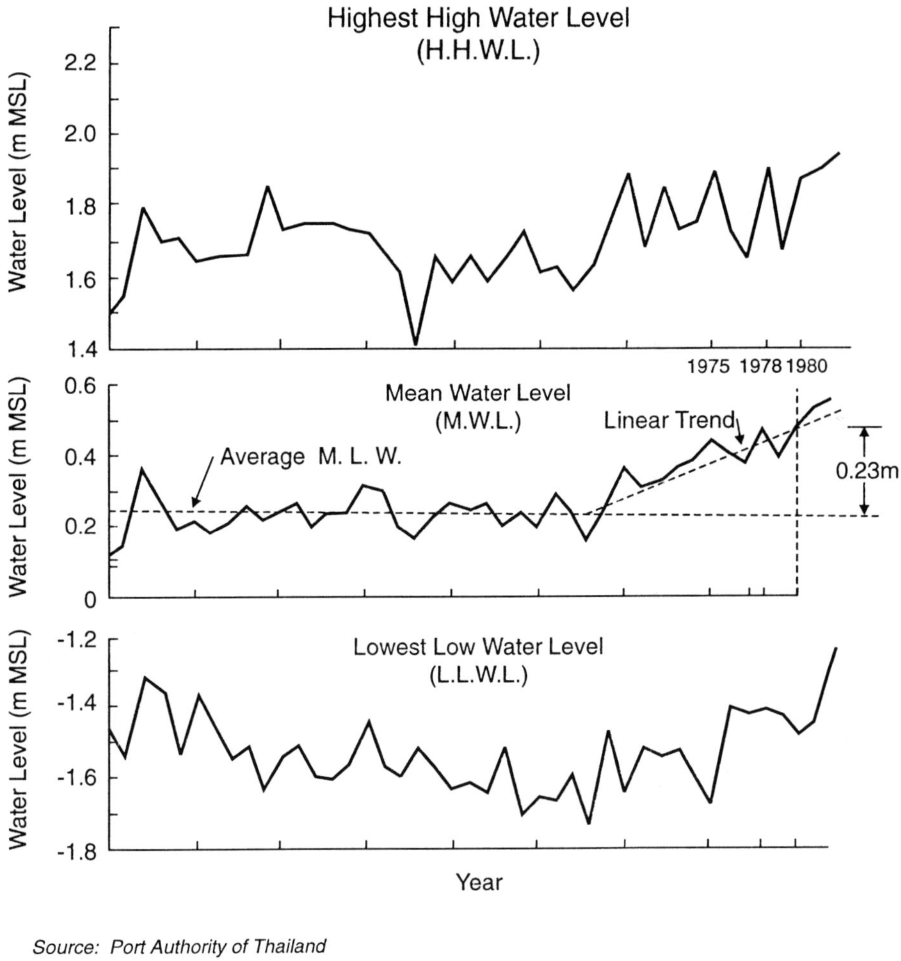

Fig. 4. Mean annual water levels at Bangkok Port.

project (NEDECO, 1985) was estimated at 4,000 million Baht. A priority package for 2,000 million Baht was formulated including the following components:

flood barrier	225 million Baht
canal improvement	880 million Baht
pumping stations	220 million Baht
box conduits and culverts	300 million Baht

The remaining amount was for land acquisition, administration, supervision, etc.

Flood control schemes after 1980 led to the monitoring of land subsidence, limitation of the existing drainage system against the increasing run-off rate of the paved urban areas. The need for non-structural measures had been first brought to attention by the JICA (Japan International Cooperation Agency) study on flood protection and drainage in eastern suburban Bangkok. The assignment included a preliminary study, preparation of a master plan and a feasibility study. The preliminary study covered the eastern suburbs as far as the BMA territory up to the Green Belt which had been earlier warded off by a 70-km long perimeter earth dike.

The total project which covered 360 km^2 was estimated to cost 11,000 million Baht or 30.6 million Baht per km^2. The main expenditures for these works are:

pumping stations	1500 million Baht
canal improvements	2000 million Baht
main pipes	2000 million Baht
gates	300 million Baht
embankment and dikes	5200 million Baht

The pumping cost at 1 million Baht per km^2/year and the structural investment cost of about 30 million Baht each km^2 indicate the high economic burden of the structural measure, and consequentially the importance of a non-structural flood-loss prevention system.

River Diversion

The flood routing and control alternatives (AIT, 1985) offer another approach of the structural measure affecting the river engineering. The National Economic and Social Development Board (NESDB) initiated a study to select the most effective flood control schemes in reducing flood water levels in the Chao Praya in and near Bangkok. The study involved provision of schematic hydraulic designs and cost estimates together with assessments of hydraulic, economic and social effects of alternative flood control schemes. The study focused on the following selected measures:

1) Bypassing river meanders.

2) Diversion of part of the peak flow of the river through a canal along the eastern perimeter of the city.

3) Embankment along the river north of Bangkok to accommodate the back-water rise during flood season.

4) Improvement of the river by deepening or widening of shallow or narrow reaches.

However, none of the investigated alternatives was reported to significantly lower the peak stages of water flow in the river. This was attributed to the uncontrolled effect of the diurnal tides.

Another study continuing the river engineering approach followed almost immediately in 1986 by the Thai Austrian Consortium (ACE-Consultco-CAE). The AIT also had been employed by the BMA to cooperate in this subsequent study. The new study was made on a new alignment west of the Chao Praya River (TAC, 1986). The study recommended excavating a 2,000 m^3/sec capacity diversion canal for the Chao Praya river from the northern boundary of the metropolitan area to the sea. In addition to the diversion, two river barriers were proposed; one in the north to divert the flow into the diversion canal, and one in the south to close the mouth of the river against diurnal high tides. Navigation gates were to be provided

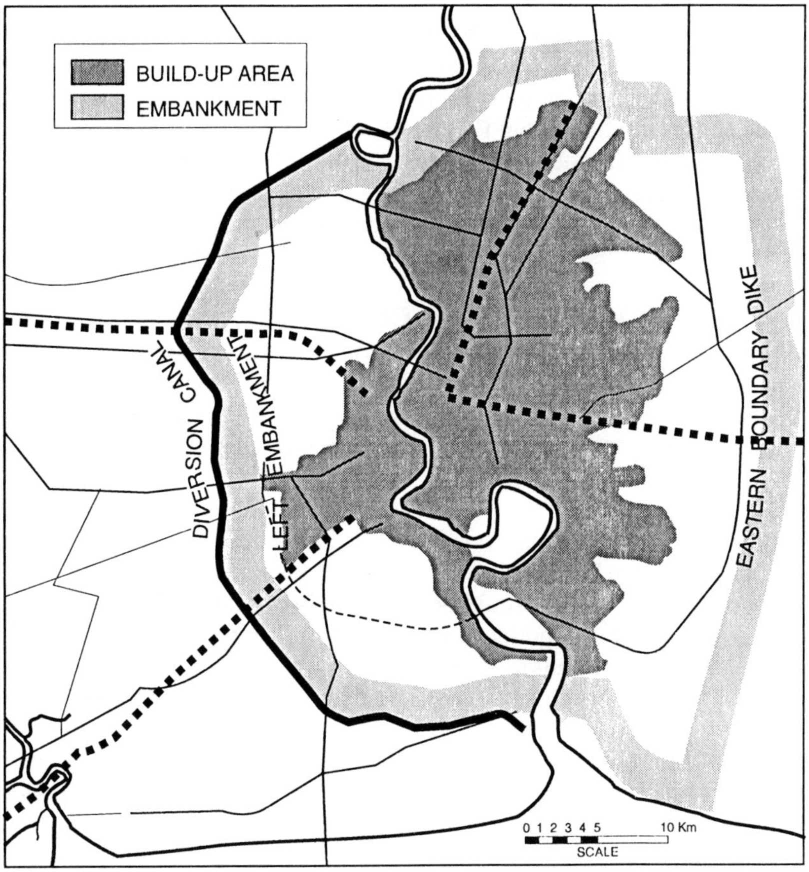

Fig. 5. The Chao Praya II Diversion Project.

at both barriers. A pumping station was to be built in the river mouth barrier to relieve the rainfall runoff from the inner closure (Fig. 5).

This project is unique in lowering water level in the Chao Praya River in the vicinity of Bangkok, which should improve the efficiency of drainage in the city without necessitating construction of raised embankments to protect the city against the seasonal rise of water level. Protection against sea-level rise would be provided by a sea-barrier structure at the river mouth. The sea barrier completes a ring of dikes along the east and west of the city encompassing a protected area of about 1,500 km^2 on both sides of the Chao Praya. The existing King's dike on the eastern boundary across the river would be extended by another hydraulic structure called "Diversion Structure". This structure would contain provisions to regulate both the flow and the navigation traffic through the river. The excess flows would be diverted through the periphery canal "Chao Praya 2" outside the west bank dike around the protected area and return to the mouth of the river.

The design criteria of this scheme have been laid-out for:

Discharge from the north: 100-year return period; 3,600 m^3/sec from Bangsai

Diversion channel capacity : 2,000 m^3/sec.

The ancillary flood protection facilities would include:

1) A diversion channel on the west bank with 135 m bed width and 220 m top width. The channel would start at Pak Kret and run along the west bank to the sea at the Chao Praya River mouth. Total length would be about 60 km.

2) Control structures located immediately downstream of the diversion intake would divert flood water to the diversion channel and control the amount of flood water flowing to the Chao Praya River. The structures would include weirs and a navigation lock.

3) A tidal barrier located across the Chao Praya River mouth would protect against high tide intrusion into the city. Navigation locks are also provided at tidal barrier. A 1,600 m^3/sec-capacity pump would be required to evacuate flood water released from control structures and drained from the city. The salinity gates would protect against salt intrusion into the diversion channel and control flow in the diversion channel.

This scheme would lower the water level in the Chao Praya River to +0 m MSL at Pak Kret and –0.8 m MSL at Fort Chula for floods less than 100-year extreme (approximately 3,600 m^3/sec). In such conditions, the existing flood risk at +1.2 m MSL in the east bank polders would be completely eliminated. The effectiveness of the CP2 is even more pronounced northward of Bangkok where it would help the farmers near the river banks.

A general conclusion can be drawn that the polder-dikes approach would provide effective solution to the impoundment type of flooding within the protected area, but it would increase flood risk outside the dikes. The water approach schemes eliminate this negative effect and also reduce the peak water level inside the protection areas by allowing gravity flow inside the control basin.

BMA Flood Plain Management

The Bangkok Metropolitan Administration acknowledges the need for flood plain management. The proposal drafted by the Policy Department acknowledged that the metropolitan area lies below the mean sea level and that urban areas needed to be protected by dikes, and the drainage capacity of the existing canals and pumping stations needed to be enhanced. It was urged that the district offices request the government to assign ponds and parking lots for early rainfall detention. Private houses, schools and large building were also proposed to follow the practice. The advisory report also contains a description of the hydraulics of the drainage system and examples of construction of local ponds in Bangkok and retention facilities in Japan.

The implementation plan proposed by the Department of Drainage and Sewerage of the BMA to this committee includes the following targets in the five year plan:

1) Production of a flood risk map, to be implemented by the Department of Drainage and Sewerage of the BMA. The initial step would be the training of personnel on surveying and data collection.

2) Establishment of permanent marks of previous floods. These levels would be marked clearly so that they would be observable by the general public.

3) Conservation of degrading areas. Initially the implementation was to be on public lands while the suitable for privately owned lands. Basic criteria mentioned were: empty spaces; low lying in relation to the surroundings; suitability for conservation. (The BMA located 177 potential locations for degradation in a reconnaissance survey).

4) Draft regulations for land-fill practices and compensation for the increase in run-off rates by providing a storage capacity. The working group for this subject was to make a draft based on experiences in other countries.

5) Set up a public relation program for flood-risk information.

Additional long-term policy also has been laid out as follows:

1) To establish a legal requirement for each household to provide for an adequate storage volume of rainwater in the hope to maintain the original runoff rate before land development.

2) Set-up the measures to include the observation of important impacts of land-use development to the urban flood phenomena.

3) Set up a tax system that relates to the benefit level of land use and also allows an effective regulation of land-use development.

Impact of Upstream Agricultural Development

For many decades the upper catchment area of the Chao Praya River has been managed by authorities in charge of agricultural promotion, irrigation and hydro-electric production. Two large storage reservoirs were created on main tributaries of the Chao Praya River, with a total capacity of 23,000 million m^3 and operated by the Electricity Generation Authority for power generation and irrigation. Releases are adjusted to the requirements of salinity control in the lower Chao Praya. Flood control is not a major objective of the reservoirs, but the conditions and operation appear to be favourable for flood control as the average annual peak discharges below the dams are reported to have been reduced by 34% after completion of the two dams and reservoirs.

Possible negative effects of the agricultural development in the upper flood plain, however, is a matter for increased concern by Bangkok. The riverside dikes between Ayudhya and Bangkok would literally "funnel" the overspills directly towards the city. A simulation by NEDECO (1987) indicates that the maximum daily peak water level could rise from +1.5 m to +3 m MSL in the city centre.

Land Subsidence

Land subsidence is the most significant cause of the relative sea-level rise in Bangkok and vicinity, which had been observed since 1969. Results of field measurements from 1978–1981 showed that the major cause of subsidence was the rapid drop in piezometric levels caused by the ground-water removal (AIT, 1981; Ramnarong, 1983). Land subsidence has occurred over the central, eastern and

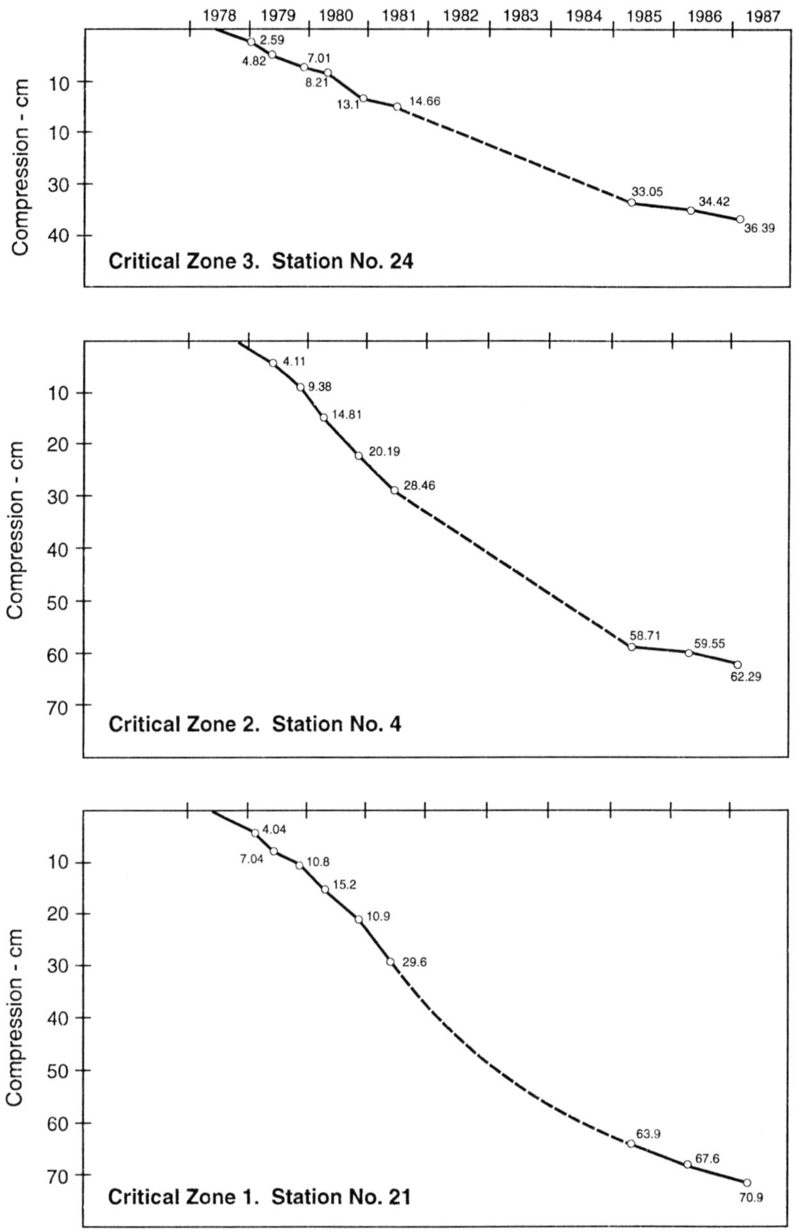

Fig. 6. Ten-year record of subsidence in three critical areas of Bangkok (1978–87).

southeastern parts of Bangkok. Since the present altitude of Bangkok is very close to the mean sea level, flooding of low lying areas is a significant problem. The flooding of urban and suburban Bangkok has caused many millions of Baht of loss annually, even though the depth of inundation only has been between 0.2–1.0 m.

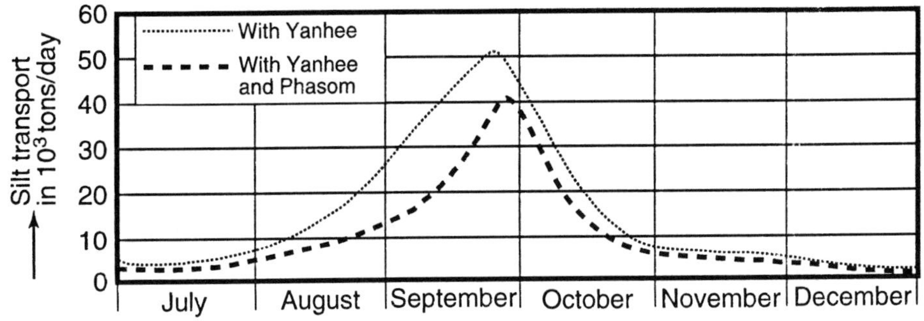

Fig. 7. Calculated sediment transport of the Chao Praya River at Bang Sai.

Maximum subsidence occurred in the eastern part of Bangkok, with as much as 40–70 cm between 1978 and 1987 (Fig. 6).

Sedimention due to the Chao Praya River

The most comprehensive study of sedimentation in the Chao Praya Estuary, made 30 years ago (NEDECO, 1965), led to the following conclusions:

1) In view of the overall flat nature of the upper plain and the regular flooding of large areas during the rainy season, the upper plain probably does not contribute to sediment transport. The upper plain may even be an area of slight aggradation, although negligible.

2) Assuming that the sediment yield of the basins is directly proportional to their respective runoff, the annual sediment transport of the Chao Praya River amounts to 5.8 million tons per year. Reservoirs act as sediment traps even for the fine suspended particles. Thus much of the sediment that passes the dam sites will be contained in the reservoirs. It also was concluded that possible siltation or scouring of the rivers below the dams was negligible.

3) These two assumptions mean that the sediment transports at Nakorn Sawan will decrease to 3.7 million tons with Phumipol Dam in operation, and to 1.6 million tons when both the Yanhee and Phasom Projects are completed (Table 3).

However, it was anticipated that more water would be diverted above the Chao Praya Dam, reducing the flooding in the Delta. As a result, the water that overflows the rivers would carry less sediment. The net result would be that sediment deposited in the inundated areas in the future would be considerably less than before.

The actual future transport may be somewhat higher than estimated because the assumptions for the above estimates were rather extreme to establish the maximum effect of the water-control work on the sediment supply. Also, future hydrological conditions may be somewhat different from those in 1952–1962. However, deviations would be greatest in the dry season and would not affect the sediment supplies. Taking everything into consideration, the transport curves of Figure 7 could provide a suitable basis to conclude that there will be less future sedimentation in the the Chao Praya River Estuary than in the past.

TABLE 3

Chao Praya sediment transport between July 1–December 31, 1962 (in million tons) at Yanhee and Phasom dams.

	1962	Yanhee	Yanhee & Phasom
Supply at Nakorn Sawan	5.6	3.5	1.4
Diverted above Chao Praya Dam	0.6	0.9	0.4
Released by Chao Praya Dam	5.0	2.6	1.0
Lost by flooding	1.3	0.5	0.1
Resulting supply Chao Praya	3.7	2.1	0.9
Supply by Pasak	0.5	0.7	0.7
Transport at Bang Sai	4.2	2.8	1.6

References

Asian Institute of Technology, 1981. Investigation of land subsidence caused by deep well pumping in the Bangkok area. Comprehensive Report, submitted to National Environment Board, Bangkok.

Asian Institute of Technology, 1982. Groundwater resources in Bangkok area. Development and management study. Comprehensive Report, submitted to National Environment Board, Bangkok.

Asian Institute of Technology, 1985. Flood routing and control alternatives of Chao Phraya River for Bangkok. Research Report No.164, submitted to the NESDB, Bangkok, Thailand.

Camp Dresser and McKee (CDM), 1968. Master plan, sewerage, drainage and flood protection systems, Bangkok, Thonburi, Thailand. Boston, Mass., USA.

Japan International Cooperation Agency (JICA), 1985. Master plan for flood Protection and drainage project in eastern suburban-Bangkok.

Litchfield Whiting and Associate/Adams, Howard and Greenley, 1960. Greater Bangkok Plan,1990. for The Ministry of Interior, Bangkok, Thailand.

NEDECO,1965. A study on the siltation of the Bangkok Port Channel. v. 2.

NEDECO, 1987. Master plan for flood protection and drainage for Thonburi and Samutprakarn West. Master Plan Report, Bangkok.

Piancharoen, C. and Chuamthaisong, C, 1976. Groundwater of Bangkok metropolis, Thailand. Inter.Hydrogeological Conference,Budapest.

Ramnaraong, V, 1983. Environmental impacts of heavy groundwater development in Bangkok, Thailand. International Ground Water and Man, Sydney, Australia.

TAC, 1986. Bangkok flood protection Chao Phraya 2. Feasibility Study Final Report. ACE Consultco. CAE and AIT.

Thai Institute of Science and Technological Research, 1988. Master Plan for Flood Control and Drainage for East Smuthprakarn. No. 2.

Coastal Vulnerability: Hazards and Strategies

Bilal U. Haq and John D. Milliman

Introduction

The change in sea level at any location represents the combination of global sea-level change (e.g., due global warming) and local/regional sea-level change due to subsidence or uplift. Because of the wide variations in the latter component, the rate of future sea-level rise and its implications must be assessed separately for each coastal zone. To plan a meaningful response strategy, risk assessment for shoreline regression must be based on local empirical data, preferably as an integral part of the coastal zone management activities.

Once the local threat has been identified and evaluated, a clear response strategy can be instituted, taking into consideration physiographic and socio-economic factors of the affected coastal zone. Mitigation responses on the global scale (e.g., limiting greenhouse warming by reducing emissions and/or augmenting carbon sinks and reservoirs) and on local/regional level (e.g., limiting the factors that accelerate subsidence and environmental degradation) must go hand in hand if adaptive strategies are to be successful.

The Intergovernmental Panel of Climate Change (IPCC) in their recent report (1990) underscored the urgency for maritime nations to undertake early planning of adaptive strategies in order to avoid detrimental impacts. Unfortunately, Unfortunately, such opportunities are often lost through inaction.

Impacts of Accelerated Sea-Level Rise on Coastal Areas

Many uncertainties remain in existing climate prediction models (Raper et al., this volume), but the trends in climate change are becoming increasingly apparent. The climatic record of the recent past contains evidence of steady increase of greenhouse gases in the atmosphere, rising mean temperatures, diminishing mountain glaciers and enhanced cloudiness in mid latitudes. When the trend of increasing greenhouse gases is projected into the future it implies a doubling of pre-industrial levels of CO_2 by the year 2065. It has been estimated that this will elevate global mean temperatures by 1.5–4.5°C. The responding global (eustatic) sea-level rise is projected to be between 20 and 70 cm over the same period, and between 30 and 110 cm by the end of the 21st century. Thus, the present rate of average sea-level

J.D. Milliman and B.U. Haq (eds.), Sea-Level Rise and Coastal Subsidence, 357-364.

rise of about 1–2 mm/year could amplify up to 6 mm/year in the next century (IPCC, 1990).

Increased vulnerability of the coastal areas must be recognized in view of this potential accelerated sea-level rise due to human activities that threaten to enhance the efficacy of natural hazards along the coasts. A more rapid pace of global sea-level rise will increase the rates of transgression of wetlands, submergence of coral islands, erosion of the shoreline, and salt-water intrusion in coastal aquifers.

Increase in mean sea-surface temperature in the tropical ocean also could lead to greater frequency of hurricanes that already cause significant damage to property and human life. As discussed in several papers in this volume, accelerated subsidence in many coastal areas (such as the highly- populated river deltas and coastal cities) due to excessive withdrawal of water and/or hydrocarbons also promises to exacerbate the situation by amplifying the magnitude of relative sea-level rise locally.

Impacts of Sea-Level Rise

The impacts of a more rapid advance of the sea on the coastal areas can be placed in three main categories (Gornitz, 1991): 1) intensified flooding, 2) increased erosion, and 3) greater intrusion of saline marine water into estuaries and coastal aquifers.

Intensified Flooding

Most coastal features, such as low beaches, lagoons, bays, estuaries and delta plains, are more vulnerable to accelerated sea-level advance due to their relatively low gradient. Initially the coasts may suffer episodic inundation, but later this may become permanent. Wetlands, including marshes, bogs, mangrove swamps, as well as many estuarine areas, normally can keep pace with slowly rising sea level. When vertical accretion is limited by lack of sediment input or fresh-water flux, or when wetlands cannot migrate landward due to artificially-protected coastlines, a significant loss of these ecosystems may occur. A rapidly advancing sea, as projected for the next century, also could jeopardize the ability of the wetlands to migrate; instead they may drown. According to one estimate, some 40% of all US wetlands may be vulnerable to drowning in the next century, and less than half are expected to be able to migrate with the more rapidly advancing seas (Gornitz, 1991). The economically important mangroves along many tropical coasts may be especially threatened due to the impeded capacity to migrate.

Deltas and large coastal urban centers where subsidence has been accelerating may be prone to more rapid and permanent inundation. Some examples of vulnerable areas include the deltas of the Bengal, Amazon, Orinoco, Niger, Indus and Chao Pharya Rivers, coastal cities such as Shanghai, Tianjin, and low-lying areas such as, the eastern coast of Sumatra, the coast of Borneo, the low countries of Europe, southern Baltic and the coast of Louisiana (Milliman et al., 1989; Gornitz, 1991; Haq, 1994).

Tropical coral islands are a special case. Most of these islands are only 1-3 m above sea level and may be particularly vulnerable to episodic and/or permanent flooding. Normally, a healthy coral reef can keep pace with rising sea level. However, damaged or unhealthy reef may not be able to produce enough carbonate limestone or sediment to aggrade at the rate of sea-level rise and thus may drown. Also, coral growth may be impeded if sea surface temperatures in the tropics rise much above the present 30°C. As recent exceptionally warm years have witnessed, corals respond by "bleaching", which impedes their growth or kills them (Bunkley-Williams and Williams, 1990). Thus, higher eustatic sea levels and elevated temperatures may combine to cause even more rapid flooding of coral islands. Such areas include the Great Barrier Reef of Australia, the Laccadive and Maldives off India, Marshall Islands in the Pacific, and many islands in the Caribbean (Gornitz, 1991).

INCREASED EROSION

A majority of the worlds sandy beaches are in retreat due to erosion. When offshore sand sources are depleted, erosion can exceed accretion. Often human activities upstream as well as along the coast have added to the problem of rapid erosion. Upstream damming, as in the case of Indus and Nile Rivers, reduced or eliminated sediment flux to the coastal areas and has encouraged erosional processes (Milliman et al., 1989; Haq, 1994). Coastline retreat is also taking place more rapidly where sediment supply is impeded or eliminated. Along the mouth of the Nile, for example, the promontories are being eroded rapidly so that the coastline will be increasingly wave-dominated and straighter (El Sayed, this volume). The damming of the Niger and Benue Rivers has also choked sediment flux to the river mouths, and the barrier islands that front the ocean have been retreating at the rate of a few tens of meters every year (Ibe, this volume). The practices of building breakwaters and of sand mining also add to the increased erosion.

During a slowly rising sea level, as in the case of the transgression that followed the last deglaciation, the beach profile maintains an equilibrium. Initially, a landward migration of sand barriers takes place, followed by accretion of the dunes. However, during a rapid advance of the sea the barrier islands may not be able to keep pace and may drown, especially if sediment flux to the coast is low. An accelerated sea-level rise combined with a low rate of sediment supply (e.g., away from sediment sources or at the mouths of the dammed rivers) could intensify coastal erosion, at the expense of accretion. This is already taking place along the coast of Louisiana where relative sea-level advance is rapid due to high local subsidence rates and sediment flux from the Mississippi River has been more than halved in the past 60 years (Gornitz, 1991; Wells, this volume).

INCREASED SALT WATER INTRUSION

Salt water intrusion is another serious problem along coasts that could intensify with accelerated sea-level rise. The combined effect of impeded flow in rivers due to human activities and rapidly rising sea level could mean greater intrusion of saline waters into estuaries and streams and contamination coastal aquifers. This is now a common scenario in some deltas and large urban areas. In the Bengal Delta salt-water intrusion in the estuaries has been observed up to 240 km upstream during dry season when the river flow is impeded due to sequestration of water upstream (Milliman et al., 1989; Alam, this volume). Nile and Niger Deltas are also experiencing increased salt-water intrusion. In Bangkok aquifer contamination has been encouraged by excessive and rapid ground-water withdrawal (Nutalaya et al., this volume). With the projected rise in sea level, tides will penetrate further inland and will be able to contaminate greater areas. Seepage of salt into soil via groundwater is already a serious problem in low-lying areas that also could be exacerbated. Rising sea level would tend to steepen the salt water-fresh water gradient and lead to enhanced seepage. These threats underscore the need for expert management of the coastal zone.

Adaptive Strategies

The IPCC (1990) identified three adaptive strategies to combat rising sea level at local levels: retreat, accommodate and protect. The retreat response characterizes relinquishing structures in existing developed areas threatened by advancing seas. It may entail significant resettlement of local population further inland, beyond the area projected to be affected. The second strategy accommodates the advancing sea by accepting the fact of future flooding, and by scheduled conversion to alternative economic activities, e.g., switching from agriculture to aquaculture. This response, characteristically, is more inclined to preserve a balance between nature and human needs. The protective response embodies a defensive strategy, where economically-prized structures and resources are shielded from the advancing threat. Building sea walls, dams, and locks are common protective strategies.

The exigency of taking early action also needs to be underscored, because some responses may take decades to be fully effective. Failure to consider responses with long lead-times at an early stage will limit the range of adaptive options later on, perhaps to the considerable detriment of the nation's economic and environmental well being. It is clear the while some activities require only short decision-making time, other may require decades of lead time to be effective (IPCC, 1990).

Examples of Action and Consequences of Inaction

There are numerous examples where mitigation activities, instituted at the eleventh hour, have made less impact than if they had been attempted earlier. The city of Venice is a celebrated example, where the threat of relative sea-level rise due to accelerated subsidence was finally heeded, and belatedly, in the late 1960s steps were taken to reverse the trend. The accelerated rate of sea-level rise in the middle part of this century led to several corrective measures, including building locks on the inlets to the Venetian Lagoon to control inflow from the open ocean (Sestini, this volume). More importantly, the practice of using artesian water was reserved. A rapid decline in extraction of ground water since the 1970s has slowed the subsidence, and parts of the city have actually rebounded by as much as 20 mm.

Experience in other areas have shown that total rebound is never much more than about 20% of the cumulative subsidence, and thus, Venice is expected to rebound by a total of only about 30 mm (Emery and Aubrey, 1991).

Mitigation and reversal of adverse consequences of previous developmental activities often involve a complete rethinking of long-practiced traditions. A rare example of such a change in policy concerns the northwest coast of the Netherlands. On the Wadden Sea coast past practices had accelerated the subsidence rate to a high 13 mm/year in the last two decades. According to previous practice there would have been an attempt to reclaim the land for agricultural use. Instead, a revised government policy designated the Wadden Sea area as a type of "Biosphere Reserve" park, allowing only those activities that avoid conflict with the natural processes. This experiment to let the area revert to a wetland, and again become a hatchery for shrimp, fish and invertebrates, also makes better economic sense in view of the agricultural surpluses in the countries of European Economic Community (Hekstra, 1988).

What mitigation strategies can be instituted for some of the case studies described in this volume? Another IPCC (1992) report on dealing with the challenge of rising sea level summarizes many of the necessary mitigation approaches for the most vulnerable coasts of the world. Selected examples of adaptive strategies for some areas are discussed below.

For Bangladesh a retreat from the coastal areas, prone to both riverine and severe-storm flooding, would seem like the most likely strategy. However, a total retreat in not a practical choice due to the area's high population density. A combination of all three adaptive strategies may be necessary. To prevent further acceleration of subsidence of the delta would require regional, rather than local solutions. Reduction in the rate of ground water withdrawal from the delta could further help slow down subsidence. More important, however, is whether or not the World Bank moves ahead with its plans to build extensive embankments that would prevent seasonal flooding of the deltaic plains; without knowing the exact rates of regional subsidence (Alam, this volume) or the amount of fluvial sediment load deposited on the flood plains, construction of such embankments might cause far more problems than they solve. Protection is not feasible for much of Bangladesh's coastline, but more developed areas and ports with high economic activities could be protected from the imminent threat of advancing sea.

For the Nile Delta, on the other hand, protection may be provide the only logical solution, given the fact that the Nile River already is completely dammed, thus preventing the influx of any new sediment to the coastal area. Sea walls and other structures could be erected to protect developed areas and urban industrial centers such as Alexandria. Similarly, in Nigeria protecting the entire coastline may not be economically feasible, but in the Niger Delta, shielding the economically-vital hydrocarbon infrastructure and ports may be the only attainable solution. A relocation of people from the lesser developed parts of the delta and the coast might be necessary in a selective protective program.

For the coastal urban centers such as Bangkok, Shanghai and Tianjin, the response must be along several fronts, including protection and retreat strategies. First, past practices, such as rapid ground-water withdrawal, that are responsible for accelerated subsidence and relative sea-level rise have to reversed to slow down the rates of sinking. This almost certainly will require searching out new sources of potable water and the recycling of used water. Second, the economically-important civil and industrial areas may have to be selectively shielded by building protective structures that can withstand the projected rise in sea level. To protect urban centers such as Shanghai and Tianjin would require large investments amounting to several hundred million dollars. In addition, inhabitants of some areas may have to be relocated. In all these cases substantial external assistance, both financial and technical, will be needed for such high-investment mitigation activities.

In case of small island nations such as Maldives, a mixture of retreat and protection also may be the preferred adaptive response. Since all islands cannot be protected, scheduled evacuation and resettlement of people from the lower-lying islands may be an important adaptive consideration. Further investment into building protective structures on the seaward sides of economically-important lagoons should begin at an early stage. A protective strategy for Marshall Islands may be more difficult because of a general lack of suitable in-fill material on the islands. Any excavation for such material would also exacerbate the already considerable problems of fresh-water supply. Even if protective structures could be built around some of the main urbanized islands, it could mean considerable loss of land and dramatic changes in the nations socio-cultural and economic life (Crawford et al., 1992). Ultimately, however, atoll islands do have the capacity to accrete naturally if the source of the sands (offshore coral reefs) are preserved; in part, then, the survival of these low-lying atoll islands depends on the conservation practices of the islanders themselves.

Concluding Statement

Mitigation actions can only be planned effectively when the scope of the problem is fully understood. Most of these problems are best approached at the national or local level, although many cases further research may require international funding.

Recognizing potential environmental impacts of engineering projects in coastal areas and the hinterland does not necessarily mean that projects should be scrapped. Rather than questioning if the project should be done, one might question how the project should be done so that downstream impact is minimized. For example, local sea-level rise can be predicted accurately only if local rate of subsidence is known, and effect of river damming or channelizing can only be predicted if the flux and fate of river sediment is accurately known. If the hydrological cycle is not fully understood, for example, the impact of projects that affect downstream water usage cannot be fully evaluated. The effect of river diversion or damming, for example, can have great impact on the coastal environment as well as the river and flood plains downstream from the project; witness the aftermath of the Aswan High Dam in Egypt.

Building and utilizing coastal defenses against rising sea level, subsidence and erosion should be viewed with particular caution. Such defenses (e.g., jetties and breakwaters) often simply pass the problem elsewhere (i.e. downdrift) without actually solving it. Adapting or relocating may be economically wiser, but addressing the causes of increased subsidence or erosion in the long run may be the wisest course of action.

The problem of sea-level rise perhaps has been best stated by Robert M. White (1990): "Our global environment is under attack on many fronts. Climate warming is but one, perhaps the most complex, of these issues. If the changes occurring in our atmosphere are likely to cause consequences, we must understand the problems and promote sensible policies to remedy them. What would be unwise is to lapse into apocalyptic thinking or ostrich-like denial."

Acknowledgments

We thank SCOPE for funding the initial Bangkok workshop in 1988. Any opinions, findings or conclusions are those of the authors and do not necessarily reflect the views of the National Science Foundation.

References

Alam, M., 1995. Geotectonics and subsidence of the Ganges-Brahmaputra Delta of Bangladesh, and Associated drainage, sedimentation, and salinity problems. (This volume).

Bunkley-Williams, L. and Williams, E.H. Jr., 1990. Global assault on coral reefs. *Natural History*, April 1990: 47-54.

Crawford, M., Holthus, P., Makroro, C., Nakasaki, E., and Sullivan, S., 1992. Vulnerability assessment to accelerated sea-level rise. Case study: Majuro Atoll, Republic of the Marshall Islands. Executive Summary of the Report to US NOAA. 44 pp.

El-Sayed, M.Kh., 1995. Rising sea level and subsidence of the northern Nile Delta: A case study. (This volume). Emery K.O. and Aubrey, D.G., 1991. *Sea Levels, Land Levels and Tide Gauges*. Springer-Verlag, New York.

Gornitz, V., 1991. Global coastal hazards from future sea level rise. *Paleogeo., Paleoclimat., Paleoecol.* (Global and Planetary Change), 89:379-398.

Haq, B.U., 1994. Sea level rise and coastal subsidence - rates and threats. Rept. to the Environment Department, the World Bank, 34 pp.

Hekstra, G.P., 1988. Prospects of sea-level rise and its policy consequences. Discussion paper for Symp. on Controlling and Adapting to Greenhouse Warming. Washington DC, June, 1988.

Ibe, A.C., 1995. The Niger delta and the rise in sea level. (This volume).

IPCC, 1990. *Climate Change, The IPCC Scientific Assessment*. (J.T. Houghton et al., eds.) Cambridge, Cambridge University Press, 365 pp.

IPCC, 1992. Global Climate Change and the Rising Challenge of the Sea. The Hague, Directorate General Rijkswaterstaat.

Milliman, J.D., Broadus, J.M. and Gable, F., 1989. Environmental and economic implications of rising sea level and subsiding deltas - the Nile and Bengal examples. *Ambio*, 18, 340-345.

Nutalaya, O., Yong, R.N., Chumnankit, T. and Buapeng, S., 1995. Land subsidence in Bangkok during 1978-1988. (This volume).

Raper, S.C.B., Wigley, T.M.L., and Warrick, R.A., 1995. Global sea-level rise: Past and future. (This volume).

Sestini, G. and Carbognin, L., 1995. Land subsidence and sea-level rise: The case of the Po delta region, Italy. (This volume).
Wells, J.T., 1995. Subsidence, sea-level rise, and wetland loss in the lower Mississippi River Delta. (This volume).
White, R.M., 1990. The great climate debate. *Sci. American.*, 263 (1), 36-43.

List of Contributors

Mahmood Alam
Department of Geology
Dhaka University
Dhaka-1000
Bangladesh

Eric C.F. Bird
Department of Geography
University of Melbourne
Melbourne
Australia

James M. Broadus (†)
Woods Hole Oceanographic Institution
Woods Hole, MA
USA

Somkid Buapeng
Department of Mineral Resources
Bangkok
Thailand

S. Charulukkana
Bangkok
Thailand

Thongchai Chumnankit
National Environment Board
Bangkok
Thailand

J.G. de Ronde
National Institute for Coastal
and Marine Management
P.O. Box 20904
The Hague
The Netherlands

Mahmoud Kh. El-Sayed
Department of Oceanography
Faculty of Science
Alexandria University
Alexandria
Egypt

Bilal U. Haq
Bethesda, MD 20816
USA

A. Chidi Ibe
Nigerian Institute for Oceanography Research
Victoria Island
P.M.B. 12729
Lagos
Nigeria
(present address: UNESCO, Paris, France)

S. Jelgersma
Geological Survey of The Netherlands
Haarlem
The Netherlands

C. Manotham
Bangkok
Thailand

Robert H. Meade
U.S. Geological Survey
MS 413, Box 25046
Denver, CO 80225
USA

John D. Milliman
School of Marine Science
College of William and Mary
Gloucester Point, VA 23062
USA

Prinya Nutalaya
Asian Institute of Technology
Bangkok
Thailand

R. Polsi
Bangkok
Thailand

H. Postma
Netherlands Institute for Sea Research
Texel
The Netherlands

S.C.B. Raper
Climatic Research Unit
University of East Anglia
Norwich NR4 7TJ
UK

A.L. Ramanathan
School of Environmental Sciences
Jawaharlal Nehru University
New Delhi - 110 067
India

Ren Mei-e
Department of Geo and Ocean Sciences
State Laboratory for Coatal and
Island Exploitation Nanjing University
Nanjing
People's Republic of China

Sanga Sabhasri
Faculty of Forestry
Kasesart University
Bangkok 10900
Thailand

G. Sestini
London
England

D. Srisaengthong
Bangkok
Thailand

K.R. Subrahmanya
Department of Marine Geology
Mangalore University
Mangalagangothri 574199
India

V. Subramanian
School of Environmental Sciences
Jawaharlal Nehru University
New Delhi - 110 067
India

Ksemsan Suwarnarat
Faculty of Forestry
Kasesart University
Bangkok 10900
Thailand

S. Vongvisessomjai
Bangkok
Thailand

R.A. Warrick
Centre for Environmental and Research Stuc
University of Waikato
Hamilton
New Zealand

John T. Wells
Institute of Marine Sciences
University of North Carolina-Chapel Hill
Morehead City, NC 28557
USA

T.M.L. Wigley
University Corporation for
Atmospheric Research
P.O. Box 3000
Boulder CO 80307-3000
USA

R.N. Yong
McGill University
Montreal, Quebec
Canada

Geographical Index

Coastal Systems and Continental Margins

1. B.U. Haq (ed.): *Sequence Stratigraphy and Depositional Response to Eustatic, Tectonic and Climatic Forcing*. 1995 ISBN 0-7923-3780-8
2. J.D. Milliman and B.U. Haq (eds.): *Sea-Level Rise and Coastal Subsidence.* Causes, Consequences, and Strategies. 1996 ISBN 0-7923-3933-9

KLUWER ACADEMIC PUBLISHERS – DORDRECHT / BOSTON / LONDON